KT-227-358

ECONOMIC GEOGRAPHY

A Contemporary Introduction

Neil M. Coe, Philip F. Kelly and
Henry W.C. Yeung

Blackwell
Publishing

© 2007 by Neil M. Coe, Philip F. Kelly, and Henry W.C. Yeung
'The River' by Bruce Springsteen. Copyright © 1980 Bruce Springsteen (ASCAP).
Reprinted by permission. International copyright secured. All rights reserved.

BLACKWELL PUBLISHING
350 Main Street, Malden, MA 02148-5020, USA
9600 Garsington Road, Oxford OX4 2DQ, UK
550 Swanston Street, Carlton, Victoria 3053, Australia

The right of Neil M. Coe, Philip F. Kelly, and Henry W.C. Yeung to be identified as the Authors of this Work has been asserted in accordance with the UK Copyright, Designs, and Patents Act 1988.

All rights reserved. No part of this publication may be reproduced, stored in a retrieval system, or transmitted, in any form or by any means, electronic, mechanical, photocopying, recording or otherwise, except as permitted by the UK Copyright, Designs, and Patents Act 1988, without the prior permission of the publisher.

First published 2007 by Blackwell Publishing Ltd

1 2007

Library of Congress Cataloging-in-Publication Data

Coe, Neil M.
Economic geography : a contemporary introduction / Neil M. Coe, Philip F. Kelly, and Henry W.C. Yeung.
p. cm.
Includes bibliographical references and index.
ISBN 978-1-4051-3215-2 (hardback : alk. paper)
ISBN 978-1-4051-3219-0 (paperback : alk. paper)
1. Economic geography. 2. Economic development. I. Kelly, Philip F., 1970– II. Yeung, Henry Wai-Chung. III. Title.
HF1025.C73 2007
330.9—dc22

2006037361

A catalogue record for this title is available from the British Library.

Set in 10/13pt Sabon
by Graphicraft Limited, Hong Kong
Printed and bound in Singapore
by COS Printers Ltd

The publisher's policy is to use permanent paper from mills that operate a sustainable forestry policy, and which has been manufactured from pulp processed using acid-free and elementary chlorine-free practices. Furthermore, the publisher ensures that the text paper and cover board used have met acceptable environmental accreditation standards.

For further information on
Blackwell Publishing, visit our website:
www.blackwellpublishing.com

University of Liverpool

Withdrawn from stock

014132071

ECONOMIC GEOGRAPHY

CONTENTS

DETAILED CONTENTS

LIST OF FIGURES

LIST OF TABLES

LIST OF BOXES

PREFACE

The world around us is powerfully shaped by economic forces. The economy, as we experience it in everyday life, is innately geographical. There is no economy 'out there', floating in the atmosphere, detached from the lived reality. Rather, the economy is a set of grounded, real-world processes, a set of complex social relations that vary enormously across, and because of, geographical space. Our argument in this book is that the set of approaches offered by the field of economic geography is best placed to help us appreciate and understand the modern economic world in all its complexity. To ignore geographical variation leads to a retreat into the unreal, hypothetical world of mainstream economics, with all its many underlying assumptions and simplifications.

In the pages that follow, we adopt a particular approach in making our case for an economic-geographical perspective. Before outlining this approach, we should be clear about what this book is not! This note is especially important for instructors and professors teaching economic geography courses. First, our book is not a statistical and factual compendium on the current geographies of the global economy. In other words, this is not an almanac for economic geography courses; there are many such books already available in the market. The Internet is now a much more effective medium to access contemporary economic data, which has an exceptionally short shelf-life. Hence, we are not primarily concerned with how much coal there is in Northeast England or Shenyang, China, or how many textile factories are located in Bangladesh or the Mexican *maquiladoras*. Moreover, we do not attempt to offer a systematic survey of all parts of the global economy, either by sector or by region. Again, this kind of economic geography text is already available.

Second, the book is not structured as an intellectual history of economic geography, systematically charting a path from the sub-discipline's origins in the commercial geographies of the late nineteenth century through to the very pluralist economic geography that exists today. Nor does it offer a series of literature reviews of work at the research frontiers of contemporary economic geography. Existing 'readers' and 'companions' offer exactly such an access to the economic geography literature (Bryson et al., 1999; Barnes et al., 2003). Our view is that many undergraduate students are initially nervous and/or ambivalent about this intellectual history approach, and that we first need to engage them fully in the *substantive issues* of economic life. By demonstrating the insights economic geography can offer, students will then be equipped to later explore the intellectual and methodological lineage of the field.

Third, the book deliberately blurs the distinction between economic geography and what has conventionally been labelled 'development geography'. In various ways, we have woven issues of development, poverty and inequality into our discussion of an economic-geographical perspective, thereby rejecting the notion that development geography is about the Global South and economic geography is about the Global North. By integrating substantive issues and empirical examples from across the globe, we adopt a more inclusive approach to economic geography.

Given these parameters, what, then, is this book really about? In its essence, this book takes the form of a series of linked chapters on *topical issues* and *contemporary debates* that draw upon, and showcase, the best of economic geography research. These issues are drawn from contemporary economic life, which is increasingly constituted at a global scale – from uneven development, space-shrinking technologies, and environmental degradation to powerful global corporations, organized labour, and ethnic economies. We see each of these as issues rather than just phenomena – that is, they are processes to be debated rather than factual realities to be described. Each chapter thus seeks to answer a significant contemporary question that a curious and well-informed undergraduate reader might reasonably be expected to ask about the world around them.

This, then, is not a conventional text: our aim is to develop well-grounded *arguments* from an economic geography perspective, not necessarily to present simplifications of multiple viewpoints or collections of facts and data. We are, however, trying to develop these arguments in straightforward and accessible ways. The book is intended to be used in introductory courses in economic geography in the first or second year of an undergraduate degree programme. The chapters should be seen as bases for discussion rather than collections of facts and truths needing to be reproduced in examination scripts. Nor are they intended to substitute for classroom lectures to elaborate on some of the theoretical themes or empirical case studies provided here. This is a book very much intended to support an introductory course in economic geography and to socialize students into the fascinating world of contemporary economic geographical research.

Notwithstanding the above comments, in this book we are seeking to advance and advocate a *certain kind* of economic geography, although this underlying position, for reasons alluded to above, is not always made explicit on a chapter-by-chapter basis. From the outset, however, it is important that readers – and particularly instructors and professors – are at least aware of how our favoured approach sits within the sub-discipline of economic geography more broadly. In short, we seek to combine the best of both the *political economy perspective* that entered economic geography from the 1970s onwards and the *new economic geographies* that are generally seen to have risen to prominence since 1990 (see Box 1.2). We are not championing and legitimizing an either/or approach to the theoretical and empirical insights afforded by these perspectives. Such an epistemological or paradigmatic debate is much better covered in more advanced texts (e.g. Hudson, 2004). Instead, we intend this book to be a celebration of *multiple* theoretical perspectives, including the new economic geographies, in contemporary economic geography research.

The political economy perspective is now well known in economic geography, but we should distinguish clearly how our conception of new economic geographies is different from the quantitative modelling version of so-called 'new economic geography' in economics (e.g. Krugman, 2000; Fujita and Krugman, 2004). In common with other economic geographers, we will refer to the latter as 'geographical economics'. 'New economic geography' is a very different proposition and describes an approach to economic geography that has been influenced by the recent 'cultural turn' in human geography (and the wider social sciences). This has created a geographical approach to the economy that contextualizes economic processes by situating them within different social, political and cultural relations.

In doing so, new economic geography is not merely concerned with the economic realm, but also with how such a realm is intertwined with other spheres of social life. While there are publications that encapsulate the various dimensions of this broad approach (e.g. Lee and Wills, 1997; Sheppard and Barnes, 2000), we believe this is the first textbook that seeks to demonstrate the benefits of such a multi-faceted approach to an undergraduate audience. Moreover, our aim in this book is to try and blend the nuanced insights of these new economic geographies about everyday economic life with the analytical rigour that a political economy approach brings to understanding the inherent logics and mechanisms of the capitalist system, and the social and spatial inequities that it actively (re)produces. We also take from the political economy perspective a critical and normative stance that leads us to question and constantly interrogate those inequities. We therefore make an explicit effort to demonstrate the value-added of such a geographical approach in relation to conventional economic analyses of these topics.

The book itself is structured around answering thirteen important questions that arise in everyday economic life. We also preface these questions, listed as each chapter heading, with an important analytical theme. Part I is entitled 'Conceptual Foundations' and explores in turn the 'geography' and 'economic' of economic geography. Chapter 1 counterposes a geographical view on the economy with that which might be adopted by a mainstream economist, and introduces the key geographical vocabulary of space, place and scale. Chapter 2 unpacks the apparently common-sense notion of 'the economy' as something 'static' and 'out there' in conventional economics to reveal how we might think about the economy in more creative and critical ways. In Part II, 'Dynamics of Economic Space', we focus on four broad dynamics inherent to the capitalist system: uneven geographical development (Chapter 3), commodity chains and their role in organizing economic space (Chapter 4), technological change and its ability to alter (albeit unevenly and partially) the geography of the economy (Chapter 5), and the commodification of nature and the environment (Chapter 6). Part III, 'Actors in Economic Space', looks at four main groups of actors who play an active role in shaping economic geographies, namely the state, in all its scalar forms (Chapter 7), transnational corporations (Chapter 8), labour/workers (Chapter 9) and consumers (Chapter 10). In Part IV, entitled 'Socializing Economic Life' – and here we draw in particular on the so-called 'new economic geographies' – we bring in the dimensions of culture (Chapter 11), gender (Chapter 12), and ethnicity (Chapter 13) to our understanding of the spatial organization of economic activity, explicitly moving beyond conventional economic analysis to incorporate consideration of how these 'non-economic' variables shape economic processes.

A few further caveats should be noted at this point. First, in a text of this type and length we cannot hope to cover every aspect of economic geography, either within individual chapters, or across the book as a whole. We do, however, feel that we have covered the most significant debates in which economic geographers have been active and offering valuable insights in recent decades. Second, the book's structure and scope make it impossible to explore all the intersections between the chapter topics: in producing a text of this kind, some simplification is inevitable. Gender and ethnicity, for example, are bracketed out into separate chapters but in reality could be part of many others, and, indeed, heavily intersect themselves. Our strategy has been to break up economic geography into manageable and relatively coherent segments for an undergraduate audience and to use the introductory chapters in Part I to offer a more integrative analysis and to clearly differentiate our field from mainstream economics. Extensive cross-referencing of chapters also helps to make explicit connections between different themes, examples, arguments and case studies in various chapters.

Each chapter in this book follows a similar structure. We open with what we call the 'hook'; a (hopefully engaging) contemporary example or issue used to introduce the key theme of the chapter. In the second section we tackle a commonly held myth or misapprehension about the topic at hand (e.g. the nation state is dead or transnational corporations are all powerful) and illustrate how these myths often rest, in large part, on an 'ageographical' (i.e. non-geographical) understanding of the world around us, particularly in mainstream economics. The main body of each chapter then serves to illustrate the necessity and effectiveness of taking an explicitly geographical approach to understanding different aspects of the economy. Our aim is to make these arguments in a clearly understandable, lightly referenced, jargon-free manner, drawing on a wide range of examples from across different sectors of the economy, and from around the world. Boxes within the text (between three to five per chapter, on average) offer further development of key concepts, case studies and examples for the reader, and the diagrams and photographs have been carefully chosen to illustrate further the various points we are making. The penultimate section of the chapter is designed to add a 'twist' to the arguments that have preceded it; or in other words, to probe somewhat more deeply into the complexity of contemporary economic geographies. Additional nuances and insights are offered in these twists. Each chapter then concludes with a deliberately short and pithy summary of the main themes covered.

What lies after the summary is also very important. First, for ease of use, the reference list is included on a chapter-by-chapter basis. For the instructor, this is also meant to facilitate use of the chapters in a more 'modular' manner that does not have to deal with topics in the order we have presented them here. Second, the further reading section guides the reader towards what we identify as the most engaging and accessible literature on the chapter's topic. To be clear, our objective here is not to bamboozle the reader with complicated and advanced texts at the cutting edge of geographical research, but rather to select readings that further explicate and develop our arguments in a digestible manner for an undergraduate audience. Some of these readings identify the sources of well-known case studies we have drawn from the geographical literature, enabling students to 'flesh out' the inevitably brief summaries we have been able to offer in the text. Third, we identify up to five online resources per chapter that can also be used to supplement the chapters. Overall, our intention is to offer an exploration of economic geography rich in examples and case studies that can, on the one hand, open students' eyes to economic life and practices in various parts of the world, and, at the same time, introduce concepts that can be 'put to work' in local contexts. Hence the text can be used alongside local literature and case studies wherever the book is used.

References

Barnes, T.J., Peck, J., Sheppard, E. and Tickell, A. (eds) (2003) *Reading Economic Geography*, Oxford: Blackwell.

Bryson, J., Henry, N., Keeble, D. and Martin, R. (eds) (1999) *The Economic Geography Reader: Producing and Consuming Global Capitalism*, Chichester: John Wiley & Sons, Ltd.

Fujita, M. and Krugman, P. (2004) The new economic geography: past, present and the future, *Papers in Regional Science*, 83(1): 139–64.

Hudson, R. (2004) *Economic Geographies*, London: Sage.

Krugman, P. (2000) Where in the world is the 'new economic geography'?, in G.L. Clark, M.A. Feldman and M.S. Gertler (eds) *The Oxford Handbook of Economic Geography*, Oxford: Oxford University Press, pp. 49–60.

Lee, R. and Wills, J. (eds) (1997) *Geographies of Economies*, London: Arnold.

Sheppard, E. and Barnes, T. (eds) (2000) *A Companion to Economic Geography*, Oxford: Blackwell.

ACKNOWLEDGEMENTS

In an undertaking of this scale and scope, we have certainly benefited from the generous help and assistance of various people and institutions and we would like to acknowledge them here. Our editor at Blackwell, Justin Vaughan, has been patient and supportive as this project has slowly and steadily moved towards completion. We are very grateful for his ongoing encouragement and confidence in our project, and for the excellent editorial assistance of Ben Thatcher and Kelvin Matthews at Blackwell. We would also like to thank the many anonymous reviewers who helpfully and constructively commented on the book proposal in its various iterations. Trevor Barnes, in particular, deserves special mention. He has been most generous and helpful on many different occasions, and his detailed comments on the earlier version of the full manuscript were extremely important in guiding our revisions. Peter Dicken has been our primary inspiration in trying to become better economic geographers and in striving to improve the accessibility and visibility of the sub-discipline. Gavin Bridge, Tim Bunnell and Peter Dicken kindly commented on individual chapters. Clive Agnew, Gavin Bridge and Martin Hess gave us permission to reproduce their photos in the book. Graham Bowden did a fantastic job of producing all the figures in the book, sometimes at extremely short notice, and often from almost unintelligible scribblings! None of these individuals, however, are responsible for any errors or mistakes that remain.

In order to overcome the tyranny of geographical distance, we met three times, in Singapore (December 2004), Manchester (July 2005), and Bellagio (February 2006) to discuss face-to-face the book proposal, detailed structure, chapter drafts, and so on. These intensive meetings were supplemented by many other brief exchanges when the two or three of us got together during professional meetings or while on research trips. Various institutions provided financial support for these meetings, either directly or indirectly. In particular, we would like to thank the Rockefeller Foundation for its Team Residency Award that enabled us to work together for almost ten days in February 2006 in the tranquil setting of its Conference Center located in Bellagio, Italy. We are enormously grateful to our referees for their support: Ann Markusen, Jamie Peck, and Trevor Barnes. Various resident fellows at the Bellagio Center, and Tom Bassett in particular, shared many interesting insights and suggestions for the project. We are also grateful to the staff at Bellagio, particularly Pilar Palacia and Laura Podio, who took care of us during our stay. The following institutions also funded our travels and/or hosted some parts of the manuscript writing: Isaac Manasseh Meyer Fellowship, Department of Geography, National University of Singapore (Neil); the Social Science and Humanities Research Council of Canada, the Asian Institute at the University of Toronto, the Department of Geography at Queen Mary, University of London, and the Asian Metacentre at the National University of Singapore (Philip); and, the Faculty Research Support Scheme of the National University of Singapore, and the International Centre for the Study of East Asian Development, Kitakyushu (Henry).

At the more personal level, we all would like to thank our respective colleagues, friends and families for their support and understanding. Neil would like to thank, in Manchester, his fellow economic geographers – Gavin Bridge, Noel Castree, Martin Hess, Jennifer Johns and Kevin Ward – for creating a stimulating, collegial and fun working environment; the golf gang – Tim Allott, Martin Evans and Chris Perkins – for occasional, but much needed distraction; and Clive Agnew, Michael Bradford and Fiona Smyth for their support and encouragement. More broadly, he is extremely grateful for the ongoing support offered by Trevor Barnes, Ray Hudson, Lily Kong, Jamie Peck, Roger Lee, David Sadler and Neil Wrigley. Closer to home, heartfelt thanks go to Laura and Adam for generally heeding the cries of 'Don't go down the basement, Daddy's working down there', and even more so for occasionally ignoring them! And Emma, the rest is down to you. Your support is much appreciated.

Philip is grateful to colleagues and family in Toronto who have supported this endeavour, knowingly or unknowingly, along the way – either on the 'production line' of research assistance, in the 'industrial milieu' of academic life, or in the 'domestic sphere' of a larger life. They are: Keith Barney, Ranu Basu, John and Susan Britton, Simon Chilvers, Raju Das, Anne-Marie Debbané, Don Freeman, Sutama Ghosh, William Jenkins, Lucia Lo, Tom Lusis, Glen Norcliffe, Linda Peake, Cesar Polvorosa, Valerie Preston, John Radford, Esther Rootham, Robin Roth, Peter Vandergeest, and Junjia Ye. For all sorts of reasons, Hayley, Alexander, Jack, and Theo also deserve acknowledgement.

Henry is immensely grateful to Weiyu, Kay, and Lucas for their tolerance of an 'absentee' husband/father during those writing trips. The earlier-than-expected arrival of Lucas in November 2005 also inadvertently delayed our Bellagio meeting originally scheduled for early December. He is thankful to Neil and Phil and Susan Garfield at Rockefeller for reworking their schedules to accommodate the postponed Bellagio meeting. To Lucas, he would like to offer this book as his first big present from Dad.

NC, PK and HY
Glossop, Toronto and Singapore
July 2006

PART I

CONCEPTUAL FOUNDATIONS

CHAPTER 1

A GEOGRAPHICAL APPROACH TO THE ECONOMY

Aims

- To understand the assumptions used by economists in understanding the economy
- To recognize the limitations of economic approaches to the economy
- To appreciate key concepts in economic geography.

1.1 Introduction

In 2005, Niger in West Africa seemed to be on the brink of widespread famine (Figure 1.1). More than three million people (around one-third of the country's population) were suffering from severe hunger. While crops had been planted, by the middle of 2005 they were still a few months away from being harvested and food stocks from the previous harvest were running dangerously low. Those stocks were smaller than usual because of a drought and locust infestation in 2004, which resulted in reduced yields.

International media organizations and developed country governments started to take notice, and pictures of skeletal children and women lining up for food aid, along with desperate pleas for further aid, were broadcast around the world. In the United States, the Public Broadcasting Service (PBS) aired a segment on the crisis in Niger on 4 August 2005 during *The NewsHour* with Jim Lehrer. PBS is one of the country's most respected broadcasters, and *The NewsHour* represents perhaps the most cerebral of all newscasts available in the USA. Conscious, however, of its viewers' need to match a picture and a headline with the story, PBS had a logo designed for the occasion (Figure 1.2). Against the backdrop of the African continent, the logo featured women clutching emaciated children, one of them plaintively holding out a bowl, presumably to receive a

Figure 1.1 Map of Niger in West Africa

ration of food aid. We will return to this image later, but first we will hear from the show's reporter in Niger, who set the scene for the broadcast. His commentary reveals some important insights into who is affected by famine, but it also reproduces a number of commonplace assumptions about how economic processes work and how we should understand them:

> At first glance, you wouldn't think there was anything wrong in Masochi [a village in Niger]. There's poverty, yes, but then that's a way of life here. It's only when you come across children like this little boy that you realize the village has a problem: He's two years old and has chronic conjunctivitis; he's struggling to see as flies feast on the discharge from his eyes. Suddenly, we were presented with

Figure 1.2 The PBS logo for its coverage of famine in Niger, 2005
Source: PBS Newshour, with permission.

at least a dozen children with similar problems. Classic signs of malnutrition: Distended bellies and discolored hair. Their mothers showed me what they were feeding their children: Two ladles of watery-looking porridge per day . . .

These children are of course so vulnerable because they've had nowhere near enough to eat. But although it might look like it, the disaster taking place here is not the result of famine. And this is why: [in the] market about ten minutes walk from that intensive care unit . . . there's plenty of produce on sale. [S]acks are stuffed full of onions and sweet potatoes. The problem is that millions of people in this country just can't afford this food.

Aid convoys are slowly making their way across this vast country, but there's food being pushed around here by the cartload. Up country there are areas where it's difficult to find any really young children. In this tiny village, six have died in the last two months. It's weeks until the next harvest, so in the meantime the older ones are being fed weeds.

(Geriant Vincent, *The NewsHour* with Jim Lehrer, PBS, 4 August 2005, http://www.pbs.org/newshour/bb/africa/july-dec05/niger_8-04.html, accessed 4 May 2006)

The report vividly portrays the desperation and human misery of food shortages and disease. But it also points out that famines are not necessarily the result of absolute food shortages and environmental calamities – they often have more to do with the (market) *distribution* and *allocation* of food than the amount of food being grown. While climatic and environmental factors do play a part, they do not fully explain why some people in particular places go hungry. They are hungry, the reporter notes, because they do not have enough money to buy food. Both within the village, and globally, we are therefore seeing a phenomenon that is not so much about food production, but about how we construct economic arrangements to share our planet's resources.

We start this chapter with a profound economic problem in order to highlight what is at stake in understanding our economic world. But we have also deliberately drawn attention to international media coverage of that problem because a key purpose of this book is to think carefully and critically about *how* we understand economic processes. The economic world around us is usually explained by economists. As an academic discipline, economics has achieved a notable dominance over popular understandings of how the economy works and, more importantly, how economies should be managed and economic problems such as poverty and famine should be solved. Economics, however, approaches the economy with a great many simplifying assumptions, designed to render economic processes knowable and manageable in quantitative and technical terms. We elaborate on these assumptions in Section 1.2.

Through most of this chapter, we will pursue the example of Niger as a grounded case study of the contrasts between an economist's and a geographer's approach to the same problem. In Section 1.3, we suggest how economic geography would offer something different in understanding the specific phenomenon of the 2005 famine in Niger. We do not provide comprehensive explanations for the events in Niger – that would be another book in itself – but we do outline some of the questions that economic geographers would ask and seek answers to. For example, a geographical analysis would carefully examine the positioning of Niger in a global pattern of uneven development and consider the ways in which the country is integrated into processes of globalization. It would also examine the spatial patterns of famine *within* Niger, thereby uncovering the uneven impact of food shortages and the difficulties of traversing the country's landscapes. Beyond spatial patterns, geographers are also concerned with the specific characteristics of particular places. In this way, a rich understanding of famine in the local context can be built, based on analyses of environmental degradation, Niger's social structures, the characteristics and capabilities of the national government, and many other localized features. At its core, then, a geographical approach is about *understanding patterns and processes in space and the particularities of places*. This represents a contrast to the abstract and generalizing tendencies of economists. At the end of the chapter we again highlight these contrasts, but this time using a very different set of economic processes an ocean away – the workings of the financial markets in New York City.

1.2 Poverty and Economics: Explaining What Went Wrong

We pick up the story in Niger again by returning to the PBS broadcast. After hearing from its reporter on the ground, the programme then turned to a guest expert, an *economist*, for further insight into the looming famine. We will quote at length from the exchange that ensued as it illustrates how the crisis was understood (and how the media sought to construct that understanding). The studio anchor, Margaret Warner, conducted the interview, which is reproduced in Box 1.1.

Box 1.1 An economist explaining Niger's famine

Margaret Warner: Niger, a poor country twice the size of Texas, is said to be losing now 15 people a day to hunger, and nearly three million of its 11-plus million people are at risk. For more on this unfolding crisis we're joined by Christopher Barrett, Professor of Economics and Management at Cornell University, and co-director of Cornell's African Food Security and Natural Resources Management Program. Professor Barrett, welcome. What would you say is the root of this hunger crisis in Niger?

Christopher Barrett: Well, the root is chronic poverty, Margaret. There's been, as your correspondent just mentioned, a bit of a shock over the past year, low rains, a locust infestation, all of which knocked the harvest down. But the core problem is that there's no margin here, that these are such desperately poor people that even the slightest shock can cause irreparable damage. Prices have spiked for food in Niger and the real problem here isn't no food; the problem is people can't afford food when they're so poor.

Margaret Warner: What else beside those two factors? I mean, can aid get around the country, can food get around the country, or are there problems of transportation and infrastructure?

Christopher Barrett: Well, as the correspondent mentioned, Niger is a country that is more than twice the size of Texas, with a population about the same as that of Ohio, and paved road infrastructure about the same as that of Dayton, Ohio. So once you spread such limited road networks across such a big area trying to serve a fairly large population, it does become very expensive and very logistically tricky to reach them all. So this is a real problem . . .

Margaret Warner: Now, Niger over the past few years has certainly received millions of dollars in different kinds of aid. Has it just not been spent wisely? Has it been mostly spent on paying off their debt? What's been the problem there?

Christopher Barrett: Well, it's true that there has been a fair amount of aid flowing into Niger, on average a bit more than US$300 million a year, only about 5 per cent of that from the United States, by the way. But the majority of that aid has been on debt relief; it has been emergency assistance; it has been a variety of things that don't really get at those fundamental issues of improving the productivity and the health and the education of this population. As a result, they simply can't be very productive and they remain desperately poor. So while it's true that there has been aid flowing into Niger, and it's an open question how well it's all been used, I think one needs to be very careful about assuming that there's this great generosity of flow to Niger. I mean, on average, Americans are giving Niger about a nickel per person, about a nickel per American per year goes to Niger, that's not exactly a generous flow of aid . . .

Margaret Warner: But briefly, what you're saying is the long-run answer is not this emergency relief aid, but it's the kind of aid and management of the aid that actually builds the infrastructure that addresses the problems you were talking about earlier?

Christopher Barrett: Exactly right, Margaret. We absolutely must do something now. It's the humanitarian imperative. But we also need to anticipate that this problem will recur if we fail to address the underlying structural causes. If we don't improve agricultural productivity and we don't improve the marketing infrastructure so that people can earn a living so they can secure their own families.

Margaret Warner: All right, Professor Christopher Barrett, thank you so much.

Christopher Barrett: Thank you very much, Margaret.

Source: PBS, *The NewsHour* with Jim Lehrer, 4 August 2005. http://www.pbs.org/newshour/bb/africa/july-dec05/niger_8-04.html, accessed 4 May 2006.

Where does this exchange leave our understanding of famine in Niger? Is it simply an economic problem of food shortage and high food prices? The issues of drought and locusts are again raised to explain poor harvests – the role of natural disaster is never far away. Nevertheless, the correspondent's earlier point about the famine being a problem of poverty rather than food production is reiterated. Ultimately, then, this is an economic problem rather than an environmental one, but the root causes of poverty itself are not actually addressed in this exchange. We are told that people starve because they are poor, but we are not told, ultimately, *why* they are poor in the first place. Instead, the discussion moves on to address two issues – the response to the crisis and the longer-term solutions that might be found. With respect to crisis response, the difficulties of getting food aid into the more remote parts of the country are acknowledged, but the abilities of the Niger government to manage crisis and use aid efficiently are also questioned. On the longer-term solutions to the problem, we are told that the core problem is a lack of productivity in Niger's economy. Such productivity increases, we are told, will come from investment in people and infrastructure. In short, what Niger needs is capital investment, and such investment has to come from the 'international community'.

This is not a perspective that is unusual. Jeffrey Sachs, a high-profile economist who advises the Secretary-General of the United Nations (UN) on development issues, has argued that 'ending poverty is a grand moral task, and a geopolitical imperative, but at the core, it is a relatively straightforward *investment proposition*' (Sachs, 2005a; our emphasis). Sachs goes on to suggest that if the USA and its rich allies fulfil their long-standing pledge to provide 0.7 per cent of their national income to finance development aid, we can win the war against extreme poverty. For Sachs, however, the root causes of poverty are 'geographical'. In his book, *The End of Poverty* (2005b), Sachs argues that 'geography is destiny'. A country will be poor if it has a location that is inaccessible, an environment that is prone to disease, an extreme climate, and fragile soils:

> In all corners of the world, the poor face structural challenges that keep them from getting even their first foot on the ladder of development. Most societies with the right ingredients – good harbors, close contacts with the rich world, favorable climates, adequate energy sources and freedom from epidemic disease – have escaped extreme poverty.
>
> (Sachs, 2005c: 47)

The required investment, then, is to overcome the natural disadvantages dealt by geography.

International financial institutions are inclined to agree. In a diagnosis that has direct implications for the investment that is available to a country such as Niger, the International Monetary Fund (IMF) and the World Bank provide a specific list of factors that explain the country's poverty: 'limited resources, the climatic conditions, the weak development of income-generating activities in rural areas, strong demographic growth, the scarcity of arable land and environmental degradation, as well as the inadequacy of basic infrastructures' (IMF, 2005: 8). Thus, climate, natural resources, environmental degradation and scarcity of land are all seen as natural 'limits' on growth. In fact, in the list provided in this quote, only demographic growth and basic infrastructure can actually be addressed. The answer, therefore, is to control population growth and to build infrastructure to attract capital investment. Once again, we are back to the need to invest capital in order to grow, and the sources of such capital are international investors like the World Bank.

All these diagnoses represent the application of prevailing economic orthodoxy to the specific case of Niger. It is a powerful, yet simplifying, orthodoxy that tends to homogenize the economic world in a way that economic geographers try to avoid. We identify four components to the economic orthodoxy.

The first is *universalism*, which implies a one-size-fits-all approach to poverty, development and other economic processes. It is represented by the belief that economic processes will work the same in every context and that growth will result from the right stimuli. Basic health, education and infrastructure are seen as the precursors to inevitable growth based on whatever natural advantages a country might have (which in the case of Niger, might relate to mineral resources and a labour force that is plentiful and very inexpensive). Furthermore, growth is assumed to move along a trajectory of development in which countries gradually become more and more like those industrialized countries from which such diagnoses are made. Hence, for example, the notion in the interview quoted above that Niger is not a democracy *like us* 'just yet', but is on the path towards that goal.

The second assumption evident here is that economic *rationality* always prevails. In many ways, this is an extension of the idea of universalism. For economic principles to apply everywhere, it is necessary to assume that people will respond in predictable and rational ways to 'market signals'. Individuals are treated as isolated actors who make judgements in order to maximize their economic gains. This is often known as the *homo economicus* ('economic man') assumption. Under this assumption, it is, for example, seen as 'natural' and expected that food traders in the market should demand massively inflated prices for food even while some people go hungry – self-interest is 'normal'. This assumption thus makes the concept of famine in the context of food availability a notable but perfectly understandable phenomenon. The problem is seen as being located in the inability of certain people to pay for the food, rather than the artificially high cost of the food.

The third assumption concerns *competition* and *equilibrium* – the notion that the market mechanism will always find the greatest efficiency and productivity. It is a basic principle of economics that equilibrium can be achieved through perfect competition in the market. This equilibrium produces a situation of maximum economic efficiency when supply meets demand at a particular price. Thus, by opening up to international investment, Niger will be incorporated into a global economic system that will naturally find the sources of growth that are appropriate to that context – exporting raw materials from the uranium mining industry, for example. As Niger competes in the global economy, we are told that it will inevitably benefit.

The final assumption is that economic processes are based on certain *laws* and *principles*. This relates closely to the previous three points, but it also draws attention to the assumption that practitioners of economic 'science' have insights into the predictability and operation of certain fundamental processes. These processes are often, therefore, reduced to formal statistical models in which only quantifiable processes can be accommodated. The insights of economics are often taken, however, to be the equivalent of the expertise of natural scientists concerning processes in physics and chemistry. Thus, while PBS sought a description of the situation from its reporter on the ground, it turned to an economist for expert analysis of the reasons behind the crisis.

1.3 Geographical Perspectives on the Economy

Having established the ways in which economists 'read' the situation in Niger, in this section we introduce the distinctive set of geographical sensibilities that can be applied to understand the situation there. While 'geography' enters the vocabulary of some economists, notably Jeffrey Sachs, what we have in mind is very different from the environmental and locational determinants of poverty that he spells out. Instead we argue that a geographical approach puts such spatial concepts as *space*, *place* and *scale* at the centre of analysis, while also remaining conscious of how these concepts are *represented*. They form part of the common language that is shared among professional geographers. We will introduce each of these concepts with reference to the example of Niger, but it is also worth noting their relationship to broader intellectual traditions that have developed over time in the field of economic geography. These are described in Box 1.2.

Space

In its most fundamental sense, space refers to physical distance and area. Every economic process must exist 'on the ground' in a bounded area and at some definable distance from other activities. This definition of space entered scholarly work in economic geography during the 1950s and 1960s when location theory held sway (Box 1.2). In Figure 1.3, this is represented by the flat plane that forms the surface on which places are located. It would seem to be common sense that all human and environmental processes happen *in space* in this way, and yet it is not unusual to find that abstract processes are discussed without any sense of how relative distance and location might play a part. A prime example was described above, whereby an understanding of Niger's economic problems was based on an abstract theory of how the economy 'should' work, without any sense of *where* Niger is located in global space.

We can elaborate on this notion of space by defining four interrelated elements of the concept. The first relates to *territoriality and form*. Niger is a territory defined by lines on a map and jurisdiction on the ground (see Figure 1.1)

Box 1.2 Major theoretical perspectives in economic geography since the 1960s

During the second half of the twentieth century, economic geography witnessed the rise and fall of different and often competing theoretical traditions (Scott, 2000). Broadly speaking, we can identify a series of overlapping trends in economic geography over this period:

1 *Location theory and the neoclassical approach* represented an attempt by economic geographers to emulate the scientific methods and philosophies of the natural sciences (and economics). Starting with the German sociologist Alfred Weber's industrial location theory, published in 1909, and the translation of German economist August Lösch's work on the economics of location into English in 1954, location theory and analysis gathered pace throughout the 1950s and the 1960s. During the 1960s, the theory was imported into economic geography primarily through the classic work of Brian Berry and William Garrison in the US and Peter Haggett in the UK. This genre of economic geography was particularly interested in establishing and explaining patterns and order in the distribution of economic activities across space. In methodological terms, locational analysis adopted mathematical forms of geometrical modelling to describe and explain spatial patterns. This tradition in economic geography has been picked up by some economists since the 1990s, who have developed sophisticated models of economic activities across space – thereby appropriating the term 'new economic geography' to describe techniques that are quite different from those now used by most economic geographers (see below).

2 Between the late 1960s and the early 1980s, some economic geographers experimented with a *behavioural approach*. This approach questioned the rationality assumption that underpins location theory and neoclassical economics (general equilibrium analysis). By adopting Herbert Simon's idea of *bounded rationality*, behavioural economic geographers examined the role of cognitive information and human choices in determining decision-making and locational outcomes. While the analytical focus remained on locational issues and spatial behaviour, this genre of economic-geographical research tended to shy away from the mathematical modelling that dominated location theory. Instead, large-scale surveys were used to investigate the economic decision-making of human actors in various situations.

3 *Marxist political economy* emerged in the early 1970s as a reaction to the failure of locational analysis to adequately address the social and spatial inequities in economic development and wealth that were

emerging. Beginning with David Harvey's classic work on urban social justice in 1973, Marxism opened up entirely new horizons in economic geography that explored social relations of production and the geography of capitalist accumulation. The main foci of analysis were not spatial patterns or location decisions, but rather the structures of social relations that underpinned capitalism. Spatial patterns of industrial location, urban form, or economic development in general were seen as outcomes of struggles between capital and labour (see Chapters 3 and 9). During the late 1980s and the early 1990s, the political economy approach manifested itself in the *post-Fordism* debate. This theme focused in part on the ways in which capitalist economies are regulated through institutions, but also on the transactional relationships between firms in particular *industrial districts* and high-growth regions (see Chapter 5).

4 Since the mid-1990s, *new economic geography* has moved away from viewing economic processes as separate from social, cultural and political contexts. Instead, social, cultural, and institutional factors now tend to be seen as key factors in understanding economic dynamics. Unlike previous genres, the new economic geography is not represented by a particular theoretical perspective or methodological practice. Rather, it is characterized by an eclectic collection of philosophical standpoints and social theories ranging from poststructuralism and postmodernism to institutionalism and feminism (see the chapters in Part IV).

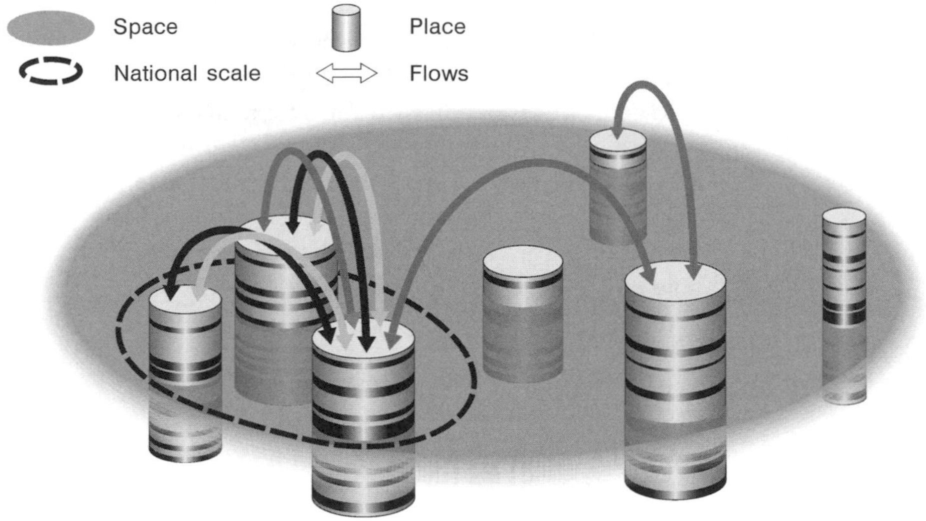

Figure 1.3 Place, space, and scale

– jurisdiction that is clearly bounded in certain locations, for example, at border crossings. Importantly, Niger's territory takes on a particular form – a contiguous and landlocked territory. Furthermore, this spatial expansiveness is marked by unevenness in infrastructure, resources, and even the reach and effectiveness of the government. The country is not a point on a map – it is, as the TV interviewer quoted in Box 1.1 reminded viewers, a country twice the size of Texas. The first question that might arise in relation to famine in Niger, then, is 'Where in Niger?' A sense of Niger's territoriality thus straightaway provides a more nuanced set of questions about what is happening there.

A second dimension of space, beyond its form, is *location* within space. Indeed, long before most of us would ask 'Where *in* Niger?', we would most likely be asking 'Where on earth *is* Niger?' This notion of space concerns the coordinates of a location in relation to others. The location of Niger in global space is important, as we shall show below. With geographical location come the environmental characteristics emphasized by Jeffrey Sachs – climate, ecological features, and so forth. No one would deny that the country's location in belts of Saharan desert and savannah grasslands is materially important in understanding the resource base on which its economy rests (although the *use* of these resources is a separate and more important issue, as we will discuss later).

Location is also important in terms of the country's significance (or rather insignificance) to global geopolitics. Unlike other countries that have developed rapidly over the past 50 years, little is at stake in Niger for the great powers of the world. In contrast, it would be impossible to understand the processes that led to the economic successes of countries such as South Korea and Japan without acknowledging their significant geopolitical location in a world region that was deeply divided during the Cold War between the Soviet bloc and the West. Their economies were supported, and their successes tolerated, by Western allies in a way that would have been unimaginable without their geopolitical significance, located adjacent to communist China and the Soviet Union. Niger, meanwhile, has no such geopolitical significance and so it tends to require scenes of profound human misery before any interest in its fate is mustered by Western governments and media. As noted by the economist in Box 1.1, helping Niger is a 'humanitarian imperative'. Its priority in the eyes of wealthy nations such as the United States, however, may be easily replaced by other strategic initiatives. We thus need to know where Niger is located – both physically and geopolitically – in order to understand its position in the priorities of the global political and economic order.

A third dimension of space helps us examine the situation of Niger in relation to *flows across space* (see Figure 1.3). A careful reading of the diagnoses of Niger's plight offered by the expert economist quoted in Box 1.1 and the IMF report described earlier would suggest that the causes behind its problems are essentially *inside* the country. Poor resource endowments, low productivity, natural disasters, environmental degradation, inadequate infrastructure, overpopulation, inefficient government – all relate to processes contained within Niger's borders. None of them recognize the significance of Niger's connections and relationships with the outside world. The historical roots of Niger's economy (and, indeed, its definition as a national entity) in the French colonial period that ended in 1960 have created an economy and a set of dependent international linkages that continue to the present. In 2004, 41 per cent of the country's exports went to France. In the same year, the only direct long-haul flight out of the airport in the capital city Niamey went to Paris.

Another dimension of Niger's positioning in international flows is its trade balance. The country imports approximately US$360 million more than it exports – a shortfall that is greater than the total budget of the national government. In reality, then, the country's international economic linkages constitute a yearly extraction of wealth from what is already one of the poorest countries in the world. This trade deficit is only just offset by annual aid and development assistance contributions (and as a staunchly Islamic nation, much of this aid to Niger comes from other Islamic countries). A recognition of the importance of flows across space can also be applied to relationships *within* the country. The PBS reporter quoted earlier noted that 'up country' there were villages with almost no young children because of famine-related deaths. Clearly we would therefore want to ask what kinds of resource flows exist across Niger's national space – flows that leave certain people in rural food-producing areas with less food than their urban counterparts.

The fourth and final role of space in our understanding of the poverty that afflicts Niger is rather more abstract and draws upon the Marxist tradition in economic geography (see Box 1.2). In short, it is possible to construct an argument that unevenness of wealth and development across space is not an anomaly that economic processes will eventually iron out – as assumed in the general equilibrium models in economics. Instead, uneven development is a fundamental and necessary characteristic of the capitalist system through which our global capitalist economy is organized. Thus, rather than seeing equilibrium as a natural state of affairs, some would argue that unevenness is an inevitable outcome of capitalism. If we understand the global capitalist system in this way, then Niger's poverty and hunger are necessary parts of the system. Only through such destitution are the raw materials that the country exports kept inexpensive. And only then are the lifestyles of those of us in wealthy countries sustained. For example, without Niger's uranium, the nuclear power plants supplying electricity in many parts of the developed world could not operate. In a sense, then, *we need* the poverty of others in order to keep ourselves in affordable comfort. We will explore this idea in some detail in Chapter 3, but it is worth highlighting here as it opens up a rather different notion of space, one in which spatial unevenness is a necessary part of our global economic system.

Place

A second set of geographical concepts relates to the *specificity* of particular places. We noted earlier how mainstream economics tends to create universal principles that are assumed to apply equally in all contexts. Geographers, in contrast, tend to focus on the specificity and uniqueness of places. This goes back to the tradition of geography that sought to provide detailed descriptions of regions that integrated both physical features and social processes, but it is also a feature of the so-called 'new economic geography' (see Box 1.2). In both cases, the emphasis on place draws upon a strong geographical tradition of empirical fieldwork – requiring a detailed understanding and lived experience of a place as necessary conditions for generating knowledge about it.

Nevertheless, 'place' is a very vague notion. When we use the word, it can refer to anything from a shop or street, to a region, a country, or a continent. We will return to these multiple scales later on, but for now the important point is that the concept of place relates to somewhere *in particular*, whatever the scale involved. This might seem very obvious, but the imperative to think about economic and social processes in particular places forces us to consider them not as vague and abstract forces somewhere 'in the air', but as real and lived experiences on the ground. Thus, while economists are seeking universally applicable generalizations, in some ways geographers are going in the opposite direction – trying to understand why certain things happen in specific places in the context of all the richness and complexity of that place.

Clearly, this is a massive and complicated undertaking. It means that in trying to understand famine in Niger, we are not just arriving with a set of assumptions for which we will collect evidence. Instead, we are seeking to understand how economic processes are deeply embedded in an environmental, social, cultural, institutional and political context. First, we might start by examining what exactly we mean by 'the economy' in Niger. We have seen an economist's explanation that food shortages occurred because people could not afford to buy food. But this indicates that a whole range of other sources of food were exhausted too. Imagine, for a moment, what *you* would do if you did not have enough money to buy food. Most likely you would be fed by a relative or friend, or you would borrow some money. But you might also have access to free food through a workplace or through a charitable community organization, or you could have something edible growing in your garden, or have access to someone else's crops by hopping over a fence into a field or an orchard.

In any given place, people will have a variety of ways in which they acquire food – not all of it, perhaps not even most of it, would come from buying produce at the market. We could start, therefore, by trying to understand what comprises this bundle of sources for the most vulnerable people in Niger, and examine why all of their various means of coping broke down. In other words, we would take a more comprehensive view of 'the economy' than the one based on formal monetary exchange and the meeting of supply and demand in the marketplace that is the foundation for conventional economic analyses. Once we know the diverse forms of the economy, we might work towards supporting them, rather than applying a one-size-fits-all approach to poverty based on 'big' capital investment in infrastructure. This is precisely the approach to the economy that we will explore in Chapter 2.

Another way of understanding the place-based specificity of Niger would be to ask about the structure of its *society* and *government*. There are a variety of questions we might ask. We know that not everyone was affected by food shortage, so how exactly were its effects distributed across the population in relation to various dimensions of social identity such as age, gender, ethnicity, and so on? If specific groups appear to have been most affected, such as rural landless labourers, or women and children, would the kinds of solutions being proposed – such as infrastructure and agricultural productivity – help them directly? In fact, it might be that inequalities in the society would simply be accentuated by these solutions – intended or not. It would also be important to know about the relationship between large landholders and politicians and the role that the government plays in economic processes, such as cropping decisions by farmers, irrigation schemes, and the marketing of produce. All these questions require detailed and place-specific answers – they require going beyond the headline story that trumpets 'another' famine in Africa, to actually understanding the complexity of a place and its relationships with the rest of the world.

In short, the characteristics of places are unique assemblages of a huge array of political, economic, social, cultural and environmental factors. These features are built up gradually and incrementally, so the characteristics of Niger (and places within it) are the products of historical layers of change over time. Thus, colonialism, for example, left an array of economic legacies and bequeathed a set of political institutions and practices. But these, in turn, were superimposed upon a pre-existing set of cultural, political, and ecological practices. And the legacies of colonialism have, in turn, been taken in certain directions by Niger's post-independence history. This continual accretion of place-based historical legacies will be discussed further in Chapter 3.

Finally, while we have, in this section, primarily focused on the internal characteristics of a place, references to colonialism and contemporary flows of trade and aid make it clear that places are as much created by their relationships with the 'outside' as they are by things that happen internally.

Niger, as a place, is a unique intersection of various relationships and flows that intersect there. The same could be said for specific places within Niger. Each place is, then, the unique intersection of its past and present engagements with broader flows. For those reasons, in Figure 1.3 we show places as having a depth, a layering, and a uniqueness – depicted, in the diagram, as a bar code.

A geographical approach to the economy seeks to understand and problematize the elements of uniqueness that arise in particular places. Niger has issues that cannot be encompassed by generalized and abstract ways of thinking prevalent in economic analysis. The universal rules and assumptions that characterize economics tend to gloss over the distinctive contextual circumstances that shape the causes and the solutions to famine and poverty: ecological systems and adaptations to them, including farming and pastoral practices; political structures and power relationships; gender relations; ethnic identities; various legacies of colonialism, and many more. A geographical approach takes most, if not all, of these issues as analytical problems to be studied, understood, and eventually resolved.

Scale

In the last subsection we focused on the analysis of difference and uniqueness as a core geographical approach. In this sense, a place is simply a defined site for exploring idiosyncrasies that have developed in relation to other places and over historical time. But we have not yet defined what a place actually *is*. Indeed, we noted that places may take all shapes and sizes – from a street or village, to a nation, and even a continent – just so long as it is a specific piece of territory that is being discussed. This brings us to a third geographical concept that helps us to organize the world of space and place around us, namely geographical *scale*. Clearly there are differently sized units that might constitute places carved out of space and we provide an organizing framework for understanding these units by using the language of scale. This provides us with a typology for describing the various scales at which we might find some coherence in the processes that are at work (Figure 1.4).

At a *global* scale, we have already noted the kinds of unequal economic exchanges in which Niger is engaged. More specifically, however, we could consider the case of the global uranium industry. As a mineral that is needed for nuclear power generation and weapons manufacture, uranium is much in demand. Yet, just ten countries control 96 per cent of the world's reserves, with Niger among them. However, the process of enriching uranium, essential for its subsequent use, is carried out almost entirely in the United States because of its military and geopolitical implications. The highly restricted nature of this prized commodity at the global scale, and the lack of an internationally accessible market for it, mean that the potential for Niger to benefit greatly from its uranium endowment is limited.

Another scale at which we might understand Niger's economy is at the level of the Sub-Saharan *macro-region*, where common environmental and colonial experiences exist. The similar circumstances experienced by neighbouring countries also explain why food was scarce across the region in 2005 – neighbouring Burkino Faso, for example, banned food exports and thereby exacerbated the market shortages in Niger.

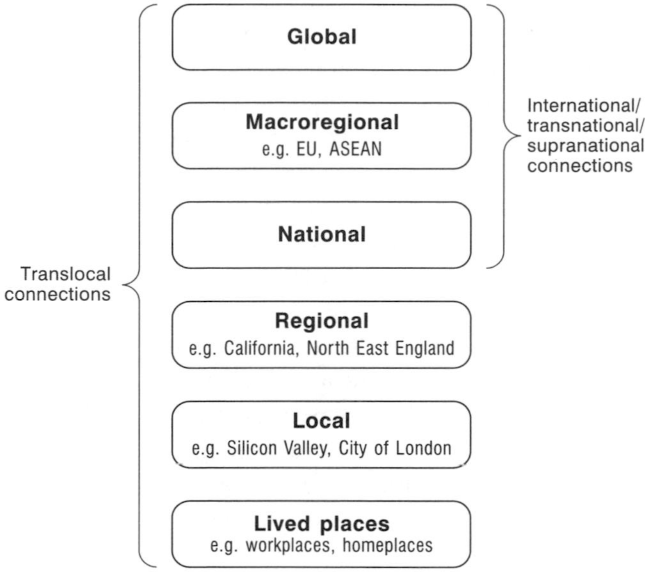

Figure 1.4 Spatial scales
Source: Adapted from Castree et al. (2004), Figure 0.1.

Other processes, however, can only be understood at the *national* scale defined by Niger as a territorial state. During the food shortage in 2005, for example, the national government initially refused to provide food hand-outs, preferring instead to offer subsidized grain supplies to areas affected by the shortages. Clearly, such policies have to be understood in relation to a national political system. The national scale is also important in a variety of other ways. As the first two sections of this chapter illustrated, the challenge of food shortage gets defined by outside experts and media pundits as 'famine *in* Niger' (i.e. a national issue). The solutions that are then presented by economists and aid or lending agencies are *national* level policies: signing on to trade agreements, providing investment incentives, implementing birth control, and developing infrastructure.

But we have also noted that hunger in Niger was spatially as well as socially uneven. Different people and places were affected in various ways. We do, therefore, need to look at *sub-national* or *regional* scales in order to understand this unevenness. The Tahoua and Maradi districts of Niger in Figure 1.1 were especially afflicted by hunger in 2005, as livelihoods there were largely dependent on agricultural crops. Also, the famine was largely a rural, rather than an urban, phenomenon.

Finally, we can examine the nature of famine in towns or villages that we often refer to as the *local* scale. Access to agricultural land or grazing rights, money lending, and retailing are all parts of the system through which food access is determined and are all, in part, constructed at the local level. The PBS reporter quoted at the start of this chapter focused on experiences of famine in the village of Masochi and pointed out that starving children in an intensive care unit were just a few minutes walk away from food being carted around in large quantities at a local market. Uneven access to food is clearly created in part through the way in which society operates at the village level.

There are, then, a variety of scales at which we might construct an analysis of famine in Niger. The important point, however, is that none of them in isolation is adequate to develop a comprehensive understanding. In fact, the spatial relationships and place-based processes that we have noted are operating within and across all of these scales. An important part of a geographical approach is therefore the awareness of how economic processes are constituted at *multiple* scales simultaneously. This was notably absent in the analyses of economists and international financial institutions quoted earlier: each located the causes of famine squarely within the national context alone.

Representations of space, place, and scale

So far, we have focused on geographical language and categories as *descriptive* of the spatial characteristics of the processes we are analysing. But the ways in which space, place, and scale are discussed in everyday life (as well as in academic and political contexts) means that these terms are not just descriptive. Instead, a type of space, or a particular place or scale, becomes soaked in connotations and implied meanings. In other words, they are not simply neutral tools for describing the world; they are also *representations* of the world.

We can start illustrating this idea of representations by exploring the ways in which types of spaces become represented in particular ways. It is common, for example, to see poverty in Africa as closely associated with desert and grassland zones. These ecosystems are represented as dusty, desolate and unsuitable for human habitation. In reality, adaptations to such environments have made them homes to many people for a long time, but the Eurocentric association between a type of space and its habitability makes deprivation in such a space seem *inevitable* or *natural*. We might make similar points about spaces such as 'the village' or the 'Third World'. Both are generic types of space but ones that conjure up particular notions of what happens there and assumptions about those who live there.

While these representations concern generic types of space, a closely related form of representation is associated with specific places. This can best be illustrated with some examples that have been left un-analyzed since the beginning of this chapter. In the account from the PBS reporter, he notes that 'there's poverty, yes, but then that's a way of life here'. We also noted that PBS provided its viewers with a logo depicting an African woman holding an emaciated child, with an arm stretched out as if asking for food (Figure 1.2). Both of these examples act as representations of a place – namely Niger – but they also tie into representations of Africa as a whole. In short, Africa is represented in these textual and graphical descriptions as a place of poverty, hunger, drought, locusts, and deserts. In this place, poverty is represented as 'natural' there – it is 'a way of life'.

Ultimately, this is a very powerful representation. It provides a narrative for understanding suffering that somehow makes it comprehensible and perhaps even ignorable by the public and their politicians in wealthier countries. If hunger is natural in Africa, then no one can be responsible. If poverty is a way of life, then why would we intervene? It does not seem to be our 'business' because it's a way of life 'over there'. But this representation is, of course, obscuring the reasons for poverty and hunger. Not everyone in Niger was starving, nor is Africa a continent defined by desert and drought. By representing places in this way we cast a veil over inequalities and differences *within* those places and more importantly we ignore the relationships *between* such places and the world in which we *all* live. By colouring a place with a particular representation, all of the subtleties of a place such as Niger that we discussed earlier are blocked out.

Scale too can be constructed in ways that are misleading. If we represent just one scale as the defining container for understanding a phenomenon, then we ignore the processes that operate at other scales. Hence an emphasis on the national scale serves to ignore the global relationships that shape a country's poverty. But equally, an emphasis on the global scale would obscure national systems of inequality that cause or exacerbate hunger. It is, therefore, important to remain conscious of the ways in which scale is being represented or ignored in accounts of economic processes.

Representation, then, can have important implications for the way in which we understand a problem, and, by extension, for the way in which we imagine solutions. Alongside a geographical understanding of economic processes based on space, place and scale, it provides an important means of correcting the limited or partial analyses that we often hear. In this section we have barely scratched the surface of an explanation for, or solution to, African poverty, but we have laid out some of the important questions that a geographical approach would ask. In this way, we can at least start the task of making better diagnoses of such problems, and hence avoiding bad prescriptions in which the medicine may at best be ineffective, and at worst have severe side-effects.

1.4 A World of Difference: From Masochi to Manhattan

The previous two sections have highlighted what can be missed in conventional economic analyses of poverty and how a geographically conscious approach may do a better job of explaining what is a highly complex, real-world problem.

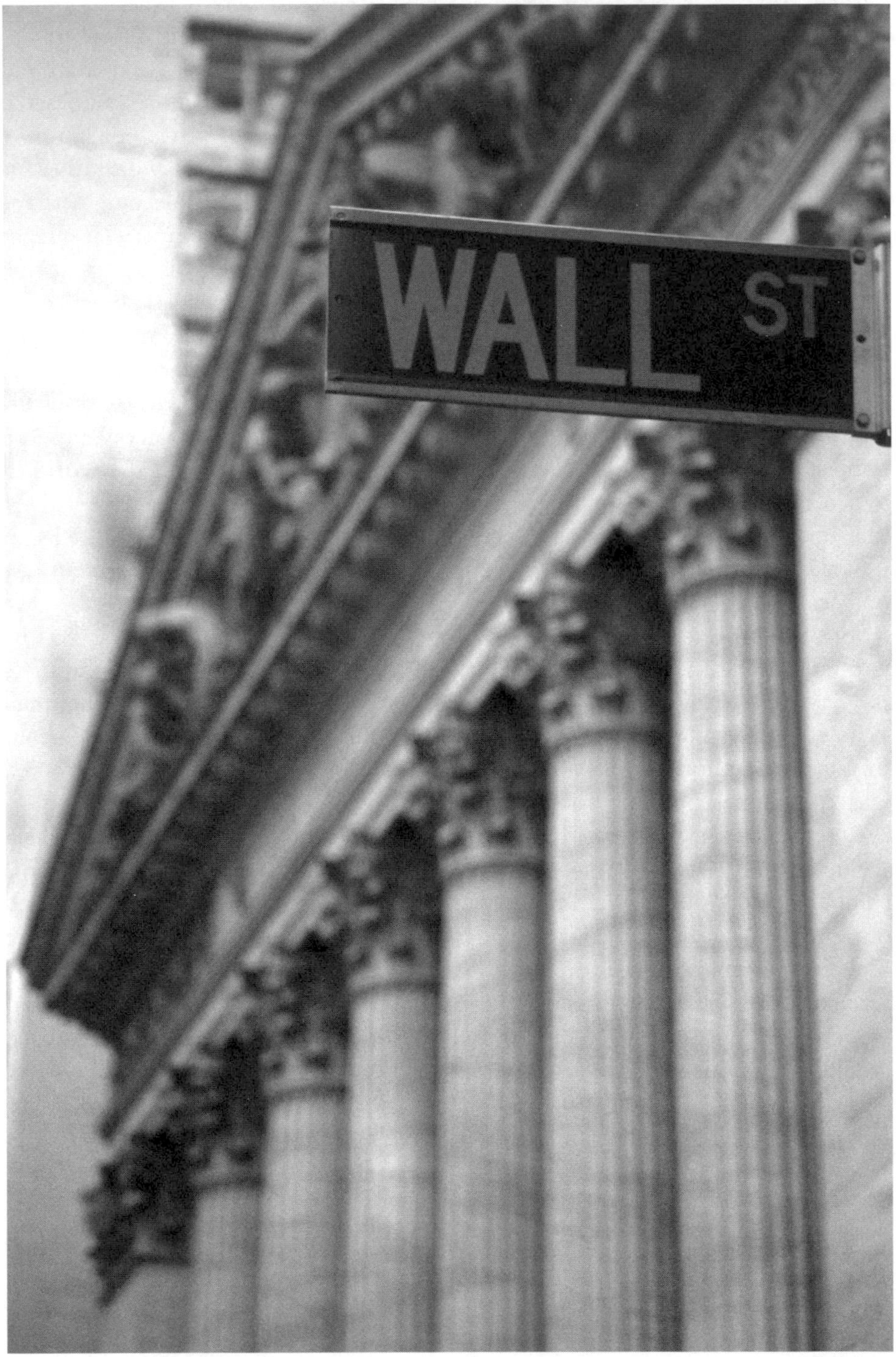

Figure 1.5 Wall Street, New York City
Source: © Getty Images.

In this section we further develop the idea of a distinctive economic-geographical perspective by looking at a completely different example – the financial power of Wall Street in New York City – and considering the contrasting ways in which economists and economic geographers might approach that topic. Asking different research questions about the same economic phenomenon can produce very different analytical insights. Our purpose here is not to denigrate economics, but rather to demonstrate the potential benefits of the insights that can be derived from adopting an economic geography perspective.

As is well known, Wall Street, a simple street in New York City (Figure 1.5), plays a dominant role in the global financial market. Wall Street is at the heart of lower Manhattan, America's leading financial district and, together with Broad Street, defines the boundary of an area that hosts the world's most powerful financial institutions. These institutions range from the New York Stock Exchange (NYSE) to the headquarters of many large financial groups such as Goldman Sachs and Merrill Lynch. The NYSE in Wall Street is particularly significant, with its website claiming that 'the world puts its stocks in us' (http://www.nyse.com). Its stock market indices literally *dictate* the global price movements of stocks in numerous other stock exchanges; its daily trade of stocks represents more than 80 per cent of all listed US firms and over 60 per cent of all listed firms in the world. In 2004, its average *daily* value of trade was \$46.1 billion, a figure exceeding the *annual* GDP of many countries (in the same year, Niger's entire economic output was about \$3 billion).

How might an economist approach try to understand this incredible confluence of financial power? A financial economist might immediately think of money and banking as defining Wall Street's central importance. The research questions that then arise from this viewpoint would relate to several key concepts, including price, equity, rate of return, growth, volatility, and risk. A universal law of financial economics – the *efficient market hypothesis* – puts these concepts into practice, depicting the financial market as a self-regulating machine whose primary function is to clear (i.e. balance) the demand and supply for particular financial instruments such as currencies, shares and bonds at a certain equilibrium price. Alternatively nicknamed the *random walk theory* after Burton Malkiel's (1985) bestseller, this hypothesis has been 'proven' through numerous empirical studies in financial economics with the result that it has become seen as a universal law, akin to the law of gravity in physics. In this law of finance, the financial market is assumed to be so efficient that virtually nobody can gain exceptional profits. This happens because everybody has access to 'perfect' information and the market is fully competitive. Taken to its logical extreme, Malkiel (1985: 16) argues that 'a blindfolded monkey throwing darts at a newspaper's financial pages could select a portfolio that would do just as well as one carefully selected by the experts'.

Armed with this law of efficient markets, financial economics would approach Wall Street as an efficient market for settling the demand and supply of different financial instruments. As the law is assumed to prevail in all circumstances, the critical research question for financial economists is to determine the functioning of different financial markets in Wall Street, to improve its economic efficiency, and to predict its behaviour. By adopting such a financial approach to the understanding of economic activities in Wall Street, we would argue that financial economists have often ignored or indeed completely missed some important aspects of the Wall Street phenomenon.

An economic geographer, by contrast, armed with the conceptual toolkit of space, place, scale and representation, might seek to ask the following important questions:

- Where does the capital, people, and technology in Wall Street come from? How is Wall Street 'plugged into' the global economy through different connections across *space*? How are these connections organized at different spatial *scales*?
- Why do these different elements in the financial industry coalesce together in a particular *place*, i.e. Wall Street? What are unique attributes of Wall Street and its environs that have led to the continued concentration of global financial power in such a small area?
- Is it just economic factors that shape decision-making and market movements in Wall Street? What are the non-economic dimensions of these economic processes?

To be clear, we are not arguing that financial economists must necessarily raise and answer these thorny questions. But we believe these are important questions that economic geographers can productively seek to address.

Unlike the law of efficient markets in financial economics, there is no single theory that explains everything in economic geography. Instead, as we saw in section 1.3 in the case of Niger, we rely on several interrelated geographical concepts to tackle these research questions. First, we would seek to explain how Wall Street serves as a key *node* in the global financial system by exploring how New York City has come to dominate the global financial economy since the nineteenth century. It would demonstrate how Wall Street institutions are connected to financial circuits operating at the *global* scale. This line of geographical inquiry requires deep historical analysis of both the growth and role of New York City in an expanding hierarchy of urban centres in the United States, and the rise of American power and global finance in the post-Cold War era of financial globalization (see more in Chapters 3 and 7). Equally, it necessitates detailed analysis on the contemporary flows of people, capital and knowledge into, and out of, New York City that are part of the global financial system. These kinds of detailed analyses clearly do not lend themselves to the law-making approach of financial economics, as they are not readily amenable to generalization, statistical analysis and mathematical modelling.

The second set of research questions concern the 'coming together' of different financial institutions in one particular *place*. For over a century now, economic geography has sought to explain the spatial clustering and co-location of economic activities (see Box 1.2 and more in Chapter 5). Part of the reason may be to do with the cost savings of locating together and doing business with one another. And yet why, with the recent quantum leaps in information and communications technologies, do institutions still want to trade *on* Wall Street when presumably they can do so from anywhere in the supposedly 'borderless world'? After all, property in Manhattan is some of the most expensive to rent or own in the world. As our third set of questions suggest, to tackle this conundrum effectively we need to move beyond purely economic logics to bring in social and cultural factors (indeed, concern with these aspects is central to the 'new' kind of economic geography outlined in Box 1.2). Over time, certain norms, values and ways of doing business have developed in and around New York's financial district, offering a different kind of 'glue' that binds firms and other institutions together in a particular place. This alludes to a local world of bars, cafés, clubs, gossip and rumour in which vital financial knowledges are exchanged between bankers, stockbrokers, business analysts, journalists and so on. Crucially, these intense networks are primarily organized at the *local* scale (Thrift, 1994). Equally importantly, these rather intangible, but highly significant elements of Wall Street's uniqueness and continued vitality are beyond the radar of most economists' models.

We could make a similar argument about many, many other economic phenomena. What we have tried to show in this section, through looking at the example of the Wall Street financial district, is that employing our geographical vocabulary leads us to approach the topic in a distinctive manner, and one which, we would argue, leads to richer and more nuanced analyses of grounded, real-world economic processes than might otherwise be the case.

1.5 Overview of the Book

This chapter has introduced four key concepts – space, place, scale and representations thereof – that will be used throughout this book and which we see as defining a geographical approach to the economy. In Figure 1.6, we gather these concepts together graphically in order to illustrate how the various chapters of this book connect to form a coherent geographical narrative. In short, it shows both the comprehensiveness and complexity of an economic-geographical perspective on the global economy.

We first focused on *space*, noting in particular the importance of location (that is, patterns *in* space) and flows (movement *across* space). *Location* in space will be a key issue in Chapter 3, as we explore why economic development appears to occur unevenly, but throughout the book the question of why certain

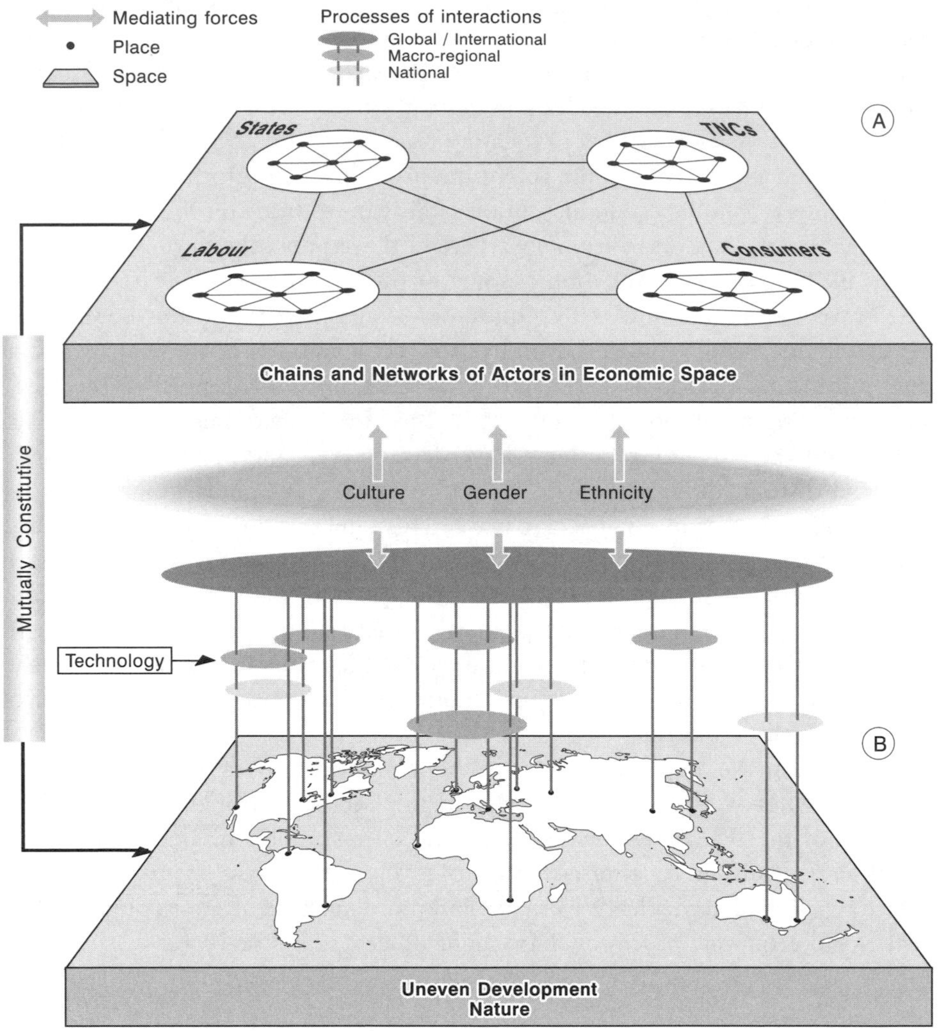

Figure 1.6 An economic-geographical perspective on the global economy
Source: Inspired by Dicken (2004), Figure 2.

economic activities happen in certain locations will underpin many of our discussions. In Figure 1.6 (B), we show different locations in space and patterns of uneven development.

Flows across space will also be a common theme in this book. In Figure 1.6, we show how space both serves as the backdrop through which these flows occur *and* affects how these flows are organized. There is a *mutually constitutive* relationship between the networks linking states, firms, workers and consumers in economic space (A) and the placed-based attributes in the space of uneven development and the natural environment (B). Chapter 4 will explain how commodity flows across space connect the different groups of actors depicted in the figure. Chapter 5, on technology, will establish some of the space-shrinking technologies that have facilitated and accelerated flows of capital, people, and commodities in different processes of spatial interactions shaded in Figure 1.6. At the same time, however, it will emphasize the continued importance of physical proximity in space. Flows of commodities are discussed further in Chapter 6 when we examine the ways in which natural resources located in different parts of the world map are commodified into economic products.

In Part III, each of the key actors involved in shaping these economic flows (illustrated in the upper plate in Figure 1.6) will be explored in more detail. Chapter 7 will unpack how the state is related to other actors in the global economic space. Flows of investment by transnational corporations (TNCs) will be explored in Chapter 8, where we will ask how TNCs organize their global activities and integrate complex networks across space. Flows of migrants form a part of Chapter 9 where we will also discuss how space is actively used by employers and workers when they confront each other. Chapter 10 will examine how consumers perform their role as the 'end users' of the commodities that flow across space through complicated chains and networks.

Equally important to us is the geographical imperative to understand the specificities of *place* featured throughout this book. Such places are denoted in the lower plate in Figure 1.6 and our role is to elucidate their inherent complexity and diversity. The uneven development discussed in Chapter 3 is meaningless without acknowledging that places are distinctive, while the 'space-shrinking' technologies described in Chapter 5 have brought places closer together without necessarily making them the same. Indeed, it is their differences that make them worth connecting! Technology merely serves as a mediating force that facilitates or enables these processes of spatial interactions.

Relatedly, Chapter 9 (on labour) will note how workplaces are distinctive places with complex internal processes and how a labour control regime is locally constituted in specific places. Similarly, Chapter 10 will focus on particular places of consumption such as high streets and shopping malls.

Part IV emphasizes the ways in which economic activities become embedded in the particular socio-cultural characteristics of places. We consider the ways in which firms are reflections of the places in which they originate and/or operate (Chapter 11), the importance of gender relations in shaping economic life (Chapter 12), and the formation of ethnic business clusters (Chapter 13). We will also describe the place-based formation of gender- or ethnicity-specific labour markets. Taken together, these social-cultural processes play a *mediating role* in shaping economic processes, as shown in Figure 1.6. Place is important here because it provides the site for these processes to play themselves out.

While the importance of place is a theme that recurs in almost every chapter of this book, an attention to *scale* is also explicit throughout – from the workplace and home-place through to the global economy. While the understanding of a single geographical scale may in some cases be pertinent, the approach we champion in this book is inherently *multi-scalar*. We emphasize the need to develop a scalar understanding of *each* economic process and its spatial outcomes, defined in Figure 1.6 as 'processes of interactions' at different spatial scales. Although it is the global reach of transnational corporations that forms the basis for Chapter 8, for example, we will also point to the importance of national regulations and local characteristics in shaping their operations. Chapter 7, meanwhile, which focuses on the state, will consider the way in which governmental functions are shifting from the national scale to local (e.g. city) and supra-national (e.g. European Union) scales. Sensitivity to multiple scales of analysis is also critical in Chapter 11, as the culture of economic practices can be interpreted as either a reflection of firm-by-firm differences in corporate culture, regional business cultures, or national business systems.

Finally, *representation* is a key issue in understanding how we diagnose and 'treat' economic problems, and how we label spaces, places and scales in certain ways. In Chapter 4, the ways in which we represent nature affect the ways in which it becomes commodified and part of economic processes. A conception of nature as separate from society often leads to its inappropriate and unsustainable exploitation. In Chapter 12, we will explore how domestic (i.e. home) spaces are represented as feminine, with profound implications for the kinds of employment men and women find in the waged labour market. And in Chapter 13, we see how the representation of ethnic neighbourhoods, and even national spaces, can have implications for the economic fortunes of ethnic minority immigrant groups in major world cities.

But representation is also important in an even more fundamental way. As we noted with reference to Niger, it is often assumed in discussions about poverty and development that the economy is the fundamental 'thing' that we are studying. Livelihoods are, however, usually created from resources both inside and outside of the economy that is formally acknowledged and measured. What exactly the economy is, then, remains far from clear. While we have established in this chapter that there are multiple geographies of the economy to be studied, we will turn in our next chapter to explore how 'the economy' itself is a far from unproblematic category.

Further reading

- For a classic geographical analysis of poverty in Africa, see Watts (1983).
- Bryson et al. (1999) offer a good collection of important articles in economic geography.
- Barnes et al. (2003) bring together the best collection of research contributions by economic geographers with a very helpful introduction at the beginning of each section.

- Hudson (2004) provides a challenging geographical conceptualization of the spaces, flows, and circuits of economic activities.
- Lee and Wills (1997) offer the first major statement on the 'new economic geography' and herald the rise of the 'cultural turn' in economic geography.
- Sheppard and Barnes (2000) bring together student-friendly and yet state-of-the-art contributions by leading economic geographers on a wide range of important themes and issues.
- Several textbooks in industrial or economic geography are currently available. Many of them are mostly devoted to locational analysis and/or *industrial* production, e.g. Dicken and Lloyd (1990), Healey and Ilbery (1990), Hayter (1997), and Hudson (2000).

Sample essay questions

- What are the distinctive elements of an economic-geographical approach to the economy?
- How do economic geographers and economists approach similar economic phenomenon in different ways?
- What are the problems with the underlying assumptions in most economic models of the real world?

Resources for further learning

- http://www.irinnews.org: a valuable archive of material concerning humanitarian issues in Niger is provided by the UN Office for the Coordination of Humanitarian Affairs.
- http://www.econgeog.org.uk/: the website of the UK's Economic Geography Research Group provides a window on current economic geography activity in the UK context.
- http://www.geography.uconn.edu/aag-econ/: does likewise for the US Economic Geography Specialty Group.
- http://www.intute.ac.uk/socialsciences/cgi-bin/browse.pl?id=120497: this gateway provides searchable access to many websites relating to economic geography.

References

Barnes, T.J., Peck, J., Sheppard, E. and Tickell, A. (eds) (2003) *Reading Economic Geography*, Oxford: Blackwell.

Bryson, J., Henry, N., Keeble, D. and Martin, R. (eds) (1999) *The Economic Geography Reader: Producing and Consuming Global Capitalism*, Chichester: John Wiley.

Castree, N., Coe, N.M., Ward, K. and Samers, M. (2004) *Spaces of Work: Global Capitalism and Geographies of Labour*, London: Sage.

Dicken, P. (2004) Geographers and 'globalization': (yet) another missed boat?, *Transactions of the Institute of British Geographers*, 29(1): 5–26.

Dicken, P. and Lloyd, P.E. (1990) *Location in Space: Theoretical Perspectives in Economic Geography*, 3rd edn, New York: Harper & Row.

Harvey, D. (1973) *Social Justice and the City*, London: Arnold.

Hayter, R. (1997) *The Dynamics of Industrial Location: The Factory, the Firm and the Production System*, Chichester: John Wiley.

Healey, M.J. and Ilbery, B.W. (1990) *Location and Change: Perspectives on Economic Geography*, Oxford: Oxford University Press.

Hudson, R. (2000) *Production, Place and Environment: Changing Perspectives in Economic Geography*, Harlow: Prentice Hall.

Hudson, R. (2004) *Economic Geographies*, London: Sage.

International Monetary Fund (2005) *Niger: Poverty Reduction Strategy Paper Progress Report*, IMF Country Report Paper No.05/101, Washington, DC: International Monetary Fund.

Lee, R. and Wills, J. (eds) (1997) *Geographies of Economies*, London: Arnold.

Malkiel, B. (1985) *A Random Walk Down Wall Street*, 4th edn, New York: W.W. Norton.

Sachs, J.D. (2005a) The end of the world as we know it, *The Guardian*, 5 April.

Sachs, J.D. (2005b) *The End of Poverty: Economic Possibilities for Our Time*, New York: Penguin Press.

Sachs, J.D. (2005c) How to end poverty, *TIME Magazine*, 14 March, pp. 42–54 (extracts from *The End of Poverty*).

Scott, A.J. (2000) Economic geography: the great half-century, in G.L. Clark, M.A. Feldman and M.S. Gertler (eds) *The Oxford Handbook of Economic Geography*, Oxford: Oxford University Press, pp. 18–44.

Sheppard, E. and Barnes, T.J. (eds) (2000) *A Companion to Economic Geography*, Oxford: Blackwell.

Storper, M. (1997) *The Regional World: Territorial Development in a Global Economy*, New York: Guilford Press.

Thrift, N. (1994) On the social and cultural determinants of international financial centres: the case of the City of London, in S. Corbridge, R. Martin and N. Thrift (eds) *Money, Power and Space*, Oxford: Basil Blackwell, pp. 327–55.

Watts, M. (1983) *Silent Violence*, Berkeley, CA: University of California Press.

CHAPTER 2

ECONOMIC DISCOURSE

Does 'the economy' really exist?

Aims

- To show how 'the economy' is an historically and socially constructed idea
- To examine how a broader understanding of the economy blurs the distinction between economic and non-economic processes
- To understand how economic processes are represented in powerful ways.

2.1 Introduction

I got a job working construction, for the Johnstown company
But lately there ain't been much work, on account of the economy.
(Bruce Springsteen, The River, 1980)

Tonight, with a healthy, growing economy, with more Americans going back to work, with our nation an active force for good in the world – the state of our union is confident and strong.
(George W. Bush, State of the Union Address, 2005)

Bruce Springsteen and George W. Bush have about as little in common as any two wealthy, middle-aged, white men in America possibly could. There is, however, a subtext to the two quotes above which suggests that they share, along with most other people, a particular understanding of the power of 'the economy'. In Springsteen's ballad about growing up in a working-class town in America's rust belt, the economy is something outside of daily life, a force that has brought recession and despair to a locality and its people. Bush makes the same links between the economy and daily life, but also adds a set of adjectives about the economy – such as 'healthy', 'growing' and 'strong' – that are quite revealing. Each one implies that the economy is a body that can become stronger or weaker over time, can grow or shrink, and can be healthy or sick. In a sense, the economy is an 'entity' – perhaps organic, perhaps mechanical – but clearly something separate, distinct and identifiable. The sense that 'the economy' is 'out there' is common to both Bush and Springsteen, and to almost all other academic, political or popular discussion that we hear. What does it mean to understand the economy in this way? Why should we separate out particular aspects of our lives and label them 'economic'? And in doing so, what are we excluding? These are some of the questions that we will examine in this chapter. Therefore, just as Chapter 1 sought to unpack and explain what we mean by geography, this chapter examines what is implied by the economy.

The chapter starts by considering the conventional ways in which we understand the economy (Section 2.2). In particular, we note the assumptions that are commonly made about what the economy is, what it includes and excludes, and how it works. These assumptions lead to an understanding of the economy and economic processes that is quite partial and incomplete. In the next section (2.3) we show, however, that the way in which the economy is now understood is a surprisingly recent development. Even in the early twentieth century, just a few generations ago, the notion of an economic sphere was used quite differently. We will see that it was during a few crucial years in the mid-century that the economy as we know it today became embedded in the academic and popular imagination. Section 2.4 then examines what is excluded or ignored in conventional notions of the economy. This forms the basis for a broader interpretation of 'the economic' to include a variety of cultural, political and social processes, many of which cannot be given a monetary value and are conventionally ignored by mainstream economic analysis. The implication, in short, is that we need to have a less 'economic' understanding of the economy. This idea will underpin much of this book, as we explore the geographies of economic life as it is actually lived, rather than how it *should* be lived 'in theory'.

Underpinning this chapter is the idea that understanding the economy in the conventional way is a representation that is rooted in relationships of power. Power is needed to construct an imagination of the world in a particular way, and that imagination in turn perpetuates an existing set of power relations. In the final section of the chapter (Section 2.5) we illustrate this rather abstract idea through several examples of how economic processes of development and globalization are represented in the service of the powerful.

2.2 The Taken-for-granted Economy

It is a staple of our daily media diet to be informed about the state of the economy at various geographical scales – the local economy, the national

Figure 2.1 The economy as an organic entity
Source: http://www.inkcinct.com.au/, reproduced with permission.

economy, or the global economy. Political leaders describe their plans for the economy and take credit for making it grow. In 1992, Bill Clinton was elected President of the United States with the campaign slogan 'It's the Economy, Stupid!' – playing on a widespread perception that the (first) Bush administration had not competently managed the US economy and was failing to recognize the hardships faced by so many. But even beyond the rhetoric of electoral politics, 'the economy' is a notion that we constantly face as we digest the daily news. Major events – terrorist attacks, natural disasters, political tensions, and so on – are assessed in relation to their impact upon 'the economy'. A country's economy is assessed for signs of health or weakness, almost like a patient being diagnosed. Figure 2.1 illustrates this idea – an obese 'Mr Economy' is on the weighing scales and looks likely to push up interest rates due to his increasing size. Closely related, although a somewhat narrower concept, is the idea of 'the market' – most often used in relation to financial exchanges for bonds, shares, loans or foreign currency. The reaction of 'the market' to the news of the day, the prevailing sentiment or mood of 'the market', and the direction in which 'the market' is heading are all so commonly invoked in news reports that we barely stop to think about what the market, and its broader context, 'the economy', actually mean.

The collective lack of questioning that exists concerning 'the economy' perhaps reflects an assumption that it is such a solid and tangible 'thing' that its existence could hardly be in doubt. This sense of a tangible 'thing' out there is reinforced by the various ways in which the economy is defined and measured. Most commonly, this takes the form of gross domestic product (GDP), which adds together the total market value of production in a particular economy. The usual way of calculating this total is to combine recorded levels of expenditure of various kinds. Figure 2.2 summarizes the components of GDP. GDP is most often calculated for national economies, but figures are also sometimes quoted

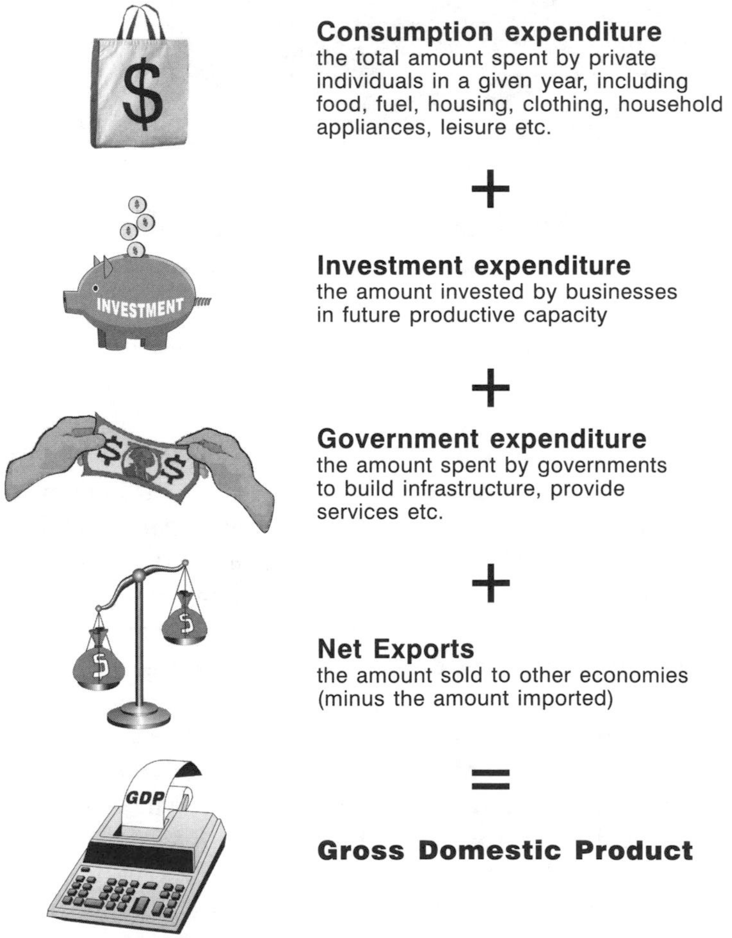

Consumption expenditure
the total amount spent by private individuals in a given year, including food, fuel, housing, clothing, household appliances, leisure etc.
+
INVESTMENT
Investment expenditure
the amount invested by businesses in future productive capacity
+
Government expenditure
the amount spent by governments to build infrastructure, provide services etc.
+
Net Exports
the amount sold to other economies (minus the amount imported)
=
GDP
Gross Domestic Product

Figure 2.2 The components of gross domestic product

for sub-national units, such as states, provinces or regions, and also for supra-national entities like the European Union.

Together, then, the components of GDP quantify all forms of production occurring within an economy in terms of the monetary value they are creating. This would appear to provide a degree of certainty about what constitutes the economy – it clearly consists of definable and real processes that can be seen, experienced, measured and perhaps even managed. But what does this semblance of certainty take for granted? What is left assumed and unexamined in this common-sense view of the economy?

In the first place we assume that we can unproblematically define what constitutes an economic 'act' or process, and what does not. This can lead to some odd conclusions. For example, if you drive or take a bus to your campus or to a place of work, then you have engaged in an economic act; but if you walk or ride a bicycle, then you have not. In one case, fuel bills and parking costs, or the price of a public transit ticket, have changed hands, but in the latter case the work of walking or cycling has been done outside of the formal monetary economy. Thus, in our accepted ways of understanding the economy, an economic act has not taken place. The conceptual boundaries we tend to place on the economy are therefore rather arbitrary. The 'economic', as it is usually understood, does not relate to the nature or purpose of the act being undertaken, but whether or not money changes hands. As we will see later in this book, the things that are excluded from the measurable economy often follow a pattern. In particular, unpaid work that is commonly done by women in the home is excluded, while work done for wages is counted in full. At the outset then, the economy as it is commonly understood could be viewed as a gendered concept (we will elaborate on this idea in Chapter 12).

The second assumption regarding the economy (one that will be explored further in Section 2.3) is that it lies outside of other dimensions of our lives. In this way, a lot is excluded: cultural processes in which we assign meanings to things and people; social relations in which we relate to family, friends, co-workers or neighbours; political processes in which we allocate and challenge power; or, environmental processes operating in the natural world. All of them are treated as having their own and quite separate operating logics. They are seen as connected with the economy, but only as external 'plug-ins' rather than as fundamental internal mechanisms. Geographers have increasingly questioned the separation of the economic from other dimensions of real life. In fact, a defining characteristic of recent approaches in economic geography, and much of what we discuss in this book, is a blurring of the boundaries between these various dimensions of lived experience and the economic processes that they intersect with.

A third taken-for-granted assumption is that the economy is some kind of collective mechanism outside of our individual control. This notion clearly underpins the words of Bruce Springsteen and George W. Bush in the quotes above. We know we are part of the economy, as workers or as consumers, but it is treated as an organism or a machine that is much bigger than any of us. Just as the cells in our bodies cannot *individually* affect, or even understand, the functioning of our bodies as whole, likewise individual economic actors cannot fully understand or affect the economy. Indeed, this metaphor, depicting the economy as some kind of organism, is quite widespread in popular understandings (hence the notion of a 'healthy', 'sick', or 'growing' economy). The implication, of course, is that while we are components of the economy, it is bigger than all of us. This is an important point, as it can easily lead to the assumption that the economy is outside of any kind of collective control – that is, it is something that controls *us* rather than the other way around. A classic *Time* Magazine cover, used several times from the 1950s to the 1980s illustrates this sense of uncontrollability, with 'the market' depicted as a rampaging bull – interestingly, however, in this case 'the market' appears to be divorced from 'the economy' (Figure 2.3). In reality, economic processes are the product of collective decisions that we make as economic actors (whether employees, employers or consumers), and which our political leaders make as economic managers – there is nothing mysterious or uncontrollable about them.

The fourth assumption is that the mechanisms keeping the economy functioning are ordered and rational. Individual producers and consumers are thought to behave in ways that maximize their satisfaction, wealth or enjoyment. Thus the individual is taken to be economically rational – behaving like the *homo economicus* described in Chapter 1. By aggregating all of these actions together it is assumed that we can identify the fundamental 'laws' that make the whole thing work. The most basic of these is the 'law' of supply and demand, suggesting that the market mechanism will naturally establish an equilibrium between buyers and sellers. It is assumed that, underlying economic processes, there is a set of structured logics that determine the direction that the economy will take. This is not necessarily to imply that every economic act is predictable, but it is to suggest that when aggregated together they will follow certain patterns. Importantly, it is also assumed that these processes will operate in the same way in all geographical contexts – that there will not be spatial variation in the fundamental 'laws' of economic life.

The fifth taken-for-granted assumption is that growth is good. A growing economy is considered 'healthy', 'robust', 'thriving'. Indeed, a growing economy is considered a necessity – new people, new resources, new efficiencies, new products, new wealth are all constantly needed. It would be unthinkable in most countries for a politician to mount an election campaign based on the idea that the economy is big enough or efficient enough already. Biological metaphors, of course, conveniently allow us to understand this kind of growth as an organic and natural process. But the separation of a 'healthy' economy from its environmental, cultural or social implications means that we often fail to take into account the full implications of a growing economy – not all of which are

Figure 2.3 The uncontrollable market as a raging bull
Source: *Time* Magazine © 1982 Time Inc. Reprinted by permission.

unambiguously healthy. If more people use their cars rather than walking, then the economy has grown, but the environmental, health and social consequences are far from positive. Even more bizarrely, if more people get involved in car accidents, resulting in repair bills, medical expenses and so on, then again the economy registers growth, even if the consequences are entirely negative.

Clearly, then, the taken-for-granted economy implies or assumes certain characteristics of economic life. As a result, a great deal is excluded from what we conventionally count as 'the economy'. In addition, the factors that are taken to be important in explaining economic behaviour are also framed very narrowly. The idea that economic processes are different from cultural or social processes, for example, immediately sets up a mental barrier to thinking about economic actions as simultaneously cultural or social. And yet, the act of selecting food, clothing, or any other product in a shop is clearly a *cultural* process, just as the working with other people in a place of employment is clearly a *social* performance. In short, the conventional understanding of the economy starts to unravel when we expose it to the realities of everyday 'economic' life.

In the rest of this chapter we will pursue two lines of argument that add substance to the idea that taken-for-granted notions of the economy need to be taken with a degree of scepticism. In the next section we will examine where and when, historically, our concept of the economy originates – thereby demonstrating that it is, in fact, a relatively recent idea. After that, we will explore some of the ways in which other dimensions of life are arbitrarily excluded when the economy is defined in narrow terms.

2.3 A Brief History of 'the Economy'

In the English-speaking world of the early eighteenth century, the word 'economy' would have referred to the management of a household. In the same way that today we use the word 'economize' to relate primarily to personal expenditures, two or three hundred years ago 'economy' was the practice of managing a family budget. In some contexts we still use the word in this sense. If we drive an 'economy'-sized car, or buy an 'economy' class air ticket, the implication in each case is that they are smaller and more frugal options.

The eighteenth century, however, was a period of rapid change, and its latter decades in particular marked the beginnings of the Industrial Revolution in England. Before then, when the production and use of goods and services were on a very small scale, there was little need to think about a set of economic processes that exceeded the scale of the family or household. In pre-industrial societies, most agricultural or craft production was for subsistence purposes, perhaps with some surplus paid to a feudal aristocrat or exchanged in local markets. The Industrial Revolution, however, saw Western Europe undergo significant changes in the nature of production and the social relations that surrounded it. In particular, larger-scale agricultural production emerged along with factory-based manufacturing. With production on a grander scale, and increasing levels of specialization, it was possible to think of a *division of labour* for the first time – in which different people carried out different tasks and were thereby dependent on each other for their needs (Buck-Morss, 1995). This had always been true in some senses – each village would have probably its own carpenter, blacksmith, etc., but the birth of the modern industrial era saw a greater degree of interdependence and specialization, and on a much larger scale.

It was this division of labour that Adam Smith famously identified in his book *The Wealth of Nations*, published in 1776. Using the example of a factory making pins, Smith showed that a group of people specializing in different stages of the manufacturing process could make pins far more quickly and efficiently than an individual craftsman carrying out every stage of the process himself. Smith's insights were, however, much broader than this. In his writing there was, for the first time, a sense that 'economy' concerned something larger than the management of a household. It represented an integrated whole at the scale of a nation – a whole with many individual parts that worked together, albeit unknowingly, to create greater wealth for all. This interdependence was summarized in Smith's metaphor of the 'invisible hand'. By seeking only their own enrichment and advancement, individuals unconsciously benefited society as a whole in the process. In Smith's words, each individual is 'led by an invisible hand to promote an end which was no part of his intention' (1976: 477). Smith's ideas have since been used to justify an ideology in which the market mechanism is seen as universally beneficial and the most appropriate way of organizing society in all contexts. This is a rather partial reading of his perspective. At a broader level, Smith was perceptively identifying the interdependence that was being created by an emerging modern industrial society. By seeing 'economy' as fostering the wealth of nations rather than the management of households, he was pointing to the development of an integrated 'whole' in which people were participating as economic actors.

The analysis of national production and consumption, which Adam Smith developed, was known as *political economy*. It was essentially concerned with the management of resources at a national scale. In Britain, institutions such as the Bank of England and the London Stock Exchange were emerging at about the same time, and these provided an organizational basis for understanding economic processes on a larger scale. Economic processes were, then, by the early 1800s, understood as matters of national significance and not simply practices of household management. This was a significant step, but it is still some way from the notion of 'the economy' as we use the concept today, or as we saw it used in the opening quotes from George W. Bush and Bruce Springsteen.

To find the beginnings of 'the economy' being understood as a separate and distinct entity, we have to move forward to the end of the nineteenth century. Until that time, it was political economy in the tradition of Smith, and later the more radical version developed by Karl Marx (writing in the mid-nineteenth century), that generated understandings of collective production, consumption, trade and wealth in society. The late nineteenth century was, however, a period of profound change – scientifically, technologically and intellectually. European societies were expanding into colonial territories, large industrial cities were growing rapidly, and great strides were being made in fields such as electrical engineering, medicine, and the natural sciences. It was also during this period that modern economics, as a profession and an academic discipline, was being established (Mitchell, 1998).

From about the 1870s onwards, economists started to think less in terms of political economy – that is, the marshalling and management of national wealth – and more in terms of *individual* economic decisions. These could, in turn, be aggregated together in order to develop a model of how the economy as whole would work as a system. It was no coincidence that such thinking emerged at the same time as physics and chemistry were developing models of the natural world. Economists adopted very similar approaches and even used terminology that was borrowed from the physical sciences. Underlying the new field of physics was an understanding of energy as the unifying force that connected all matter and processes in the universe together. Economists found their equivalent in the concept of 'utility' – the notion that anything 'economic' could be measured in terms of its value or its usefulness. Like energy in physics, utility represented a common element in all economic transactions and rendered them comparable and quantifiable. With this basis for a unified understanding of economic processes, individual producers and consumers could be viewed as analogous to the predictable behaviour of atoms and molecules, and the economic processes they created could be likened to the forces and dynamics studied by physicists. Indeed, as noted earlier, the concepts that economists used to understand economic behaviour drew directly on physical processes – for example, equilibrium, stability, elasticity, inflation, and friction (Mitchell, 1998). These kinds of metaphors are still very much in use by economists (and just about everyone else) today, as noted in Box 2.1.

Box 2.1 Metaphors of economy

Metaphors are required any time we need to *reduce* something that is complicated and difficult to *grasp* into a more conceptually manageable *picture*. The fact that the previous sentence contains three (*italicized*) metaphors to make its point illustrates how often we have to resort to them. But these are relatively minor metaphors. Economics employs much bigger conceptual metaphors in trying to make the economy comprehensible. Conventionally, these have involved drawing upon the scientific language and models of physics and biology to represent economic processes.

Sometimes this draws upon the language of Newtonian physics to conceptualize economic processes as interactions of objects, flows, forces and waves. Some everyday examples illustrate this point: the market provides a *mechanism* for bringing together buyers and sellers and for allocating resources; the location of raw materials exerts a *gravitational* attraction upon producers; and, distance provides a *frictional* force preventing consumers from shopping in far-off places. Participants in the economy are therefore reduced to rational economic beings, obeying the 'laws' of economics, just as atoms obey the laws of physics. More complex understandings of the economy may also resort to physical metaphors, for example, when we talk about economic *cycles*, and *waves* of investment.

An even more *fertile* source of metaphors is found in biological processes. To talk of the *health* and *growth* of an economy is to represent it in terms of an organism or body. This metaphor is also found in ideas such as the *circulation*, *reproduction*, *herd instincts*, and the *contagion* of economic crises. Larger logics are also represented through biological metaphors, for example, when the market is seen as providing a process of *natural selection* in which only the *fittest* and *strongest* (by implication, the most efficient) will survive, and when crises are understood to *weed out* the least efficient producers. *Development* too is essentially a biological metaphor for understanding growth and change as a natural process, as in the maturing of a human body – we discuss this idea further in Section 2.4.

Clearly we cannot do away with metaphors. We will always need them to facilitate our thinking. But we do need to be aware of the implications of understanding economic processes by resorting to these images, as they will inevitably only partially 'fit' the context in which they are being applied. Thus, while in some circumstances it will be useful to think about economic actors as law-abiding 'atoms', it is easy to see that this will exclude much that is interesting in the economic world. More broadly, by transposing the 'laws of nature' onto the social world, we are closing off many possibilities of alternative thinking and alternative economic arrangements.

For these metaphors, drawn from the natural sciences, to be meaningful, it was necessary to make some quite significant assumptions about the behaviour of human economic actors. These assumptions are still largely applied in economics and were described in Chapter 1. Essentially, people had to be assumed to behave in a similarly rational and consistent manner in all circumstances. This rationality was based on maximizing their own economic rewards (or utility) – rewards that could be quantified and counted. Equipped with such assumptions, the early economists of the late nineteenth century could start to predict how different stimuli would affect economic behaviour. There began to emerge an understanding of the economy as not simply the management of national wealth, but as a system of inputs, outputs and decisions that occupied a realm of rationality and predictability – entirely separate from issues of government, culture and society.

Perhaps the most graphic illustration of this new perspective on the economy was provided by the American economist Irving Fisher. In his doctoral dissertation, completed in 1892, Fisher designed a mechanical model of an economy with an intricate system of water tanks, levers, pipes, pivots and stoppers. The following year, Fisher actually built the model and used it during his lectures at Yale University for several decades. By adjusting various flows and water levels, Fisher claimed that he had developed a predictive model of an economy which could be used to experiment with, and predict, the effects of various changes in market circumstances (Brainard and Scarf, 2005). Figure 2.4 provides a schematic diagram of the apparatus used by Fisher, and Figure 2.5 shows a more sophisticated apparatus that he built in 1925, to replace the first model.

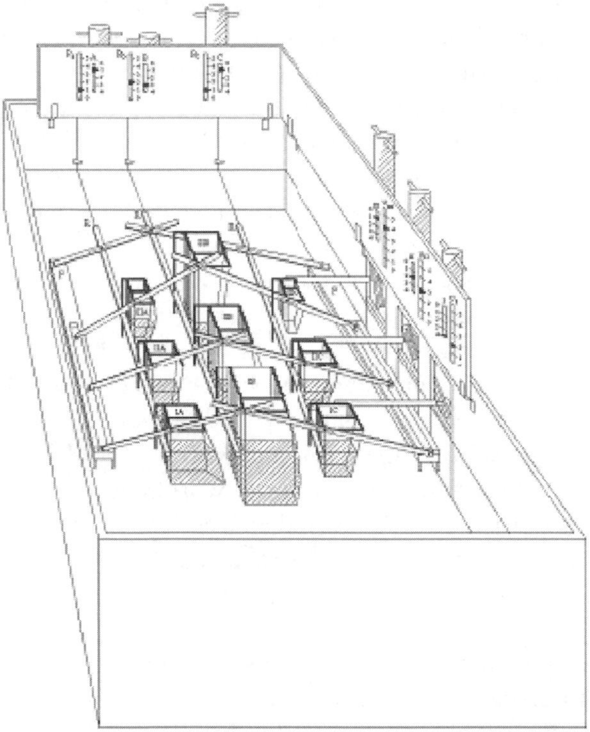

Figure 2.4 Irving Fisher's 'Economy' apparatus, as used in his lectures
Source: Brainard and Scarf (2005), Figure 5. With permission of Blackwell Publishing.

Figure 2.5 Irving Fisher's lecture hall apparatus, simulating the economy, c.1925
Source: Fisher (1925; ii).

While Fisher continued to experiment with mechanical analogies of the economy, others were developing mathematical models to study the economy as a coherent and logical entity. The 1930s, in particular, saw the emergence of *econometrics* as field of study in which complex mathematical techniques were used to capture economic processes. These models were important not just because of their sophistication and the appearance of mathematical scientific rigour that they gave to economics, but also because they enabled economists to go beyond the notion of the economy as a set of mechanical levers, tanks and pulleys. Models such as those constructed by Fisher could generate ever more complicated depictions of economic processes in the economy, but they were essentially static. They could not handle any kind of expansion in the economy as a whole, nor any kind of external change. What would happen if the water in Fisher's apparatus started to evaporate, or started to leak? Or if the tanks changed in size and number over time? Or if some kind of external shock disrupted the apparatus? Such a model could not accommodate these kinds of changes, but the mathematical versions being developed by the 1930s could analyse such dynamism. Growth and change in the economy *as a whole*, driven by both internal and external forces, could be the subject of analysis for the first time.

While econometrics delivered the analytical tools to study the economy as whole, the notion of national economic management was emerging at the same time. In his *General Theory of Employment, Interest and Money*, published in 1936, the British economist John Maynard Keynes established the idea that a national economy was a bounded and self-contained entity that could be managed using particular policy tools. These tools included controls over interest rates, price levels, and consumer demand. While Keynes developed his ideas before the Second World War, in a context of depression and unemployment, it wasn't until after the war that his ideas really took hold and governments started to engage in the kinds of economic management that he had advocated. For almost three decades from the 1940s until the early 1970s, Keynesianism was the orthodoxy of economic management in the industrialized, non-communist world.

It was, therefore, only by the 1940s, that 'the economy' as we know it today had emerged as a popular concept. The idea by that time implied several important features about economic processes, which correspond closely to the assumptions outlined in the previous section of this chapter:

1 The economy came to be seen as an *external sphere*, separate from the rest of our lives. Metaphors based on physical or biological processes greatly assisted in this conceptualization. When the economy is imagined as a machine or as an organism, it can more easily be seen as an external force bearing down upon us. In this way, it was possible to think about someone, or some place, being *affected by* the economy. Bruce Springsteen's lyrics,

which would have been unthinkable when 'economy' referred to household management or the stewardship of national wealth, could therefore only have been written in the second half of the twentieth century. Understanding *the* economy as a machine or organism also enables us to think of its health and robustness in the manner of George W. Bush's speech – again, an idea that would have seemed quite odd a century ago. The addition of the word 'the' to the word 'economy' is an important change that indicates this new meaning – while before 'economy' was an activity or an attitude, it is now a 'thing'.

2 Because of the notion that the economic sphere operates according to its own internal logics, it has become seen as *independent* of social, political and cultural processes. While political systems, ideologies and parties might vary across time and space, and while cultures and social practices might also display rich geographical variation, our understanding of the economic realm sees it as operating according to a separate and universal logic all of its own. While people may vary in their political, social and cultural arrangements and attitudes, the universality of economic rationality is seen as unchanging.
3 The economy, as it emerged in the 1940s, was primarily a concept that focused on the *national* scale. It was the national economy that was analysed and modelled using the new techniques and approaches developed in the 1930s. It was also in the 1940s that measurements of national economic well-being, such as Gross Domestic Product, were first developed. Imagining the economy as constituted at a national scale also allowed the global economy to be organized as an inter-*national* order. The institutions established in the 1940s, including the International Monetary Fund and the World Bank, were explicitly coalitions of national governments coming together in order to coordinate their economies.
4 Finally, the idea of a national economy as a machine that could be maintained and managed by government intervention introduced the notion that continual economic growth was not only possible, but it was the responsibility of *national governments* to deliver it. George W. Bush's heralding of a 'healthy, growing economy' echoes what politicians the world over are expected to deliver, but it is a relatively novel idea that is a product of the second half of the twentieth century.

We have seen, then, that much of what we take for granted in our notion of the economy is actually a set of ideas that are relatively recent. Out of the Industrial Revolution in Europe there emerged a sense of a national division of labour – a set of economic relations that constituted a sum of many parts. But it was not until the late nineteenth century that the idea of an economy as a regulated organism or mechanism started to emerge. It was as recently as the 1940s that this idea was widely adopted and started to become part of political rhetoric and popular understanding.

By examining the historical emergence of the idea of the economy in this way, we have also implicitly made an important point: *there is nothing natural or fundamental about the way in which we understand, measure and manage the economy*. Instead, current ways of thinking are a product of particular historical circumstances in particular places. Establishing this point liberates our thinking in a number of ways. First, it allows us to consider the aspects of life that are excluded from the economy and from economic analysis. Second, it allows us to think about how particular understandings of the economy, and of economic processes, might be more than just intellectual exercises. Rather, they might actually reflect and reinforce the interests of certain groups in society. In the next two sections of this chapter, we examine these possibilities.

2.4 Expanding the Economy beyond the Economic

We have seen that the economy was being re-imagined in the mid-twentieth century as a separate and identifiable entity. In the process, what counted as part of the economy, and what did not, was also being redefined. In the first instance, it was whether something *could* be counted that determined whether it *would* be counted. The rise of econometrics as the 'science' of economic processes meant that a phenomenon had to be quantifiable in order to be included in models of the economy. Thus, if an activity took place outside of the cash economy, or if a cash value could not readily be assigned to it, then it was not defined as an economic activity.

A careful examination of 'economic' processes reveals, however, that a great many take place outside of the formal, quantifiable economy. These activities have been likened to the submerged portion of an economic iceberg (Figure 2.6). Production for a market, using cash transactions, and waged labour in a capitalist firm, captures only a small part of how people actually produce and exchange resources of various kinds in their daily lives. In families, in cooperatives, in neighbourhoods, and in many other settings, people work, share and exchange in ways that are seldom recognized as formal economic processes. The effect of this neglect is not just to *mis*count the economy, but also to *dis*count the work of certain people. We noted earlier the way in which this can systematically exclude unpaid domestic work done primarily by women.

The mainstream view of the economic also has implications for groups with quite different economic cultures. For example, indigenous people in Canada, the United States, Australia and elsewhere often hold different views of property rights (Castree, 2004). While the modern Western view is that property can be

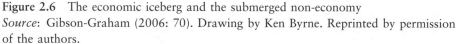

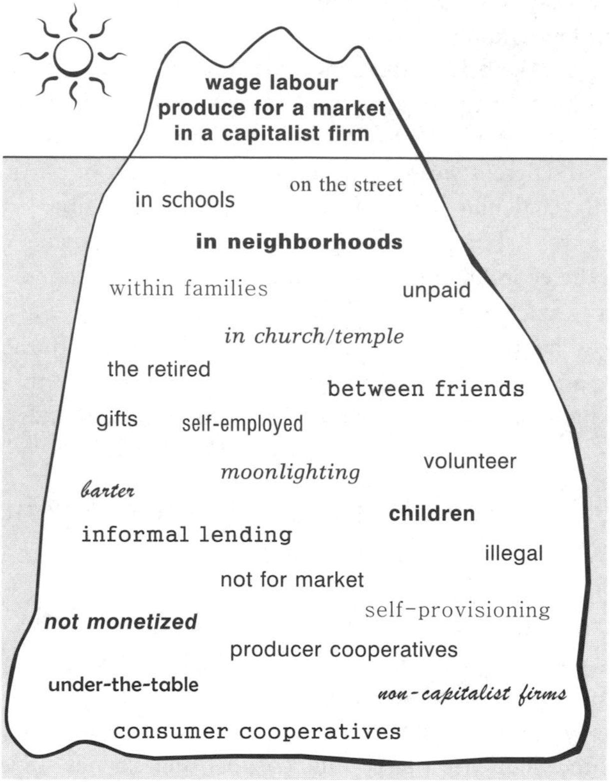

Figure 2.6 The economic iceberg and the submerged non-economy
Source: Gibson-Graham (2006: 70). Drawing by Ken Byrne. Reprinted by permission of the authors.

owned, bought and sold by individuals, in many indigenous cultures, individual ownership of this sort is an alien concept. Instead, the right to use the resources of the land or the sea are collective and no individual has the right to sell them – and yet indigenous groups have historically been pressured to do just that. Countless disputes over drilling rights, fishing licences and logging concessions around the world hinge, in part, on this difference, which Western companies have often done little to recognize.

We have also seen that beyond the need for quantification based on monetary exchange and property ownership, conventional approaches to the economy treat all other forms of human activity – such as social, cultural and political practices – as belonging to separate spheres, operating under different logics that are incommensurate with economic analysis. These other spheres might be considered as influences on the economy, or they might be seen as affected by the economy, but the idea that they are entirely separate still remains. As many of the chapters in this book will show, however, economic processes are very much embedded in other forms of human (or environmental) interaction.

Nature, for example, is usually assumed to be outside of economic processes. Most everyday definitions of nature would present it as something unaffected by human interference. As we will suggest in Chapter 6, however, nature is very much bound up with economic processes. Not only does the natural environment experience the pollutants and side-effects of economic growth, but even 'untouched' nature is given a market value – for example through the tourism industry. Furthermore, natural resources (oil, gold, water, rubber, etc.) are themselves commodified and used as inputs in economic processes. Indeed, the economy is itself, in some ways, the process of transforming nature from one form to another – from rubber plantations to car tyres, or from buried diamonds into engagement rings. At every turn, then, the economy is also an environmental process – as input, as process, and as impact.

To take another example, labour markets are often assumed to match appropriate skills with appropriate jobs in a dispassionate and objective manner. As we will show in more detail in Chapters 12 and 13, however, gender and ethnicity often have profound effects on how individuals experience waged employment. Some jobs become associated with femininity and others with masculinity. In addition, some workplaces become highly masculine or feminine environments. Similarly, many urban labour markets show very clearly that processes of ethnic 'sorting' are happening, with various jobs being disproportionately held by certain groups of people. Gender and ethnicity are not, then, 'non-economic' factors. Instead, they are forms of identity through which economic resources are allocated. To put it the other way around, economic processes, such as labour markets, are one of the venues in which gender and ethnic identities are played out.

The sphere of the state is also very difficult to divorce from the economy. As we will show in Chapter 7, governments all over the world are intimately involved in the economic activities that occur within their borders. Indeed it is quite contradictory to depict the economy as a separate and distinct sphere, on the one hand, and yet to expect governments to manage and grow 'their' economies, on the other. In reality, the economic and the political are inseparable. What we refer to as 'the economy' could not exist without the institutions of the state to establish all the pre-conditions that are required for large-scale economic interdependency – education, healthcare, transportation, currency, and the law, to name just a few. And, of course, the state itself is an entity that is defined in large part by its powers over economic activities and the territorial extent of its economy.

Even businesses and consumers themselves, engaged in the 'pure' economic acts of buying, producing and selling are often guided by factors that fall outside the bounds of economic rationality, as we will show in Chapters 10 and 11. Consumers have a very subjective judgement of what a particular product or service is worth *to them*, and looking closely at these preferences soon leads us away from motivations based on maximizing utility or profit. Instead, many consumption decisions are based, in part, on the symbolic or cultural value that is embodied in a product, or perhaps (as we will suggest in Chapter 4) its ethical acceptability. The brand and model of motor vehicle that a person might buy, for example, could be a complex combination of quantitative calculations based on reliability, resale value, fuel efficiency and utility, alongside cultural concerns about what that vehicle 'says' about its owner, environmental concerns about emissions, or social concerns about the labour that produced it. Even the coins and notes that we use for everyday economic transactions, objects that we might expect to be 'economic' in the purest sense, have a strong cultural dimension. The contentious debates in European countries over joining a single currency (the Euro), and elsewhere over 'dollarization', illustrate the point that money itself is a cultural icon (Gilbert, 2005). The converse is also true, of course: places or objects that might be considered 'cultural' – such as Disneyland, a Parisian bistro, or a music download – are in fact economic enterprises.

In a variety of ways, then, the distinctions between the economic, the cultural, the political, etc. are quite arbitrary – processes or objects that are conventionally assigned to one sphere have significant implications in another. Indeed, it is the separation of these spheres that is problematic. To understand 'economic' processes it is necessary to place them in the broader context of which they are a part. As we have shown in this section, this will be a project that underpins the rest of this book.

2.5 Representing Economic Processes

So far, we have seen how the emergence of a notion of 'the economy' reflected circumstances in a particular place and time, and how that idea has shaped the way we see the world in very selective ways. What we have done, essentially, is to see the economy as a 'discourse' – that is, a set of explanations and understandings that powerfully shape the way we think about, and act in, the world (Box 2.2). In developing a discursive analysis of the economy, we have not only shown that our taken-for-granted understanding is a product of a particular historical and geographical context, but also that it is rooted in relationships of power. Whether it is the power of Western economic interests over colonized indigenous peoples, or the patriarchal power of men over women, defining the economy in a particular way has reinforced these unequal relationships. In this section, we pursue this point further, by exploring more explicitly the ways in which discourses of the economy both reflect and reinforce power. We do so by turning our attention to two economic processes that have been defining features of the world economy since the current understanding of 'the economy' emerged in the 1940s: 'development', which dominated discussions of global

Box 2.2 What is discourse?

A discourse is concerned with the entire package of techniques that we employ to conceptualize 'things', to order things and make them comprehensible to ourselves and others. 'The Order of Things' is, in fact, the title of a key text by the French philosopher Michel Foucault who contributed a great deal to our understanding of how language affects our conception of the world and thereby affects how we act in the world. Foucault was especially interested in how people can be represented in certain ways (for example, as criminals, as insane, as abnormal) using a range of language, technologies, and institutions.

To analyse a discourse, then, is to consider how our thinking on a particular subject is being shaped by the accepted vocabulary used, the ways in which expertise in the subject is constructed, and the ways in which the analysis of a particular phenomenon is institutionalized (that is, how ideas get embedded in the institutions that organize our lives – governments, laws, customs, religions, academic disciplines, etc.). The institutionalization of discourses is important because it highlights the fact that discourses are not the conscious creations of a single author, but are collective understanding that must be constantly recreated (or 'performed') in order to continue. Not just any discourse, however, succeeds in becoming dominant. Discourses reflect, and recreate, configurations of power. They reflect them, as only the powerful can construct inclusions and exclusions, the normal and the abnormal, according to their own requirements; but they also recreate power, as a discourse powerfully 'constitutes' and naturalizes the objects that it describes.

economic interdependence from the 1950s to the 1980s; and 'globalization', which has become the dominant discourse since then. In each case we will briefly suggest how these concepts can be seen as discourses that are rooted in power relationships, and which, in turn, reinforce unequal power in the global system.

Development

The idea that economies can develop, much as a body develops through various stages of maturity, is a widespread way of understanding the problems and solutions faced by 'less developed' or 'underdeveloped' parts of the world. But, like the economy itself, it is also a relatively recent way of thinking about these areas. While 'development' has its origins in eighteenth- and nineteenth-century ideas of progress, it was only after the Second World War that societies themselves came to be labelled in relation to this notion. In a famous speech in 1949,

US President Harry Truman referred to 'undeveloped areas' of the world and the need to provide them with a 'fair deal'. This idea spread as most European and American colonies gained independence in the subsequent decades and their problems came to be seen as ones of underdevelopment rather than issues of colonial welfare. Organizations such as the International Bank for Reconstruction and Development (more commonly known as the World Bank) were established in the 1940s, thereby institutionalizing the issue of development. A field of expertise grew up around the issue that sought to identify policy solutions to poverty and underdevelopment. Development thus became a classic example of a discourse – a vocabulary, a set of institutions, and a field of technical expertise, were all established that imagined the world in a certain way, thus defining both problems and solutions (Escobar, 1994; Crush, 1995). Fostering development was turned into a technical exercise in economic planning.

Much, however, is taken for granted in this discourse. Most obviously, development discourse assumes that all economies can reach the state of being 'developed'. There is, therefore, little in orthodox development discourse to suggest that the developed world might actually *depend upon* the underdeveloped world for its comfortable standards of living (through processes of unfair trade, for example, as discussed in relation to Niger in Chapter 1). Development discourse thus obscures the interdependence of the global economy by reducing underdevelopment to a *national*-scale concern. A lack of development can then be presented as a product of poor management of the national economy. The implication is that if the correct policies were implemented, then a process of development would naturally follow. These policies have nearly always involved greater access for transnational corporations from the developed world to the resources, labour, or markets of 'underdeveloped countries'. Development discourse also assumes that Western capitalist modernity is the ultimate end-point to which all other societies should aspire. In other words, it is a Eurocentric model that celebrates one possible outcome of social change. Any deviations from that norm – for example, the alternative ideas about property rights noted earlier – are viewed as obstacles to development.

Development discourse is more than just a way of representing or conceptualizing things. It also has material impacts on people's lives. It has shaped the ways in which underdeveloped areas were 'treated' (and the medical metaphor here is quite appropriate). Solutions ranging from export processing zones to free trade agreements to infrastructure development for tourism, have all sought to pursue a particular concept of what progress and development should mean. But the discourse of development was also rooted in the geopolitics of the era in which it emerged. The Cold War of the 1950s and 1960s, which pitted the capitalist 'West' against the communist Soviet bloc, was an era in which the idea that economies could develop into wealthy capitalist societies was very useful. It isolated the problems of the Third World as internal policy deficiencies, and held out the West as a model of prosperity and stability to which such countries could realistically aspire. When the American economist Walt Rostow published his celebrated book *The Stages of Economic Growth* in 1960, which outlined the requirements for Third World countries to achieve developed status, it was pointedly subtitled *A Non-Communist Manifesto*. Development is therefore a classic example of an economic discourse that both reflects and reinforces existing global power relations.

Globalization

Although development is still very much on the agenda, we are also now constantly bombarded with the concept of globalization. It is usually taken to mean the deepening and widening of trade and investment flows around the world so that more and more places are becoming increasingly integrated into economic processes that operate beyond the scale of the locality, the region, or the nation. It is, in short, about the reformulation of the scale at which economic processes operate. Whether one sits at the bargaining table negotiating free trade agreements, or at the barricades protesting against them, there is general agreement that an identifiable process of globalization is occurring and indeed that it is inevitable. The well-known economist Lester Thurow has articulated this point of view clearly:

> The key thing about globalization is that it is not caused by a government's decision to participate. Globalization is led by businesses that are scanning the world for the best place to produce and sell their products. This process is going to continue, no matter what governments decide. If we just let globalization happen, the pluses will probably exceed the minuses. We can make the minuses smaller if we start to think about shaping globalization, rather than just letting it happen.
> (Thurow, 2004)

In Thurow's view, then, globalization is an unregulated, almost natural, process in which businesses integrate their operations across the planet. This is a view that is commonly reproduced in political rhetoric, media analysis, and academic studies.

We can, however, examine this discourse critically and try to understand whether globalization really is 'out there', and whether it is actually as inevitable as Thurow and others suggest. Several contradictions in the 'globalization as inevitable' discourse start to appear. First, globalization is often presented as being both the cause and the outcome of a particular phenomenon – China opens up to global investment flows because of the requirement of a globalizing economy, but the outcome of this process is deepening globalization. In this way, globalization is presented as having both necessitated a policy, while at the same time having resulted from it (Cameron and Palan, 2004). Second, globalization is treated as an irresistible external force that operates outside of human control, and yet it is a phenomenon that is very much socially created.

When governments suggest that flows of trade must be made freer in order to respond to a global competitive environment, they are ignoring the extent to which that global competitive environment is a product of past decisions to liberalize trade regulations. Governments are thus the authors of, and not just responding to, globalization (see Chapter 7). In short, globalization discourse in some ways creates the phenomenon that it supposedly simply describes. The problem, therefore, with Thurow's statement is that businesses don't scan the world because it is somehow natural for them to do so, but because government regulations and technology permit them to do so. Finally, globalization is not just a set of uncontrollable processes 'out there' – it is also regulated and managed by a wide array of institutions. Thurow's suggestion that globalization should be 'shaped' is undoubtedly appropriate, but in fact every aspect of globalization is already being managed and regulated in a variety of ways. From international agreements covering global trade in particular commodities (see Chapter 4 on coffee, for example), to international standards covering business practices (for example, the International Accounting Standards Committee), to certifications for organic or 'fair trade' products, the globalization of trade is already the subject of intense regulation of various kinds. The key question is whose interests these existing regulatory frameworks serve – a question we will return to several times in this book.

2.6 Summary

In this chapter we have critically examined some of the most cherished assumptions that are employed when economic issues are discussed either in the popular media or in academic analysis. We have seen that 'the economy' is not quite such a natural and unproblematic concept as we might at first assume. It is, first of all, a notion that is of relatively recent origin and one that emerged in a particular historical and geographical context. This illustrates a wider point about knowledge in the social sciences – rather than tending towards an ever more complete understanding of how the world works, knowledge is a product of its time and place. What makes sense when we discuss the economy now is a reflection of the circumstances in which we find ourselves – a few generations ago, the concept had different connotations, and perhaps a few generations from now its meaning will have shifted again.

We have also seen one direction in which our understanding of the economy could usefully move. That is, to include within our definition the various forms of production, exchange and consumption that have been excluded by narrower interpretations. As we will see in subsequent chapters, there have already been attempts to do so, for example, by examining alternative forms of labour (Chapter 9) or by including unpaid domestic work in calculations of production (Chapter 12). But the economy can also be understood more broadly if assumptions concerning the independent logic of economic processes are relaxed. When cultural, environmental, social and political processes are recognized as both a basis and context for economic processes (and vice versa), then we can understand the economy as it is actually lived and experienced. As we will show in this book, much of contemporary economic geography is concerned with reuniting these separate spheres to create a more integrated understanding of how economic processes work.

Finally, we have seen the ways in which our narrow understandings of the economy are not simply products of a particular historical period. They are also products of power, in the sense that ideas about the economy that have held sway at any point in time have been a reflection of powerful interests. Furthermore, discourses of economic processes are themselves powerful – as we saw in the cases of 'development' and 'globalization', understanding the world in a certain way forms the basis for diagnosing and treating economic ailments.

Further reading

- Timothy Mitchell (1998, 2002) has written some excellent studies tracing the emergence of the economy as a concept, and examining the consequences of development discourse in Egypt.
- For early and perceptive discussions of the interface between economy and culture, see Thrift and Olds (1996), and Crang (1997). Castree (2004) provides an updated critique of this debate.
- Barnes (1996) and McCloskey (1998) examine the role of scientific metaphors in the concepts of economics. Kelly (2001) examines metaphors of economic crisis in Southeast Asia.
- For more on discourses of development, see Escobar (1994) and Crush (1995). For an excellent account of the discourse of globalization, see Cameron and Palan (2004).
- For an innovative analysis of the political implications of seeing the economy as a discursive construct, see Gibson-Graham (1996, 2006).

Sample essay questions

- Explain how, and why, the meaning of 'the economy' has changed over time.
- When we discuss 'the economy', what activities of production, exchange or consumption are usually included and excluded?
- Is it possible to understand economic processes without also understanding other dimensions of human society and the natural environment?
- Explain how metaphors shape our understanding of economic processes.

Resources for further learning

- http://www.cepa.newschool.edu/het/: the History of Economic Thought (HET) website at the New School University in New York provides some excellent materials on key economic thinkers and concepts.
- http://www.hdr.undp.org/hd/: the website of the United Nations Development Programme which calculates a Human Development Index that provides an alternative measure of progress beyond the narrow confines of GDP.
- http://www.communityeconomies.org/index.php: the website of the Community Economies project, which includes the geographers Julie Graham and Kathy Gibson, is involved in thinking about, and enacting, alternative models of 'the economy'.

References

Barnes, T. (1996) *Logics of Dislocation: Models, Metaphors and Meanings of Economic Space*, New York: Guilford.

Brainard, W. and Scarf, H. (2005) How to compute equilibrium prices in 1891, *The American Journal of Economics and Sociology*, 64(1): 57–83.

Buck-Morss, S. (1995) Envisioning capital: political economy on display, *Critical Inquiry*, 21(2): 434–67.

Cameron, A. and Palan, R. (2004) *The Imagined Economies of Globalization*, London: Sage.

Castree, N. (2004) Economy and culture are dead! Long live economy and culture! *Progress in Human Geography*, 28(2): 204–26.

Crang, P. (1997) Introduction: cultural turns and the re(constitution) of economic geography, in R. Lee and J. Wills (eds) *Geographies of Economies*, London: Arnold, pp. 3–15.

Crush, J. (ed.) (1995) *Power of Development*, London: Routledge.

Escobar, A. (1994) *Encountering Development: The Making and Unmaking of the Third World*, Princeton, NJ: Princeton University Press.

Fisher, I. (1925) *Mathematical Investigations in the Theory of Value and Price*, New Haven, CT: Yale University Press.

Foucault, M. (1970) *The Order of Things*, London: Tavistock.

Gibson-Graham, J.K. (1996) *The End of Capitalism (As We Knew It): A Feminist Critique of Political Economy*, Oxford: Blackwell.

Gibson-Graham, J.K. (2006) *A Postcapitalist Politics*, Minneapolis: University of Minnesota Press.

Gilbert, E. (2005) The inevitability of integration? Neoliberal discourse and the proposals for a new North American economic space after September 11, *Annals of the Association of American Geographers*, 95(1): 202–22.

Kelly, P.F. (2001) Metaphors of meltdown: political representations of economic space in the Asian financial crisis, *Environment and Planning D: Society and Space*, 19: 719–42.

Keynes, J.M. (1936) *The General Theory of Employment, Interest and Money*, London: Macmillan.

McCloskey, D. (1998) *The Rhetoric of Economics*, 2nd edn, Madison, WI: University of Wisconsin Press.

Mitchell, T. (1998) Fixing the economy, *Cultural Studies*, 12: 82–101.

Mitchell, T. (2002) *Rule of Experts: Egypt, Techno-politics, Modernity*, Berkeley, CA: University of California Press.

Rostow, W.W. (1960) *The Stages of Economic Growth: A Non-communist Manifesto*, Cambridge: Cambridge University Press.

Smith, A. ([1776] 1976) *An Inquiry into the Nature and Causes of the Wealth of Nations*, Chicago: University of Chicago Press.

Thrift, N. and Olds, K. (1996) Refiguring the economic in economic geography, *Progress in Human Geography*, 20: 311–37.

Thurow, L. (2004) Globalization is inevitable, *Credit Suisse Emagazine* (http://www.emagazine.credit-suisse.com/), 9 February 2004. Accessed 1 June 2006.

PART II

DYNAMICS OF ECONOMIC SPACE

CHAPTER 3

UNEVEN DEVELOPMENT

Why is economic growth and development so uneven?

Aims

- To think structurally and systematically about uneven development processes
- To understand the fundamentals of the capitalist economy
- To consider uneven development as an outcome of systemic logics under capitalism
- To integrate an understanding of uneven development with the production of space under capitalism.

3.1 Introduction

In 2004, 24 billionaires (in UK pounds) lived in Moscow. Only New York City was thought to boast more members of this exclusive club. It is estimated that these super-rich Muscovites controlled wealth amounting to around £50 billion (or US$90 billion). Much of the wealth of such 'New Russians' has been derived from the privatization of state-owned industries and resources in the mid-1990s – many of them located far away in the vast Russian periphery. Perhaps the biggest celebrity among this group is Roman Abramovich – a man worth more than £7 billion and best known in much of the world as the owner of London's Chelsea Football Club. Abramovich's story has been presented as something of a Russian fairy tale: a poor orphaned boy, raised by his extended Jewish family in the harsh environment of Russia's North-west, who dropped out of college but went on to wield great economic and political power. Abramovich's wealth was acquired in the Russian oil industry, as shares in collectively owned state enterprises were distributed to workers in the 1990s, but quickly became concentrated in just a few hands. He became governor of the remote Russian region of Chukotka in 1999, at the age of 33, having rapidly expanded his holdings into airlines, aluminium and real estate.

Abramovich's story reveals some telling features of contemporary economic life – albeit through a somewhat larger-than-life tale. Russia has, since the collapse of communism and the emergence of market-based economic institutions, seen some dramatic concentrations of wealth in a few hands. Of even more interest here, however, is the geographical unevenness of economic development that the story reveals. While 8.6 per cent of Moscow's population was estimated to be living in poverty in 2002, in the remote Chukotka region, the equivalent figure was the 37.7 per cent (World Bank, 2005). Put simply, while some places prosper in the global capitalist economy, others are less likely to do so. In Russia, as well as many other developing countries that are dependent on primary commodities (for example, Niger, discussed in Chapter 1), impoverished peripheral regions have seen booms in economic activities, especially in mining and other resource extraction activities. But much of the wealth generated has been channelled into the bank accounts of a small and fabulously wealthy urban elite. A part of their wealth has, in turn, found its way into, or through, global cities such as London, where Abramovich has added impressive property purchases to his sporting acquisitions. On an even larger geographical scale, from a global superpower, Russia has been reduced to the status of a 'developing country', and Russians are increasingly migrating to seek work elsewhere. At the urban, national, or global scale then, we see the unfolding of a dynamic process of *uneven development*. Overall, it is a pattern of wealth concentration, rather than dispersal, but it is also a volatile process of boom and bust and boom again.

As we move on to Part II, which aims to illustrate the dynamics of economic space, this chapter provides an important conceptual foundation for thinking through *why* our everyday economic landscape seems to be changing and transformed, sometimes for the better and often for the worse. This dynamic of change is closely related to the way in which our economic system is organized and reproduced. In this chapter, we unpack this economic system, often known as *capitalism*, and show how it creates and necessitates uneven development over space and through time. We will deal with more specific dimensions of capitalism such as commodity chains, technology, and nature/resources in the next three chapters.

This chapter starts with a view of uneven development that sees it as either the result of natural (i.e. environmental) conditions, *or* as a natural state of society (Section 3.2). In many ways, this corresponds to the popular perception of why some places are richer and more developed than others. In Section 3.3, we construct an alternative view using ideas drawn from the Marxist school of political economy (see also Box 1.2). We consider how the system of creating wealth might be seen as something human-made rather than natural. These are what we will call *structural* processes, and they involve the ways in which wealth is generated, circulated, and distributed within society and across space. In other words, uneven development results from inherently *geographical processes*. We then explore in Sections 3.4 and 3.5 the mechanisms that produce uneven development in the capitalist system in particular – the system of wealth production and distribution that now prevails. Having established these fundamental mechanisms, we look at how particular places and the people living there come to reflect the shifting landscape of capitalism (Section 3.6). Finally, in Section 3.7, we note that not all capitalist happenings are necessarily the product of 'the system', but are also mediated by political and military power.

3.2 Uneven Development – Naturally!

In Chapter 2, we explored the notion of development as a discourse – a collective set of understandings, institutions, and technical expertise that diagnoses problems and prescribes solutions for the malaise of underdevelopment. But what is the development, or underdevelopment, that this discourse seeks to describe, explain, and foster? The conventional definition of development presents it as a process of economic growth, i.e. increasing collective wealth, alongside a trend towards social change. Such social change usually implies a movement in the direction of modern, urban, and industrial society – hence the criticism so often levelled against development that it is an *ethnocentric* concept being used to bring traditional 'non-Western' peoples into line with the norms of modern 'Western' life. Here, however, we are less concerned with these social and cultural implications of modern development, including Westernization. Instead, we will focus on the uneven production and distribution of wealth so intimately connected with processes of economic development.

A common approach to the unevenness of economic development is to see it as a *natural state* of affairs – natural either because of the uneven distribution of the bounty of nature (for example, oil fields can only be where oil is located), or because growth has to start somewhere and, under the right conditions, it will naturally spread. Both of these views are very much evident in contemporary debates about why economic development is spatially and socially uneven. At first glance, the notion that human societies are uneven because nature is uneven is a rather tempting explanation. After all, the Industrial Revolution was initiated in the towns of Northern England that had the appropriate endowments of water, coal, iron ore, and other raw materials. And today, nature has bestowed upon countries such as the US, Canada, and Sweden an immense bounty of mineral, agricultural, and forest resources.

But if the list of examples is expanded, this argument begins to unravel. Indonesia has a remarkable array of mining, agricultural, and forest resources, and yet its 218 million people enjoy an income per capita of just over US$3,361 (after adjustments are made for purchasing power differences). Its neighbour

Singapore, a tiny island city-state with no natural resources at all, has a per capita income of around US$24,481. Japan has negligible oil, gas, or mineral resources, and relatively little arable land, and yet it too manages to exceed Indonesia's income almost nine-fold. Likewise, Nigeria sits atop oil reserves estimated at 35 billion barrels in 2004 (3 per cent of the world's total reserves), and yet over one-third of its population lives in poverty (http://www. hdr.undp.org, accessed 6 February 2006). A territory replete with industrial raw materials is not, then, enough to guarantee the economic well-being of a nation's population – indeed there is little correlation between a list of the world's wealthiest nations and a list of those with the greatest resource endowments. Clearly we have to look for other explanations of uneven development.

An alternative argument focuses less on the natural resources of a given country and more on the *type* of society that has developed to utilize them. The characteristics of that society, for example, its political structures, technology, and modes of transport, are thought by some to be closely linked to the available ecology and resources (Diamond, 1999). Hence, over the last several thousand years, some human societies developed advanced technologies, while others continued to use 'primitive' techniques of food gathering, social organization, travel, etc., and these differences reflected the geography – the physical environment and human relationships with it – of particular places on the globe. This created initial inequalities between societies, and when advanced societies expanded and came into contact with other societies, they were able to acquire their resources. Thus, European guns and steel tools were able to conquer much of the globe and establish relationships of trade that continue to be weighted in favour of European (and European-settled) societies. The native societies of North America that Europeans encountered were complex, but they had not developed metal-working skills, nor acquired immunity to diseases that were commonplace among the more densely populated areas of Europe. This and other patterns of *colonialism* enabled Europeans to extract resources and export them to the industrializing colonial heartlands. To this day, one could argue that such a situation has continued to shape global (uneven) patterns of wealth creation and distribution.

The appeal of this argument is that it rejects the notion that there is any difference in the ingenuity, creativity, and energy of different human beings around the world. Instead, it is their societies that are differentially endowed in their physical environments. But such arguments essentially suggest that all of the unevenness in wealth and production that we now see in the world was determined by differences established between societies over the last 13,000 years, and once societies were in contact with each other (from about 500 years ago onwards), the die had been cast and the rest of history was inevitable. Clearly this is an inadequate explanation for understanding recent shifts of fortunes in the global economy. How is the wealth of Ireland and Singapore or the rise of China and India explained? And how is the unevenness of development *within* advanced industrial societies explained, including, for example, the decline of old industrial centres and the rise of new high-tech regions in Europe and North America?

A further set of arguments starts not with the causes of unevenness, but with its presumed levelling out over time through processes of development that spread wealth across societies and spaces. This has been a key argument of mainstream approaches to economic development in the past 50 years. While intellectual fashions have changed, the argument has remained essentially the same: that all economies can develop if they adopt appropriate policies and strategies, and that uneven development is merely a temporary condition that will, naturally, be overcome. One of the earliest manifestations of this perspective was in *modernization theory* – a school of economic thinking prevalent in the 1950s and the 1960s that saw largely cultural and institutional barriers to development in the 'Third World' (i.e. developing countries). Impoverished economies would develop towards the Western model of industrial production, modern society, and high mass consumption, if they first established certain pre-conditions. In early versions of the theory, these conditions included high savings and investment rates, and the removal of cultural resistance to modern science and industrial production. Such models revert to the notion that economic development will tend towards an equilibrium pattern in which differences will be smoothed out over time. This is the assumption of universalism that we discussed in Section 1.2.

More recent incarnations of this kind of economic thinking about development are found in the strong emphasis by powerful global institutions such as the International Monetary Fund, the World Bank, and the World Trade Organization on the role of free markets and democracy in generating wealth and development (see Chapter 7 for more on these institutions). What remains constant, however, is the notion that underdevelopment is an aberration – a problem that will be naturally fixed, if the capitalist system is allowed to operate successfully and more developing countries are fully democratic 'like us'. We will now turn our attention to the fundamentals of such a capitalist system in order to explore an alternative view – one that sees *unevenness* rather than equilibrium as the natural state of affairs under capitalism.

3.3 Marxian Approaches: Conceptualizing Value and Structure

The geography of value

The conventional economic approaches to explaining development in the last section are all deficient in one important respect. None of them actually seeks to explain how wealth is created, focusing instead on explanations for relative differences in wealth between regions and countries (i.e. the resource endowment approach) and how these differences might be evened out over time (i.e. modernization theory). We can, therefore, take one step backwards and think about the fundamental process through which wealth is generated and how this might have geographical inputs and outcomes.

Wealth is essentially a cumulative share of the rewards created in the economic process of adding *value*, where value is measured by the monetary worth of a good or a service traded in the market economy. The creation of value is, therefore, central to the question of economic development, and uneven development is a reflection of either a relative lack of physical or organizational resources to be used to create value (as in the explanations above), or a failure to capture and/or retain that value in the hands of the person, the household, or the community that created it.

But how is value actually created? Ultimately, value is always created by *people* – living human beings who are engaged in *labour processes* that create or change a good or service. The value of labour is thus reflected in every cost associated with producing a good or service – from building factories and offices to making machinery or transport equipment, and from extracting raw materials to developing production technologies. This is a very important point to bear in mind in order to understand the following Marxian analysis. It is not, therefore, oil companies that actually create the value that is derived from extracting oil from the earth; rather, value is created by the employees of such companies who carry out that process, or manage it, or in some other way contribute. All costs associated with creating value from oil extraction are subsumed in the labour process. The company itself is just an organizational platform that coordinates activities and determines how the rewards of that process will be distributed. Thus, the CEO of a large oil company might be allocated several million dollars per year, while a drill technician might earn a fraction of that amount. A proportion of the value will also be paid to governments in the form of tax revenue, and to shareholders as a return on their investment.

How exactly all these rewards are shared, and where their recipients are located, will affect the *geographical* pattern of development that results from a given economic activity. A key feature of explaining uneven economic development is therefore not so much where companies operate, but how different economic activities are structured spatially and which parts of their spatially dispersed operations will be allocated the most value. This is a point to which we will return later (see also Chapter 8 on the geographical organization of global companies). But for now, let us consider the process of value creation itself. Everybody in a company is involved in the creation of value, albeit some more directly and more visibly than others. The production line worker stitching together a cotton shirt in a Mexican factory is clearly adding value to the thread and pieces of fabric being used. But the advertising executive in New York who conceives a marketing campaign is also adding value. Less directly, the janitors cleaning either of those workplaces, the security staff guarding them, or the retail assistant selling the finished shirt, are all contributing to, and taking a share in, the shirt's value.

We will return to the organizational details of such commodity chains in the next chapter, but for now the important process to note is that the *distribution of value* among different points in the chain along which a commodity passes is crucial in understanding processes of development and underdevelopment. For example, since the demise of the Soviet bloc in the late 1980s, Eastern and Central Europe have seen large investments by clothing manufacturers exporting to Western Europe. By some estimates, however, less than 10 per cent of the final value of the product actually stays in the place of production. The rest is accounted for by government taxes, retail stores, brand name design and marketing firms, and transportation, i.e. nearly all of the financial benefits from production are allocated to the places where such products are consumed.

Thinking structurally

The examples above, that illustrate how value is created and where it is retained, are useful in establishing how everyday economic processes fit into a larger geographical picture. But we need to make an analytical leap that does not simply involve thinking about small processes aggregated into larger scales. We also need to think *abstractly* about how economic processes work – how they are underpinned by certain fundamental logics. This requires imagining *structures* that shape our economic lives, but which are not necessarily detectable in everyday experiences.

There are numerous ways in which economic life can be structured – peasant economies involve subsistence production so that value is almost entirely retained within the household; feudal economies involve the payment of tribute to a landlord; cooperative or collectivized economies involve the sharing of value such that the value created in productive activities is shared among the group. All of these types of activities exist in today's economic world, but the dominant form that structures contemporary economic life is capitalism.

In Chapter 2, we noted the ways in which an economy has often been presented as a mechanism or an organism with certain internal functions that are explained through economic science. Such mainstream approaches, however, tend to limit their view of these abstract structures to the functioning of the market mechanism for determining the price of goods and services. It is assumed that ultimately the competitive bargaining process of the market place will lead to the most efficient allocation of resources and that the system will find an equilibrium point at which demand equals supply. Likewise, unevenness in economic development will be 'naturally' levelled out over time. The structure in these approaches is given by the working of a market mechanism that partly defines modern capitalism.

Marxist approaches to economic processes, on the other hand, take quite a different view of economic change. Rather than seeing the magical hand of the market resolving efficiently all economic problems, and addressing developmental imbalances, Marxism sees conflict, tension, contradiction, and disequilibrium. As we will find out in the subsequent sections of this chapter, Marxist theory suggests that the unevenness of development is not only an inevitable feature of capitalism, but also a *necessary* one. It is, however, important to note that a structural approach does not seek to unearth the inevitable predictability of specific events. It is not the kind of predictive science that economics purports to be. There will, therefore, always be exceptions – an approach that predicts unevenness in economic development cannot necessarily say where it will occur and with what magnitude. We need to distinguish crucially between the identification of a 'tendency towards' and 'inevitability of' something happening. This is a key aspect of thinking structurally – recognizing that we are identifying the logics or imperatives of the system, not all of its possible manifestations everywhere in our economic life. As will be explained in Parts III and IV of this book, these possible manifestations are mediated by a whole range of different economic actors and socio-cultural processes that makes economic predictions a mostly futile exercise.

3.4 The Fundamentals of Capitalism

We have now established two key building blocks for our explanation of uneven economic development. The first is the importance of understanding the labour processes through which value is created and distributed in any given economic system. The second concerns the importance of thinking structurally about economic systems in order to understand their internal logics. Our task, then, is to explore the ways in which value is created in the labour processes and circulated within the structures of the dominant economic system of the present moment – *capitalism*.

Identifying the fundamentals of capitalism means going beyond many of the connotations it has acquired over time. These range from the implications of a triumphant system of wealth creation that has eradicated oppressive totalitarian states and has generated more wealth than any economic system in history (see Chapter 7) to those of greed and exploitation that motivates an anti-capitalist movement (closely associated with the anti-globalization movement, described in Section 8.6). In many ways, both of these contrasting perspectives correctly identify features of the capitalist system. It is true that capitalism is fundamentally a system in which value is created through exploitation. As with any other economic system, value in a capitalist system is created by human labour in the production process. But a distinctive feature of the capitalist system is that labour itself is something that can be bought and sold like any other *commodity* in the open market (see more on labour in Chapter 9).

Let us look first, then, at exactly how value gets created and distributed in a capitalist system. If labour is a commodity that is being bought and sold, then what is the price for labour? In short, it will tend towards the amount needed to keep a worker clothed, fed, housed, and raising children (future workers!) – an amount that will vary across different places and historical eras. There will, of course, also be variation across different types and sectors of work, but for the system as a whole, it will be the income necessary to sustain and reproduce the workforce that will define the average wages. The capitalist (employer), however, sells the commodity that is produced and is able to keep the difference between the market value of whatever was produced by the labour, on the one hand, and the value of the labour, on the other.

Thus, a capitalist can extract a *surplus value* from the employee – the amount of value that they produce in excess of what they are paid. In our everyday terms, the former represents revenue and the latter the total cost (bearing in mind all costs – materials and non-materials – are subsumed in Marxian approaches under the value of labour). This phenomenon of surplus value extraction is known as *exploitation* in Marxian approaches. This exploitation is only possible because the capitalist owns the *means of production* – machinery, land, raw materials, and so on. At its core, then, capitalism is about a relationship between different social classes – a capitalist class who owns the means of production, and a working class who owns the labour that they 'sell' to the capitalist. Wealth or profit is generated through the extraction of surplus value. And this wealth accumulates to those who employ workers and who have access to this surplus value – the capitalists.

This is, of course, a very simplified picture of the reality of everyday life. Some employees (such as managers, accountants, lawyers, and software specialists) are paid exceptionally high salaries, and we would hesitate to label them as 'exploited'. And the ownership of the firms (or the means of production) is often far more complicated today than it has ever been in the past – most large enterprises are now owned by diverse shareholders, and every employee with a pension plan is effectively benefiting from shares that their pension fund owns (thus the distinction between employers and employees, or between classes, is often quite difficult to establish, particularly in complex and mature capitalist economies). Nevertheless, if our purpose is to identify the structural process of a capitalist system (rather than pigeon-holing every person into a specific and singular class identity), the class relationship between capitalists and workers is an important starting point.

We have also noted that capitalism is immensely *dynamic* and *creative*. The possibilities for accumulating more and more profit mean that systemic incentives exist constantly to create or find new products, new markets, new raw materials, new ways of organizing the production process, and new ways of saving on the costs of everything. There is a fundamental logic or urge for growth in the capitalist system. This does not necessarily mean that every capitalist enterprise is buzzing with creativity and fresh thinking. But in a competitive environment,

Box 3.1 The fundamentals of capitalism

1. *Capitalism is profit-oriented* – the entire system is based upon the incentive to profit from economic transactions. To maintain profit, it is necessary for capitalism as a system to grow continually. Without growth, profits decline. If new opportunities for profit are not continually created, then existing profits will be whittled away through competition.
2. Growth in value rests on the *exploitation of labour* in the production process – that is, the difference between what labour creates and what it gets. In short, capitalism is founded on a class relation between capital and labour.
3. Capitalism is necessarily *dynamic* in technological and organizational terms. The search for profit inevitably demands that new and innovative means of extracting value from labour be devised in a competitive environment.

it is the innovators that the system will reward, while those who fail to innovate will be out-competed. Only profitable firms will survive and provide the sources of growth in a capitalist system. For example, the capitalist system as a whole has continued to design and build better and better television sets over the past 50 years. But there are no longer any major American-owned TV manufacturers, despite the fact that Americans have one of the highest levels of TV ownership in the world. Instead, East Asian electronics firms have dominated the market. American TV makers gradually went out of business as their competitors beat them with product innovations, new manufacturing technologies, and lower costs. Thus, while the capitalist system innovates and grows, some of its participants are always falling by the wayside. We have, then, a system that is simultaneously exploitative and dynamic, and we can summarize this system through the three fundamental features of capitalism in Box 3.1.

3.5 The Contradictions of Capitalism

The inevitability of crisis

A key insight of a Marxist approach to the economy is that when the characteristics of capitalism are put together, they do not create the harmonious equilibrium of commodities and prices predicted in conventional economics. Instead, they create an inherently unstable and contradictory system that is always going to move from crisis to crisis. In this section, we lay out the basis of this contradiction and crisis tendency.

Since capitalism generates surplus value, or profit, out of the labour process, it will tend constantly to grow and expand in order to accumulate even more wealth. More and more enterprises will employ more and more people and produce more and more goods. This will ultimately, over time, tend to increase the market price for labour, while at the same time dampening the price for these goods as excessive supply is building up. The first contradiction of capitalism, then, is between its internal imperative for growth and profit – growth that increases the price of labour, and profit that requires labour cost to be minimized. One way of dealing with this problem is a 'technological fix' whereby competing capitalist firms are trying to find technological ways of making their production more cost-efficient than their competitors. For these reasons, capitalists will try to push wages down or, where applicable and cost-effective, to invest in machinery in order to replace workers with machines. For individual capitalists and managers of enterprises, this is a rational thing to do.

The result of labour-saving machinery is a *reserve army* of labour (not necessarily in the same country) that is brought in and out of the workforce and always keeps wages low. There will always be someone unemployed who will fill a job if the incumbent prices him/herself out of the market. But if wages are kept low and people are kept out of work, where will the *demand* for the products come from? Workers are always producing more for the capitalists than they earn, so the aggregate demand can never keep pace with the growing supply of products – this is the second contradiction of capitalism. In short, workers just do not make enough money to provide sufficient demand for the goods they have produced.

Ultimately and collectively across the whole system, the economy is driven to what is termed a *crisis of over-accumulation*. Capitalists have more products than they can sell, or idle machinery that cannot be used to full capacity because there is insufficient market demand for the product. Idle capital and idle labour are found in the same place at the same time with no apparent way of bringing them together for socially useful tasks – wealth is accumulated, but on the basis of exploited workers who cannot then afford to buy all of the products they create.

Because of this structural tendency, the capitalist system is prone to crisis and instability. There are ways of forestalling it – such as selling to overseas markets, pressing down wages, or further investing in labour-saving machinery (we will discuss these later). But crisis is only delayed, not avoided. Therefore, capitalism contains within itself a contradiction that will keep recurring. It is not hard to find examples where wealth and surplus exist side by side with shortage and need (e.g. the food supply conditions in Niger described in Chapter 1). Indeed, booms and busts appear to be common cyclical occurrences. Over the course of the twentieth century, a number of business cycles saw the wealth of national or regional economies grow or decline. We will see this phenomenon in the case of

California towards the end of this chapter. Understanding the logic of the capitalist system puts those crises in a rather different perspective than their usual interpretations in economics. Instead of being occasional hiccups, or unpredictable storms, that hit the economy from the outside, they are in fact *inherent* to the capitalist system itself in the first place.

Coping with crisis in the capitalist system

If crises of over-accumulation are inherent features of capitalism, an obvious question to ask is why more have not occurred, and why they have not been more disastrous for the system as a whole? In the nineteenth century, Marx thought he was seeing evidence of a deepening crisis of capitalism that would eventually lead to the rise of workers (the proletariat) in a socialist revolution. Over the course of the twentieth century, it is of course true that socialist regimes rose to power, nationalizing all means of production, and implementing central planning in many parts of the world (e.g. the former Soviet Union, Eastern Europe, China, North Korea, and parts of Southeast Asia). But these socialist states varied in the extent to which they dismantled or undermined the structures of capitalism (see also Chapter 7). Also, in most cases, socialist states did not emerge in places where capitalism was most developed and therefore could hardly be viewed as outcomes of an intensified capitalist crisis. In fact, for Western Europe, North America, and the rest of the capitalist world, the twentieth century was marked by several very successful recoveries from crisis (e.g. the Great Depression in the 1930s, the oil crisis in the early 1970s, and the bursting of the dot.com bubble in 2000).

How, then, has the capitalist system managed to get beyond crises in some cases, and to contain, absorb, or delay crises in ways that do not bring about its own downfall? Here we are thinking at the level of *structures* and therefore we need to remember that we are not necessarily talking about options for particular capitalists or firms for whom crises may still be individually disastrous – leading to bankruptcy and unemployment. Instead, we focus on four ways in which the capitalist system *as a whole* restores the conditions for profitability (Harvey, 1982):

1 *Devaluation*: This process involves the destruction of value in the system – money is devalued by inflation; labour is devalued by unemployment (for example, when an important industry leaves a place); and productive capacity is devalued, in fact literally destroyed, by wars and military encounters. Recreating value gets the system moving, and capital circulating, once again, but not of course without significant political, social, and environmental costs.
2 *Macro-economic management*: This involves devising ways of bringing together idle over-accumulated capital and idle labour. This might involve heavy government spending to create jobs and stimulate demand during

Box 3.2 Regulation theory

Regulation theory was first developed in the 1970s and the 1980s by a group of French scholars, including Michel Aglietta, Robert Boyer, and Alain Lipietz. They used the word *regulation* in a broader sense than its usual meaning in English, where it is largely limited to a set of rules or procedures for governing action. In the French sense, regulation refers to the wider set of institutions, practices, norms, and habits that emerge to provide for periods of stability in the capitalist system. This mode of regulation involves both state and private sector actors, and when successful can foster a period of sustained capitalist growth and expansion. Essentially, the mode of regulation works by seeking compromises in the inherent tensions and contradictions that exist in the capitalist system. Historical periods of stability, known as regimes of accumulation do, however, eventually end in crisis and a new mode of regulation must be discovered. A regime of accumulation known as Fordism dominated in North America, Europe and Australia in the post-war decades (see Box 5.2), and featured a three-way balance between the interests of corporations, organized labour, and governments (although the precise features of the mode of regulation in each case varied). This regime crumbled in the 1970s and many see neoliberalism as a new mode of regulation in which market-based, rather than state-mediated, solutions are sought for problems of economic management (see also Box 7.1).

periods of recession. It might also involve legislation that curbs excessive labour exploitation by establishing certain standards in working conditions and wage levels. These strategies help to ensure that demand in the economy is maintained and have been conceptualized as part of a broader process of 'social regulation' (see its theoretical analysis in Box 3.2).

3 *Temporal displacement of capital*: This involves switching resources to meet future needs rather than current ones – for example by investing in new public infrastructure (such as the New Deal strategy adopted by the Roosevelt administration in the US during the 1930s), or by using over-accumulated capital as loans and thereby intensifying future production, and hence its crisis tendencies, but averting a crisis in the present.
4 *Spatial displacement of capital*: This involves opening up new spaces for capitalist production, new markets, or new sources of raw materials. Rather than expanding the time horizon of the capitalist system using credit or loans, the spatial horizons of the system are expanded. This might mean the development of entirely new production sites in newly industrializing parts of the world or the *re*-creation or rejuvenation of old spaces.

This last form of crisis-avoidance is particularly *geographical* in nature and of most interest to us here in our discussion of uneven development. It involves the ways in which capitalism *needs* space in order to function and the process through which the system values and then devalues different spaces according to its structural imperatives at a given point in time. In this sense, space is not just the container in which capitalism takes place. Rather, economic geography is fundamental and inherent to the successful operation of the system.

3.6 Placing and Scaling Capitalism

From our perspective then, capitalism produces spaces at multiple scales. At the *urban* scale, neighbourhoods and buildings can be built up as the vanguard of economic growth. In the nineteenth century, industrial towns in England grew up around the processing of natural resources, such as cotton, tea, and rubber brought in by ship from distant colonies. In the early and mid-twentieth century, large-scale manufacturing of clothing, prepared foods, cars and electrical devices formed the mainstay of urban economies in Europe and North America. In specific places, it was not just industrial landscapes that reflected these times – the small brick row houses of older industrial cities that were occupied by factory workers relatively close to their places of employment are the manifestation of a particular period of capitalist growth on the residential landscape. By the 1950s, industrial suburbs were being built to house the workforce, reflecting the growing use of private vehicles. Looking at the housing stock of any large contemporary city in an industrialized economy is, in fact, an interesting exercise in economic archaeology, as it reveals the built forms that were needed in any given period of economic growth – from the medieval alleyways of central London, just wide enough for a horse and cart, to the car-clogged twentieth-century freeways of Los Angeles.

As the imperatives of capitalist growth changed over time, many of these landscapes became redundant, only to be rediscovered at a later date when they had been thoroughly devalued and could once again be used, except for very different purposes. In some industrial cities of Europe and North America, for example, old garment factories that have been lying idle and derelict since production moved offshore in the 1960s and 1970s are now being renovated. They are prized as studio, office, or even living space by young professionals in growing industries such as graphic design, computer animation, and internet consulting. These poor and depressed neighbourhoods may become the home of the wealthy professionals – a process commonly known as *gentrification*. In this way, a devalued space of industrial capitalism has been re-valued in a new era of post-industrial capitalism.

The same economic-geographical change applies at a *regional* scale. As old industrial regions in Europe and North America succumbed to competition from

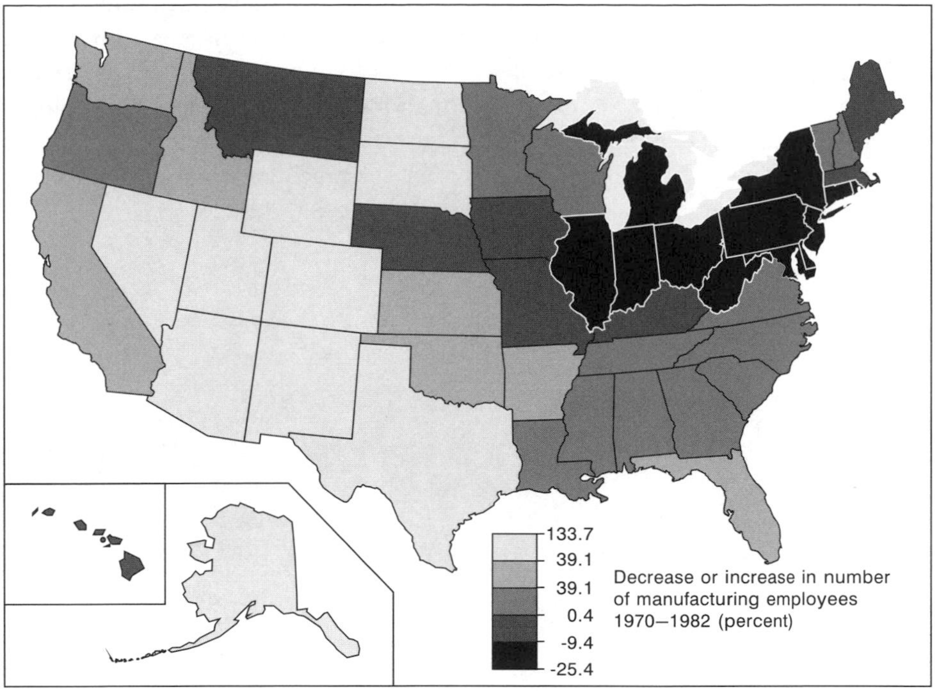

Figure 3.1 Industrial restructuring of the 1970s in the United States
Source: Peet (1983), Figure 9. Reprinted from *Economic Geography* with permission of Clark University.

Newly Industrialized Economies (in Asia in particular), places such as the Rust Belt of the US, Northeast England, and the Ruhr Valley in Germany, fell into decline – their labour forces seen as anachronistic relics of the past in just the same way as their factories, infrastructure, and towns were. Meanwhile, new growth zones were developed in the American South, London and the Southeast of England, and Baden Württemburg in Germany. Figure 3.1 graphically illustrates this process in the United States. Over the course of economic crisis years in the 1970s, manufacturing employment was dramatically reduced in Northeastern states, but at the same time growth was occurring in the South and the West (Peet, 1983).

Finally, at a *global* scale, the shift of development from region to region has been dramatic in recent years. The most striking of these shifts has been the rise of East and Southeast Asia as regions of rapid growth due to investment in manufacturing industry. A first wave of development occurred in East Asia in Hong Kong, Singapore, South Korea and Taiwan, and a second slightly later wave in Malaysia, Indonesia, and Thailand. This came to be known in the 1970s as the *new international division of labour* because labour-intensive manufacturing activities were increasingly shifting from developed countries to such economies (we will discuss this phenomenon in a later chapter, in Box 8.2). Over the last two decades, it has been the rise of China as an industrial power that has captured popular attention, although India too is rapidly establishing a presence in high-technology manufacturing, service, design, and programming (Figure 3.2). Thus, at the global scale too we see a geographical solution being sought – what might be called a *spatial fix* (Harvey, 1982).

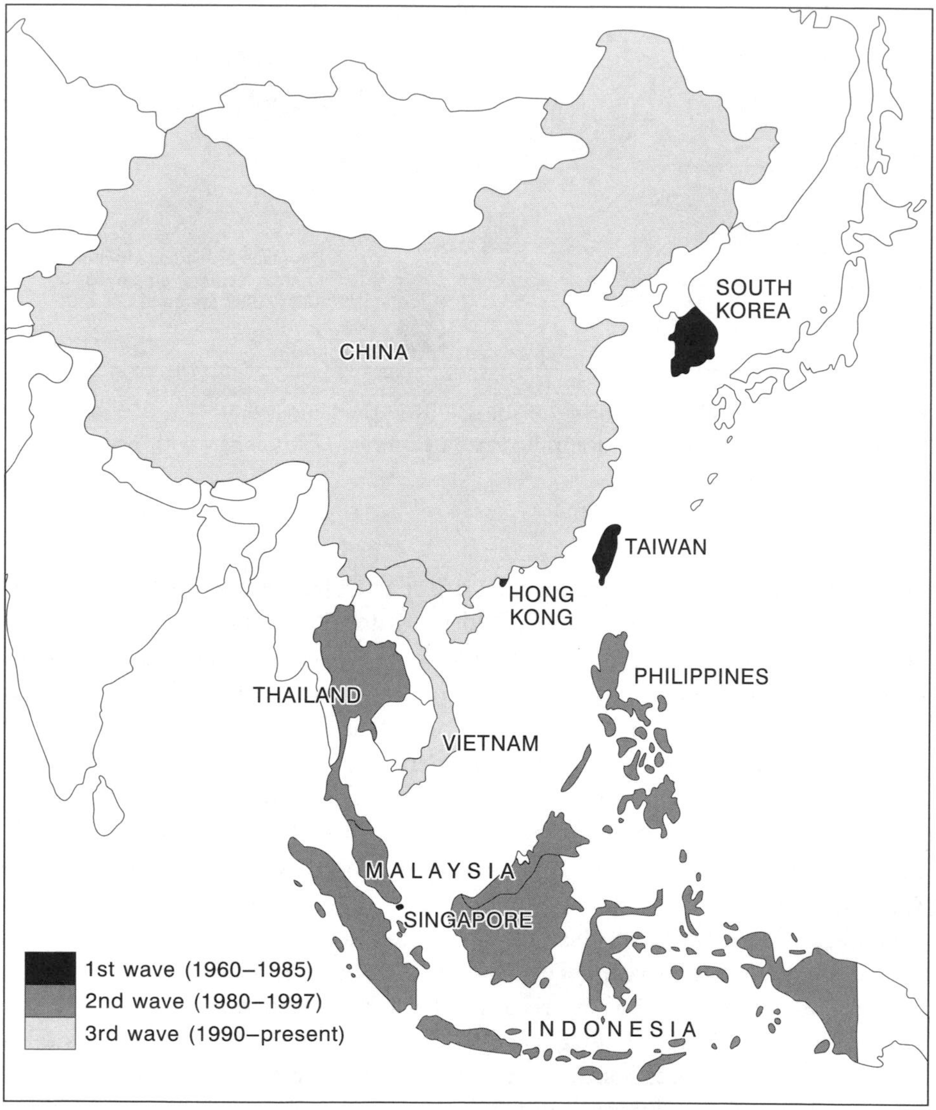

Figure 3.2 Waves of development in the Asian Newly Industrializing Economies

A global shift in development patterns is based upon a movement towards more profitable locations – either because of new sources of labour, or new markets, or, most often both. But there are also plenty of places that are incorporated into the global circulation of value in unequal ways that lead primarily to the extraction of surplus value. The fruit plantations of the Caribbean, mining activities in Africa, and the oil towns of the Russian periphery are all essentially extractive activities that suffer from a high degree of external control and dependency. The circuit of value involved in each sees wealth taken away from its source and concentrated in urban centres far away, and even in global cities in distant countries where such commodities are traded. We should therefore be conscious that for any given economic activity, the ownership, management, and actual production of value now have a complex spatial structure that stretches all the way up to a global scale (see more discussion in Chapters 4 and 8).

These different geographical scales are not, of course, unconnected, and the ebb and flow of investment at the urban scale is a product of development cycles at larger scales. Overall, we see a pattern whereby the capitalist global economy is constantly in flux: always finding new spaces to develop or old spaces to redevelop. This has an interesting implication: capitalism's inherent tensions relate to more than just the processes of capital circulation and accumulation. There is a further set of internal tensions that are related to the spaces, and especially built environments, that capitalism creates at any given moment in time. Whether it is watermills of the eighteenth century, the cotton mills of the nineteenth century, the industrial suburbs of the mid-twentieth century, and perhaps also the export processing zones and high tech parks of the late twentieth century, capitalism creates landscapes that are suited to its needs at a particular time (Figure 3.3). Then, as the system continues to grow and change, those landscapes become outdated, unprofitable, and inhibiting. They become impediments to future growth and must be devalued to make way for a new round of growth and exploitation.

To take a simple example, a city such as Liverpool in North-west England grew in the nineteenth century, reflecting its importance as a seafaring centre and as a point of transhipment for cargos from all over the world. Its prosperity reflected its appropriateness for a particular era of capitalist growth. But container ships began to increase massively in size from the 1960s onwards and to concentrate in larger and deeper ports. Air travel took the places of ocean liners around the same time, and the products of textile industries in Lancashire and the region around Liverpool were replaced with imports from abroad. Liverpool as a city, as a built environment designed to service capitalism at a given point in time, became anachronistic. The city was not just outdated in relation to newer technologies of industrial production and transportation, its whole 'set-up' was an impediment to harnessing their growth potential in a changing global economy. For much of the second half of the twentieth century, Liverpool was

Figure 3.3 A landscape of contemporary capitalism: an industrial estate in the Philippines
Source: The authors.

a city in economic decline. After it has been devalued, however, the infrastructure of the past can be re-valued in another era. Hence, the docklands of Liverpool and elsewhere, which lay derelict for so long are now being refurbished to house new art galleries, tourist attractions, and apartments (Figure 3.4).

Thus, just as capitalism must create physical forms, it must also engage in a process of *creative destruction* to break out of them as it grows and changes. And once old landscapes and spaces are broken down, they can at some future point in time be re-inhabited. This suggests that a core process in capitalism may take the forms of a 'see-saw' of uneven development in which some places are sites of rapid investment and growth, while others decline. But over time, the 'see-saw' can swing backwards so that older spaces are recreated as sites of new investments. The see-saw continues as the needs of the system to perpetuate growth and avoid crisis are satisfied (Smith, 1991).

We have arrived, then, at an argument suggesting that uneven development is far from being a reflection of the relative abundance of natural resources in different places (as suggested by some perspectives in Section 3.2). Nor is it a condition that awaits the intensification of capitalism in order to be eradicated (i.e. more growth, more investment, etc.). Instead, we can now see that uneven development is a prerequisite of the capitalist system – to continue growing and

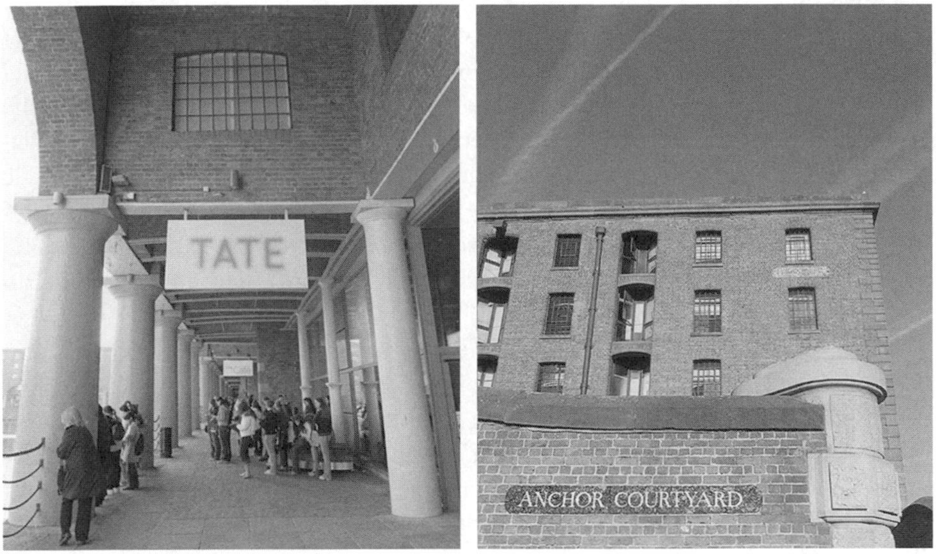

Figure 3.4 Galleries and apartments now occupy nineteenth-century industrial infrastructure in Liverpool
Source: © Mark McNulty Photography, reproduced with permission.

to avert crisis, the swinging of the see-saw from one place to another is a necessity inherent in the system. Clearly this is radically different from the perspectives that we started with in this chapter.

3.7 Putting People in the System

We have developed an argument suggesting that an outcome of the logics of the capitalist system is uneven development between places on the economic map, and the constant shifting of growth from one place to another. But with the focus on logics, fundamentals, and internal tensions in capitalist structures, we have rather lost sight of the fact that capitalism is a system inhabited by *real people* – working, playing, talking, praying, travelling, consuming and studying. Growth or underdevelopment are, after all, essentially measures of human well-being. The places created by capitalist restructuring contain real lived experiences and are therefore *social* spaces. Within these places, ways of life emerge with particular cultural practices, social bonds, class structures, traditions, labour force characteristics, and so on.

Steel towns, for example like Sheffield in the UK, Pittsburgh in the US, Hamilton in Canada, or Kitakyushu in Japan, developed working-class cultures that valued manual work, prized a collective working-class identity through trade union organizations, and were based on male-oriented cultural institutions, such as working men's clubs and professional sports teams. Gender relations were, at least in the past, based upon the assumption of a male breadwinner in each household and a role for women based on domestic work, and possibly the generation of a secondary income. There were, then, a distinctive set of social relations that emerged in association with a particular industrial sector – public institutions, gender relations, masculine and feminine identities, and class politics. The same could be said for mill towns, mining towns, export processing zones, and high tech districts. Their economic and socio-cultural relations do not determine each other, but they are inseparably interconnected.

When a particular place becomes the site for a new round of growth and investment in the ongoing dynamism of the capitalist system, it does not just develop new transport infrastructure, technology, institutions, housing, workplaces, etc. that are appropriate to that epoch. It also develops an evolving set of social relations. This economic-geographical process is nicely captured in a concept known as the *spatial division of labour* (Massey, 1995; see Box 3.3).

A further implication of the emergence of social relations associated with particular industries is that future investment decisions may be made based upon these relations. When, for example, Japanese auto manufacturers locate production facilities in North America, it is known that they seek a workforce that is flexible and open to the particular forms of work organization favoured by Japanese employers. This means seeking people who are not steeped in big-city, heavily unionized workforces with long industrial legacies (see Section 8.5). One option is to locate in small rural towns – places close to highways and air transport infrastructure, but without any historical legacy of industrial development and thus unionization (Mair et al., 1988). In other cases, though, industrial towns may be preferred where a period of decline has 'broken' the union movement, or where labour has been devalued to the point that its collective ability to make demands is greatly reduced (see also Chapter 9). Hence, Nissan's manufacturing plant near Sunderland, a former mining and shipbuilding town in the Northeast of England, established entirely new work practices, but did so in a town desperate for new investment and jobs (see also the example of central Romania in Chapter 6).

So, we have a picture of a landscape that is swept by new waves of investment as the cycles of boom and bust pass by. Each one creates particular social forms and lifestyles, and builds on past social forms and lifestyles. Different metaphors have been used to try to capture this process. Some think of it as a sequence of sedimentary layers of investment overlaid on top of each other – rather like the geological deposits left by oceans. This nicely captures the ebb and flow of investments and the record they leave behind, but it ignores the effects that past 'sediment' (or capital) has on future sediment, and the erosion of value that occurs between sedimentations (or rounds of investment). Another image is of a recurring swarm of locusts descending upon a particular area and eating

Box 3.3 Spatial divisions of labour

In a geographical account of regional development in Britain, first published in 1984, Doreen Massey (1995) describes how a contemporary capitalist economy develops a spatial structure that assigns distinct roles or functions to particular places. In Britain, coal mining towns developed in South Wales, while heavy industries emerged in the cities of the North and Midlands of England. The functions of headquarters and business services (such as law, accountancy, banking, advertising etc.), meanwhile, were concentrated in London. Massey shows how these separate functions constituted a spatial division of labour – analogous to the technical division of labour within a factory whereby different workers performed separate functions in the creation of a final product. Associated with each regional function was a structure of social relations – gender relations, social institutions, class politics, and so on – which both reflected historical rounds of growth and development and shaped the type of investment that would be attracted to the area in the future.

One example involves the reduction of coal-mining employment in South Wales during the second half of the twentieth century. This led to the dismantling of a masculine working-class culture and the social institutions that went with it. New forms of employment emerged in branch plants established by manufacturing firms, for example, in the auto sector. But the new firms sought non-unionized, female employees, and often greenfield sites outside old centres of industrial activity in the coal and steel industries. The social structures left behind and devalued by an earlier decline were effectively establishing the conditions for future investment in different industries looking for different kinds of workers.

vegetation until there is nothing more to take, and then later returning when the foliage has grown up. While these capitalist locusts provide a grimly vivid analogy of surplus extraction, and the image recognizes the distinctiveness of each place (in terms of the 'vegetation' that will eventually return), the metaphor ignores the fact that it is a very *different* round of 'locusts' that appears each time, and that regeneration is an *outcome* of capitalist development rather than its precursor (Smith, 1991).

A better metaphor for capturing the process we have described is perhaps to be found in a game of playing cards, whereby different players represent different places or regions in the capitalist space economy (Figure 3.5). Each round of investment is represented in a different suit, and the role of a place in the spatial division of labour is shown by the card itself. The characteristics of a place (the hand of cards being held) can then be seen as the product of past rounds of

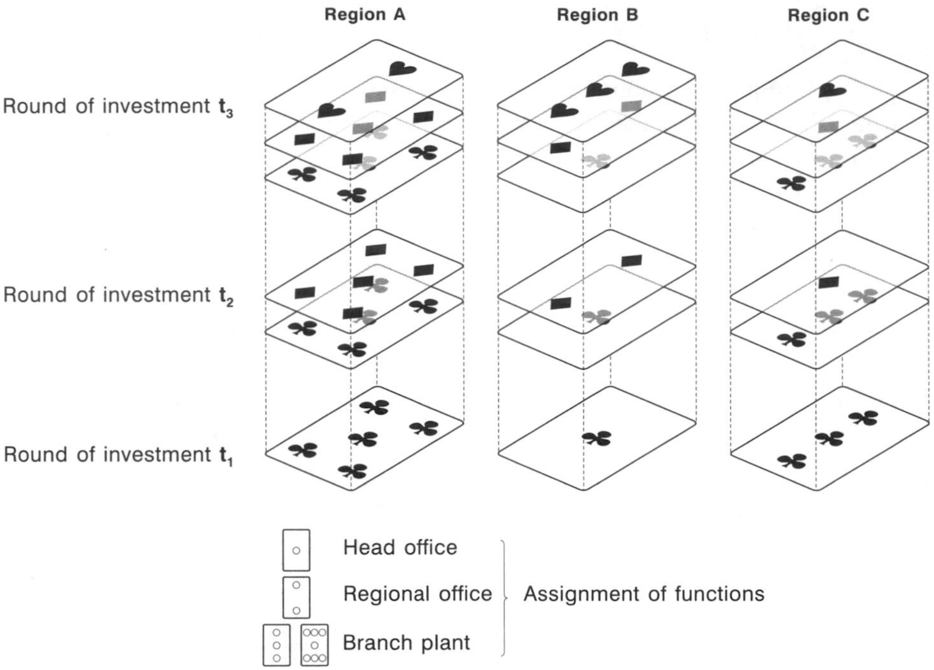

Figure 3.5 Spatial divisions of labour
Source: Adapted from Gregory (1989), Figure 1.4.2.

growth or decline, when the place was inserted into the spatial division of labour in a particular way. This metaphor also illustrates how a place can rise or decline over time in its importance. What it cannot capture, however, is the way in which cards already in a hand can affect the new card that will be dealt, which is, of course, a critical feature of the system we have described.

Whatever the metaphor used, our purpose is to highlight the fact that in order to understand the development over time of a particular place or region, we need to understand not just the present nature of the economy, but also its *historical* and *geographical* background – the previous layering of rounds of investment (Box 3.4 describes this process in the case of California's historical development). And once the social and cultural lives of people living in particular places are added into the analysis, we begin to see not just the systemic logics of uneven development under capitalism, but also the human consequences as ways of life are in turn created and dismantled. New growth will often not arrive until old social norms have been broken down – hence old industrial areas with strong working-class traditions have not seen new growth until after a period of decline that has broken the strength of unions and other embedded institutions and practices. A key part of capitalist restructuring is therefore a painful process of social adjustment and, perhaps, cultural change.

Box 3.4 Bringing it all together in California

The economic fortunes of California epitomize the ups and downs of capitalist development and the geographical unevenness that it creates (Walker, 1995; 2001). Prior to the gold rush of 1848, California was sparsely populated and relatively unproductive. The discovery of gold brought a frenzy of economic activity as prospectors and merchants arrived in droves and agricultural activities were expanded to feed a growing population. The completion of a transcontinental railroad in 1869 permitted Californian farmers to export their increasingly bountiful food crops to Eastern states and the rest of the world. The migration flow westwards, the opening up of new agricultural land, the development of growing urban centres, and the connections with all of this created by the railroad can be seen as the production of new spaces for American capitalism in the late nineteenth and early twentieth centuries. Essentially, California, and westward expansion more generally were providing a *spatial fix*, and the new opportunities for profit that the system required.

But devaluation was also going on. The workers that planted, harvested, and processed California's crops were often migrant labourers, many from the Dust Bowl states of Oklahoma, Texas, and Arkansas in the 1930s, others from China or the Philippines. Their labour was often devalued to the point of barely sustaining their survival – hence maximizing the surplus value being accumulated by California's farmers and the agro-food corporations that dominated the state. In the 1920s and 1930s, aerospace emerged as a new industry and California became a major centre for building and testing aircraft. The Second World War sparked massive federal government expenditures in this sector and the state manufacturing output tripled during the war years of 1939–45. When a new round of defence-related investment arrived, driven by the Cold War of the 1940s to the 1980s, California's already existing aerospace corporations were poised to move into missile and satellite production. Perhaps more than anywhere else in the US, California bore witness to the economic growth potential (unevenly distributed) that follows the destructive devaluation of warfare.

Aerospace and other defence-related expenditures also laid the seedbed for an innovative electronics and computer industry in latter decades of the twentieth century. Californian corporations such as Hewlett-Packard, Intel, and Apple were the driving forces behind the personal computer revolution of recent decades and Silicon Valley, south of San Francisco, remains the global epicentre of the high tech sector. Movie production also emerged in California in the early twentieth century, and following the movie industry came high fashion designers, and later clothing manufacturers. By 2000, California accounted for 20 per cent of America clothing production.

But as with agriculture, much of the actual production work was carried out by low-paid immigrants.

California has, however, also seen the realities of economic crisis. The Great Depression of the 1930s saw massive migrations of displaced farm workers and a slump in demand for California's agricultural and manufactured products. In the 1990s too, the state experienced a prolonged recession, this time itself the victim of a spatial fix that saw investment in electronics industries increasingly redirected towards the growing economies of East Asia. But on both occasions, the devaluations (of labour in particular) associated with recession have seen economic output bounce back afterwards. Over the course of its development, then, California has shown the growth potential of devaluation, spatial fixes, and government involvement in the economy. Wealth creation requires the extraction of surplus value from labour. In California, a constant stream of immigrants, lowly and highly skilled, has made this possible. It has also demonstrated the role of pre-existing patterns of economic activity in attracting future rounds of investment – a process that has historically worked in California's favour.

3.8 Going beyond Capitalism

Over the course of this chapter, we have demonstrated the way in which the internal logics of capitalist production need an uneven economic landscape – the booms and busts of development are a process of devaluation and revaluation that help the capitalist system avert crisis. But there is an anomaly here that has been quietly with us since we began the chapter with the oil wealth generated by resource extraction in the Russian periphery. Why is it that peripheral areas remain rich in resources, and alive with activity to extract them, or attractive to new manufacturing industries, and yet many are still relatively poor and underdeveloped? What has gone wrong in the 'catching up' process?

To find the answer, we have to look at the *political-economic arrangements* through which capitalism as a system of production enacts the spatial fixes described in this chapter. We must, for example, examine the exchange relationships between core country/region and periphery and the way in which value gets concentrated in some places and not others. Until around 50 years ago, the processes that determined how 'developing' countries were paid for their resources were shaped by the unequal power structures of colonialism. By taking over territories through military force, European and American colonialists asserted their ability to capture resources and establish the prices that would be paid for them. In this way, a much greater component of the value created was located in the countries of the colonizers than in those of the colonized. Think of the historical relationships between Britain and Sri Lanka in the tea trade or between Britain and the Caribbean in the banana trade. This created what has been called *dependency* or *unequal exchange*.

In most of the world, however, the formal colonialism that took hold in the nineteenth century has dissipated, but *power* in the contemporary global economy is still wielded by dominant states – it is not just capitalists who determine what, and where, economic activities will take place, and on whose terms. The wealthier nation-states of North America and Europe (in particular, the so-called G8 group of countries) are so dominant, financially and militarily, that they have ample opportunities to dictate the terms upon which less powerful countries will engage with them economically (see also Chapter 7). In particular, it is the dominance of financial centres in New York, London, and elsewhere, and their power to concentrate the accumulation of wealth among those involved in *financial* transactions, rather than production relations, that is central to the supremacy of certain core regions over peripheral ones (even when rich resources lie in the latter). This relates to global core regions and global peripheries, but also national core regions and national peripheries (as in the case of Moscow and Chukotka).

This power might be used to force peripheries to open up their economies to the activities of core-based corporations, to expose their domestic markets to imported products, to allow outsiders to operate in their financial markets, to export their natural resources while importing higher value-added good and services, to limit labour rights and working conditions, to permit environmental degradation, or to privatize common property resources such as genetic material and natural environments. Self-interest among local elites in such places will likely motivate them to collaborate in this process. The implication is that we see crises of accumulation averted not just through the *intensification* of the capitalist production process, but also through the geographical *expansion* of capitalist processes into formerly communally or socially owned spheres – for example, common property resources, citizenship rights, or national assets. In the terms described in Chapter 2, we see the *alternative* economy being encroached upon by the *capitalist* economy. The possession of communal resources (e.g. oil and gas) is transferred from local communities to global traders who concentrate their corporate wealth elsewhere. This process, through which collective resources are appropriated by private interests, has been described as *accumulation by dispossession* (Harvey, 2003).

But we also see a global situation in which this process is advanced by one capitalist *state* over another, often through direct negotiations between governments (e.g. trade agreements) or through multilateral arrangements (e.g. the structural adjustment programmes imposed by the International Monetary Fund, described in Box 7.3). This new form of political-economic domination has been dubbed the *new imperialism* (Harvey, 2003).

There are two broader points to make here. One is that the power of the financial sector (central banks, private banks, investment houses, currency traders) is such that the benefits of production are often channelled into financial centres rather than benefiting the places in which production actually takes place. The second is that the nation-state plays a key role in the process of capitalist development (see also Chapter 7). Often, the logics of the global economic system are constructed from the interests of nation-states rather than capitalist enterprises, although it is the latter that actually carry out the process of production.

3.9 Summary

We started this chapter with the great social and spatial unevenness of development in the newly capitalist economies of the former Soviet bloc. In part this represented the extractive nature of development in a peripheral area. While resources, and a manual labour force to extract them, might be located above oil fields and mines, the value that their activities create is accumulated in urban centres elsewhere in the same country or even beyond. This focus on value and its creation led us to a fundamental consideration of the processes underlying a capitalist economy – how value is created, how it is circulated, and the contradictions that exist in such a system. The result was an understanding that unevenness is not an accidental by-product of capitalist development, nor is it something that our economic system naturally levels down over time. Instead, it is quite fundamental to the workings of the capitalist system and something we should expect to continue. Indeed, if unevenness did not exist, we might expect that the see-saw of capitalist development would create it. In other words, uneven development is both a cause and an outcome of capitalist growth.

As we work through various other themes in this book, it will be worth bearing in mind the structural analysis that we have developed in this chapter. The following three chapters will explore the organization of capitalist production through commodity chains, the technological dynamism of capitalism, and the commodification of nature in more detail, along with its implications for spatial patterns of investment, wealth creation, and environmental change. Subsequent chapters on the state, transnational corporations, labour, and consumers will all, in different ways, highlight the actors involved in the contestation over the creation and distribution of value – between social groups and between places. Finally, chapters on culture, gender, and ethnicity will highlight place-based characteristics that capitalism, as a system, often incorporates and uses. Thus, while this chapter has established *why* capitalism fundamentally needs an uneven economic geography, much of this book will go on to examine precisely *how* distinct places and uneven spaces are actually created and used.

Further reading

- The single most important figure in the development of geographical approaches to studying the structure of capitalism has been David Harvey. For an introduction to, and retrospective on, his work, see Castree and Gregory (2006).
- For an early review of Marxist influences in geography, see Peet (1977). For state-of-the-art reflections at what was probably close to the peak of interest in Marxist theory in geography, see chapters by Peet and Thrift, Smith, and Lovering in the edited volumes by Peet and Thrift (1989). A more recent review is provided by Swyngedouw (2000).
- For other studies of regional economic restructuring, see Massey (1995) and Allen et al. (1998) in the UK; Gibson and Horvath (1983) in Australia, and Smith and Dennis (1987) in the United States. Soja et al. (1983) also provide a case study at the urban scale for Los Angeles.
- For a geographical review of theories related to international development, see Peet (1999).

Sample essay questions

- Explain the key differences between conventional accounts of uneven development and a Marxian approach.
- Is development inevitable for the 'developing world'?
- Why is space so integral to the survival of capitalism as an economic system?
- Using a specific example, describe the ways in which a particular era of capitalist production became imprinted upon the landscape and social characteristics of a particular place.

Resources for further learning

- Several Marxist scholars maintain extensive and informative websites with excellent introductions to the field. See, for example, http://www.ssc.wisc.edu/~wright: Erik Olin Wright's website at the University of Wisconsin; and http://www.nyu.edu/projects/ollman/index.php: Bertell Ollman's site at New York University.
- http://www.chukotka.org/root/?lang=en: the Chukotka autonomous region of Russia, mentioned in the introduction to this chapter, has an elaborate English-language website.
- International organizations such as the United Nations and World Bank have extensive websites containing studies of developing areas. See, for example,

http://www.unmillenniumproject.org: the UN Millennium Project, and http://www.worldbank.org: the Data and Research page at Development.

References

Allen, J., Massey, D. and Cochrane, A. (1998) *Rethinking the Region*, London: Routledge.

Castree, N. and Gregory, D. (2006) *David Harvey: A Critical Reader*, Oxford: Blackwell.

Diamond, J. (1999) *Guns, Germs and Steel*, New York: W.W. Norton.

Gibson, K. and Horvath, R. (1983) Global capital and the restructuring crisis in Australian manufacturing, *Economic Geography*, 59(2): 178–94.

Gregory, D. (1989) Areal differentiation and human geography, in D. Gregory and R. Walford (eds) *Horizons in Human Geography*, Totowa, NJ: Barnes & Noble Books, pp. 67–96.

Harvey, D. (1982) *Limits to Capital*, Oxford: Blackwell.

Harvey, D. (2003) *The New Imperialism*, Oxford: Oxford University Press.

Mair, A., Florida, R. and Kenney, M. (1988) The new geography of automobile production: Japanese transplants in North America, *Economic Geography*, 64(4): 352–73.

Massey, D. (1995) *Spatial Divisions of Labour*, 2nd edn, London: Macmillan.

Peet, R. (1977) *Radical Geography: Alternative Perspectives on Contemporary Social Issues*, Chicago: Maaroufa Press.

Peet, R. (1983) Relations of production and the relocation of United States manufacturing industry since 1960, *Economic Geography*, 59(2): 112–43.

Peet, R. (1999) *Theories of Development*, New York: Guildford Press.

Peet, R. and Thrift, N. (1989) *New Models in Geography*, 2 vols, London: Unwin Hyman.

Soja, E., Morales, R. and Wolff, G. (1983) Urban restructuring: an analysis of social and spatial change in Los Angeles, *Economic Geography*, 59(2): 195–230.

Smith, N. (1991) *Uneven Development*, Oxford: Blackwell.

Smith, N. and Dennis, W. (1987) The restructuring of geographical scale: coalescence and fragmentation of the northern core region, *Economic Geography*, 63(2): 160–82.

Swyngedouw, E. (2000) The Marxian alternative: historical-geographical materialism and the political economy of capitalism, in T. Barnes and E. Sheppard (eds) *A Companion to Economic Geography*, Oxford: Blackwell, pp. 41–59.

Walker, R. (1995) California rages against the dying of the light. *New Left Review*, 209: 42–74.

Walker, R. (2001) California's golden road to riches: natural resources and regional capitalism, 1848–1940, *Annals of the Association of American Geographers*, 91: 167–99.

World Bank (2005) *Reducing Poverty through Growth and Social Policy Reform*, Report No. 28923-RU, World Bank Poverty Reduction and Economic Management Unit Europe and Central Asia Region, New York: World Bank.

CHAPTER 4

COMMODITY CHAINS

Where does your breakfast come from?

Aims

- To demonstrate how capitalism serves to conceal the conditions of commodity production
- To introduce commodity chains and their basic components
- To appreciate the differentiation of commodity chains in terms of their structure and geography
- To recognize the possibilities for, and limitations of, more ethical ways of organizing commodity chains.

4.1 Introduction

Wal-Mart is the world's largest retailer by far. In 2004, the company recorded profits of just over US$10 billion on sales of US$285 billion across 4,900 stores in 10 countries, and its 1.6 million workers sold goods to some 138 million customers each week. While the huge size and ongoing rapid growth of Wal-Mart have been well documented, where do the products it sells come from? For many of the non-perishable consumer goods on the store shelves – such as toys, clothes, and electronics – the answer is increasingly likely to be *China*. Attracted by the availability of good quality, low price goods, the retailer is rapidly expanding its purchasing activities there. In 2004, Wal-Mart sourced US$18 billion worth of goods from China, representing 3 per cent of the country's total exports for the year (US$593 billion), and a remarkable 13 per cent of China's exports to the US (US$136 billion) (http://www.chinadaily.com.cn, accessed 9 September 2005). These figures placed Wal-Mart as China's eighth largest trading partner, ahead of entire national economies such as Australia, Canada and Russia.

This huge sourcing operation is run from Wal-Mart's overseas procurement office in the city of Shenzhen located in China's southern Guangdong province, from which the retailer has established ongoing supply relations with over 5,000 companies. Individual Chinese companies can do huge amounts of business with Wal-Mart. Guangdong's Yili Electronics Group, for example, started supplying hi-fi systems to the retailer in 1995, and now supplies Wal-Mart with over US$200 million worth of goods each year, accounting for half of Yili's total sales. And Wal-Mart is not alone. Other leading transnational retailers such as Carrefour and Auchan (France), Metro (Germany), Makro (the Netherlands), B&Q (the UK), Ikea (Sweden), and Home Depot (the US) also have extensive sourcing operations in the coastal provinces of China (see Chapter 10, for more on these retailers). In dynamic terms, in almost every case the share of total purchases accounted for by China is growing rapidly.

Why are these trends significant? On one level, they are indicative of the emergence of China as an economic force in the global economy, and the cost-based globalization of manufacturing production. Labour costs in China's manufacturing sector are just 4 per cent of those in the US, meaning that a product can be manufactured in China, packaged, shipped around the world, and sold to an American or European consumer and still return a decent profit for both the manufacturer and the retailer. On another level, and more importantly for this chapter, the example of Wal-Mart's Chinese sourcing opens a window on the way in which distant producers and consumers are connected together by particular commodities that travel the globe on complex journeys from their origins as raw materials, to their final destinations as items bought from the shelf of retailers and then consumed. More specifically, whether they realize it or not, the fortunes of the workers in Yili's factory in Guangdong province are intricately connected to consumers in, say, Austin, Texas – and indeed in many other thousands of locations where Wal-Mart has stores – who buy one of their hi-fi products. Equally, they are connected to workers in the many firms across China and other parts of Asia that supply components to Yili, to the shipping and logistics firms that move the goods from China to the US, to Wal-Mart's own employees, and to specialized service firms that may be called upon to fix the hi-fi should it break or need servicing once purchased. And we could easily trace this web of connections still further. The key point, however, is that one commodity – in this case a hi-fi appliance manufactured in China – connects a diverse range of people together in profoundly important ways, as it traces a chain of connections across the global economy. The simple 'Made in China' label printed on the box nowhere near does justice to the sheer complexity of these spatial and structural interdependencies.

But how can we conceptualize the complex journeys taken by commodities across the global economy? In Chapter 3 we learnt how value is produced in specific, yet interconnected, locations as part of processes of capitalist uneven development. In this chapter, we develop the notion of *commodity chains* as a way of understanding the connections and interdependencies between different workers, consumers, firms and institutions involved in the production, distribution and consumption of commodities. There are four main sections in the chapter. First, we consider how capitalism as a system hides the connections or social relations inherent to a particular commodity, and reveal the implications of this concealment (Section 4.2). Second, we explore the nature of commodity chains, revealing how they vary in terms of their structure, geography, coordination, and context (Section 4.3). Third, we look at different ways of organizing commodity chains that serve to highlight the various connections along the chain and endeavour to improve them (Section 4.4). Fourth, we evaluate the potential limitations and contradictions that such ethical interventions may generate (Section 4.5).

4.2 Capitalism, Commodities and Consumers

As explained in Chapter 3, capitalism can be thought of as a *commodity* exchange system. A commodity is simply something useful that enters the market and is available for purchase. However, commodities are much more than just material things, such as books or food. In the contemporary world, more and more areas of our everyday life have become caught up in processes of *commodification* (see also Chapter 6 on nature). Domains as varied as culture (e.g. museums and galleries), religion (e.g. celebrity preachers), knowledge (e.g. intellectual property rights), the environment (e.g. carbon credits), war (e.g. private armies) and even the human body (e.g. trade in human organs) have become commodified.

While commodities are central to the capitalist system, at the same time they may serve to hide important dimensions of how they are produced. The *exchange value* of a commodity – i.e. the price – is often indicative of how the commodity was created: the cost of the human labour that went into its production, the costs of machinery, buildings, electricity, trucks and so on that were required, and the profits extracted at various points in the process. And yet the simple price-tag itself reveals nothing of the production process the commodity has undergone and the necessary social relations that connect this production to the commodity's eventual customer. As a result, consumers in the capitalist system are largely ignorant of the geographical origins and histories of the commodities that they consume. The purchase of a commodity for money serves to *disconnect* producers and consumers, encouraging an abdication of responsibility on the part of consumers for the terms and conditions under which the commodity was made. The consumer can simply benefit from the *use value* of whatever they have purchased, i.e. the usefulness of a particular product to an individual. This poses profound challenges to both conscientious consumers who actively want to know the history of the commodities they consume, and economic geographers who want to understand connections and interdependencies within the global economy. In reality, even just drinking a coffee in a café such as Starbucks makes the consumer complicit – albeit unknowingly in many cases – in complex webs of connections across the globe (Box 4.1).

Box 4.1 Coffee, cafés, and connections

Founded in Seattle in 1971, Starbucks has become the largest chain of coffeehouses worldwide. In August 2005, Starbucks had some 6,888 locations in the US, and 2,783 across a further 34 countries, and was serving an estimated 33 million customers each week. New coffeehouse openings were proceeding at a rate of three to four stores worldwide *per day*. The company offers a range of over 30 coffees and teas in addition to a wide variety of snacks and other beverages. Through its marketing and store information strategies, Starbucks endeavours to create a 'knowledgeable' culture of coffee drinking in its cafés. The corporate website, for example, has an extensive 'Coffee Education' area, one page of which describes and contrasts the coffees from Latin America, Africa, and Southeast Asia (under the heading '*Geography is a flavour*': see Figure 4.1). Elsewhere on the website, individual coffees are described in evocative tones such as '*Sulawesi: smooth, buttery, earthy and elegant*'. The strategy is clearly to turn coffee drinking from a routine activity into a more meaningful consumption process involving certain kinds of knowledge about coffee as a commodity with a particular history and geography.

However, it is possible to offer a more critical reading of this sophisticated marketing strategy. The information on offer in Starbucks presents a highly partial interpretation of coffee and its production process (see also Figure 4.5). The structures of domination and exploitation inherent in today's global coffee industry – and indeed their colonial origins – are entirely overlooked. The story of the global coffee industry in the past decade has been one of rising production and falling prices (at least until very recently), and as a result, increasingly marginal working and living conditions for millions of farmers and farm workers in a range of poor tropical countries, many of them highly dependent on coffee exports (which make up over 50 per cent of Ethiopia's total exports, for example). Control of the industry chiefly rests with a small group of Western buying, roasting, and processing companies, the top five of which – Kraft, Nestlé, Sara Lee, Procter and Gamble, and Tchibo – account for 44 per cent of global coffee roasting. These companies have been able to maintain healthy

Figure 4.1 Geography is a flavour . . .
Source: Clive Agnew, with permission.

profit margins on their various brands, despite the price downturn since the mid-1990s. Western retailers similarly continue to do well from selling coffee. Starbucks, in short, deliberately offers a highly selective and romanticized reading of the global coffee industry in its literature and store displays. As a commodity, however, coffee also has many other less palatable stories to tell. For more, see Smith (1996), Oxfam (2003), and http://www.starbucks.com.

Moreover, the *images* we receive about commodities in our everyday life may actively serve to further conceal the origins of commodities. Advertising – a significant economic sector its own right – is extremely important here. Through the creation of various images, advertisers seek to establish time- and place-specific meanings for particular goods and services that may be a far cry from the realities of their production. Think, for example, of adverts for gold jewellery in developed country markets. Through skilfully combining pictures and words, these adverts tend to emphasize certain values and emotions that are associated with the products: love, passion, romance, commitment, and so on (for an engaging attempt to destabilize and subvert advertisements and their central messages, see http://www.adbusters.org).

But a more critical reading might ask what is missed out in this representation of gold as a commodity? A gold necklace bought in a jewellery store in a rich world city (Figure 4.2) may be the end point of a series of links that connect consumers to high-security global logistics firms, ring manufacturers in Italy, gold traders in Zurich, black male migrant miners from Lesotho working in appalling conditions in South Africa's gold mines and women left behind in

Figure 4.2 The jewellery shop window – the start or the end of a complex commodity chain?
Source: The authors.

Lesotho working long hours for negligible pay in the textile industry. In this way, 'the gold windows of Tiffany's in New York are linked to the gold widows in Lesotho' (Hartwick, 1998: 433). The harsh reality of the gold industry is that notions of love and commitment stand rather incongruously alongside the legacies of South African apartheid, slave-like working conditions in mines, and abandoned women working in unregulated factories. Curiously, in some cases, places of origin are constructed in certain ways to make the products more 'appealing' and 'positive'. Many up-market consumers are willing to pay more for products made in places well known for them – Swiss watches, Italian clothing, French wines, German cars, and Japanese digital cameras. Packaging of products often comes with labels that give a caricatured view of their places of origin. Coffee beans are often packaged and labelled to emphasize their tropical 'exotic' origins, never mind the poverty level of these severely underdeveloped countries. Advertising, then, acts as a powerful force further accentuating the disconnection of producers from consumers.

By now, the significance of the question in the chapter's title – where does your breakfast come from? – should be clearer. It alludes to the fact that even the most mundane and everyday acts of consumption tie us into these webs of connections, as the following passage makes clear:

> Consider, for example, where my breakfast came from. The coffee was from Costa Rica, the flour that made up the bread probably from Canada, the oranges in marmalade came from Spain, those in the Orange juice came from Morocco and the sugar came from Barbados. Then I think of all the things that went into making the production of those things possible – the machinery that came from West Germany, the fertiliser from the United States, the oil from Saudi Arabia . . . It takes very little investigation for the map of where my breakfast came from to become incredibly complicated. I also find that literally millions of people all over the world in all kinds of different places were involved just in the production of my breakfast! The odd thing is that I do not have to know all that in order to eat my breakfast. Nor do I have to know it when I go shopping in the supermarket. I just lay down the money and take whatever it will buy. Most of us know next to nothing about how our breakfast got on the table and even less about the conditions of life of the millions of people involved in its production.
>
> (Harvey, 1989: 3)

Commodities, then, need to be thought of as much more than just their immediate market and use values. Instead, every commodity should be seen as 'a bundle of social relations' (Watts, 1999: 307), or, put another way, as representative of the whole system of connections between different groups of people that have enabled the consumer to make a purchase. In this way, the working conditions that underlie commodity production – and which may be unacceptable to certain consumers – can be revealed, challenged and, eventually, improved. As the quotation suggests, in the contemporary era, this is increasingly about revealing

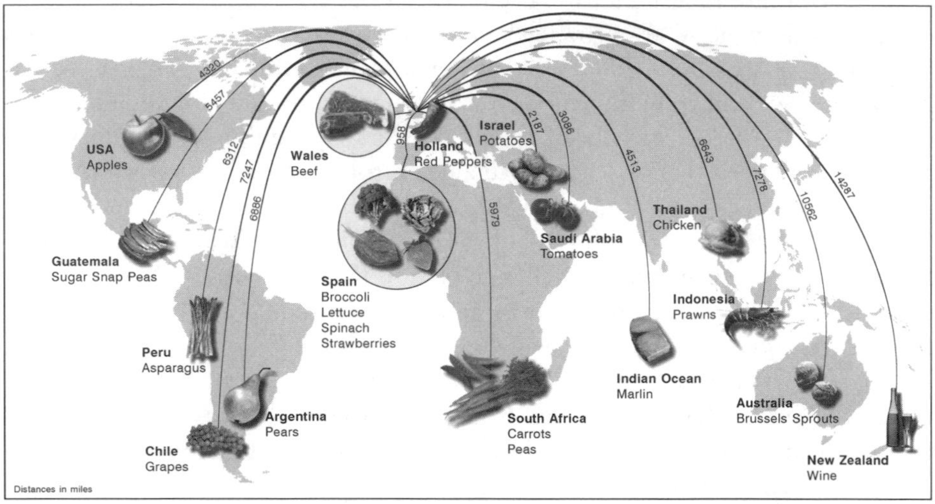

Figure 4.3 Food geographies – a basket of produce bought in a UK supermarket
Source: Adapted from *The Guardian*, 2003, p. 18.

interdependencies at the *global* scale, even in the case of relatively perishable foodstuffs. Figure 4.3, for example, illustrates how a regular basket of goods bought in the UK may have cumulatively travelled tens of thousands of miles to reach the supermarket shelves. We now move on to look at how the idea of the *commodity chain* can help us connect together the many different actors involved in producing goods and services.

4.3 Linking Producers and Consumers: The Commodity Chain Approach

How then do we bring together all the diverse actors involved in the global travels of a hi-fi system or a breakfast food? The notion of the *commodity* or *production chain* is central to conceptualizing such systems. Figure 4.4 outlines a basic commodity chain, illustrating the transformation of initial material and non-material inputs into final outputs in the form of goods and/or services. This transformation includes primary activities (e.g. production, marketing, delivery, and services) and support activities (e.g. merchandising, technology, finance, human resources and overall infrastructure). The commodity chain, then, is not simply about manufacturing processes: many of the inputs to the chain, and many of the final commodities produced, will take the form of intangible services. In producing a tangible product (e.g. a mobile phone) or a service (e.g. consumer or merchant banking), the various activities are linked together in a chain-like fashion, with each stage adding value to the process of production of

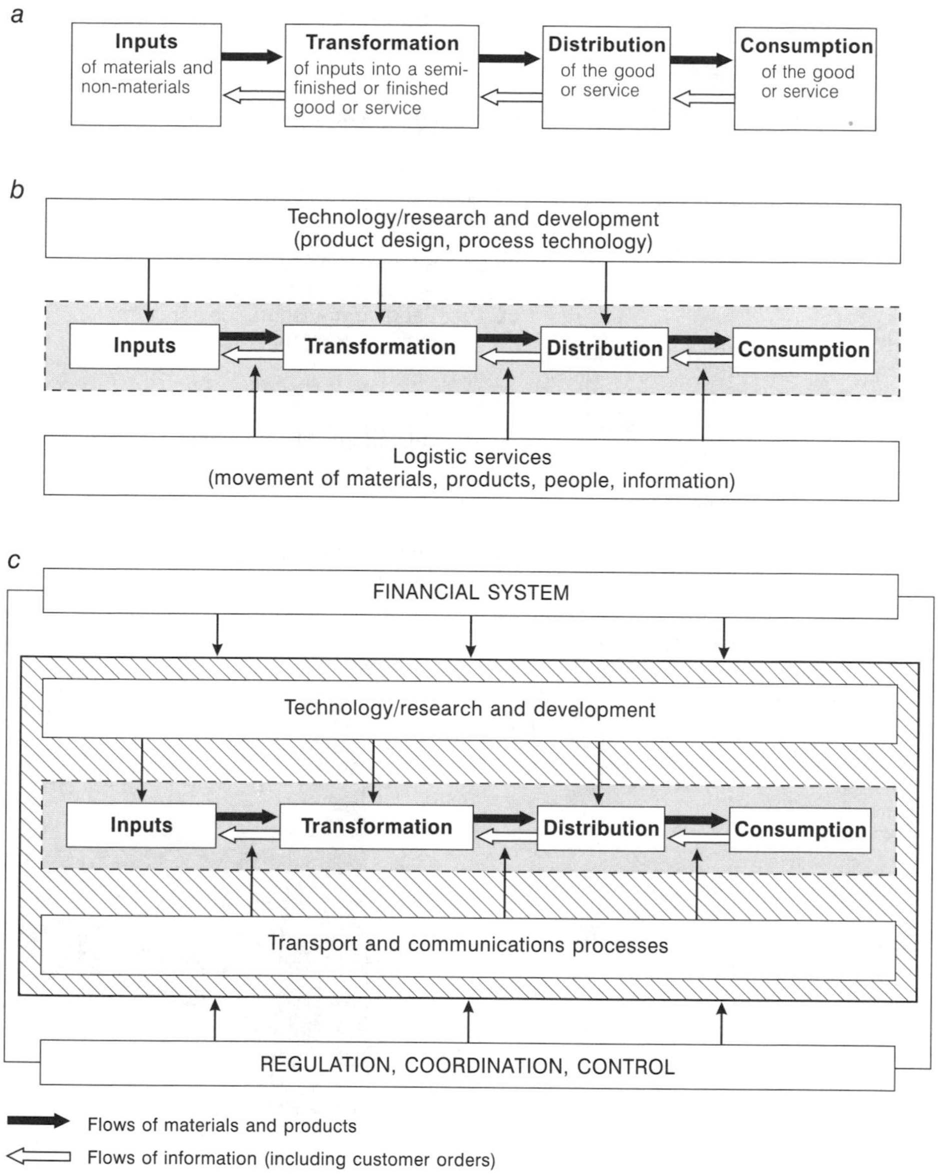

Figure 4.4 The basic commodity or production chain
Source: Dicken (2003). Figure 2.3. Reprinted by permission of Sage Publications.

the goods or services in question. One good example of such a commodity chain is coffee – a recurrent theme in this chapter. Further to the geographies of coffee beans explained in Box 4.1, Figure 4.5 shows how the market value (price traded) of a bag of coffee beans is derived and the costs and profits encountered by various producers and intermediaries.

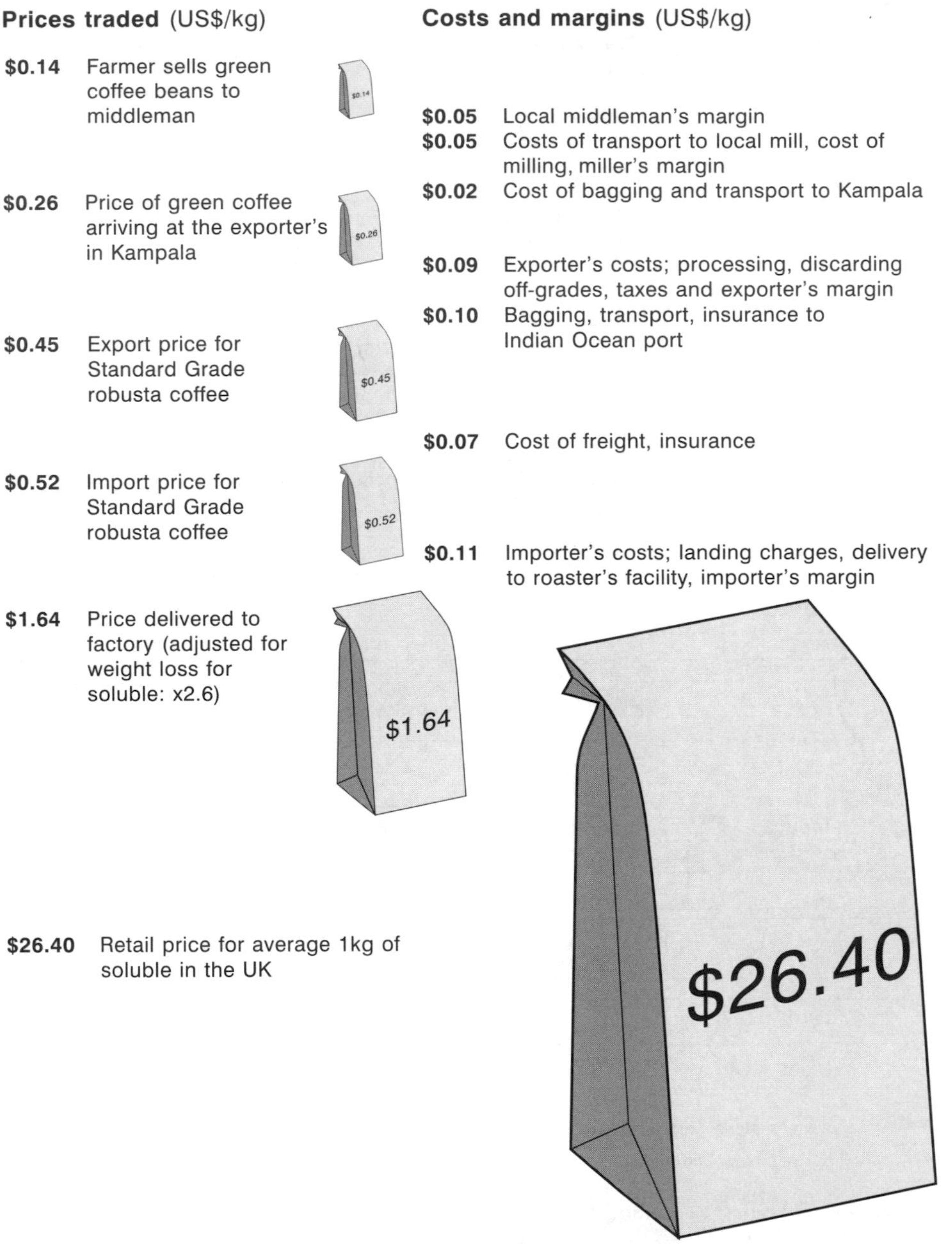

Prices traded (US$/kg)
$0.14 Farmer sells green coffee beans to middleman
$0.26 Price of green coffee arriving at the exporter's in Kampala
$0.45 Export price for Standard Grade robusta coffee
$0.52 Import price for Standard Grade robusta coffee
$1.64 Price delivered to factory (adjusted for weight loss for soluble: x2.6)
$26.40 Retail price for average 1kg of soluble in the UK
Costs and margins (US$/kg)
$0.05 Local middleman's margin
$0.05 Costs of transport to local mill, cost of milling, miller's margin
$0.02 Cost of bagging and transport to Kampala
$0.09 Exporter's costs; processing, discarding off-grades, taxes and exporter's margin
$0.10 Bagging, transport, insurance to Indian Ocean port
$0.07 Cost of freight, insurance
$0.11 Importer's costs; landing charges, delivery to roaster's facility, importer's margin
$0.14
$0.26
$0.45
$0.52
$1.64
$26.40

Figure 4.5 The coffee value chain
Source: Adapted from Oxfam (2003: 24) (no figure number).

Understanding the sequence and range of actors involved in a particular commodity chain – sometimes known as its *input–output structure* – is the first step towards developing a good understanding of commodities and their production processes. However, there are three further important dimensions to all commodity chains that we will now consider in turn: their geography or *territoriality*; the way in which they are coordinated and controlled i.e. their *governance*; and the way in which local, national and international conditions and policies shape the various elements in the chain, i.e. their *institutional frameworks* (Gereffi, 1994).

Spatial structures

In very simple terms, the geography of a commodity chain can range from being *concentrated* in one particular place, to being widely *dispersed* across a range of localities. As the earlier quotation about breakfast commodities makes vividly clear, it is hard to identify a commodity chain in the contemporary global economy that is not global to at least some degree, even if it is just seen in the sourcing of one or two inputs, or a limited export market for the final good/service. Many bring together an extensive range of international connections. *Global commodity chains*, as they have come to be called, are one of the primary organizational features of the world economy (Gereffi, 1994). This territoriality is important not only because it determines precisely which actors are connected together across the global economy, but also in revealing the unequal geographical distribution of value, and associated economic development benefits, between different points along the chain. As we saw in Chapter 3, the location of high value-added activities (e.g. design, marketing, etc.) in global cities is particularly important in the spatial inequality engendered through these commodity chains.

We can make five further arguments about the territoriality of commodity chains. First, in general, the *geographical complexity* of global commodity chains is increasing, enabled by a range of developments in transport, communication and process technologies. This complexity is illustrated in Figure 4.6, which shows the production system for hard disk drives, a key component of personal computers. The various tasks shown are distributed across a wide range of countries. Globally, the majority of disk drive assembly takes place in Southeast Asia, particularly Singapore and Thailand. In 2000, Singapore maintained a 35 per cent share of the world's hard disk drive market by volume. The leading American manufacturer Seagate, for example, splits the tasks shown in Figure 4.6 between operations in the US, China, Thailand, Malaysia, Indonesia, the Philippines and Singapore, the latter being a key hub where a lot of the final assembly and testing takes place. If the commodity chain was extended to include the production process of the whole personal computer, and incorporate its final consumption, the geographic complexity would, of course, increase still further (see examples in Chapter 8).

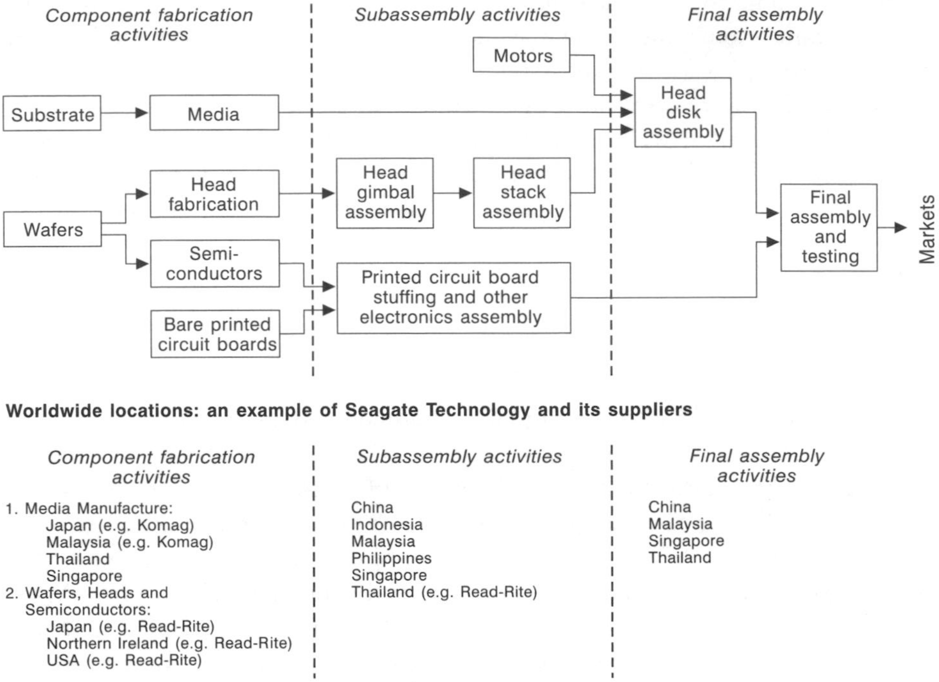

Figure 4.6 The geography of the hard disk drive commodity chain
Source: Adapted from Gourevitch et al., (2000), Figures 1 and 3.

Second, the geographic configurations of global commodity chains are becoming *more dynamic* and liable to rapid change. This flexibility derives both from the use of certain space-shrinking technologies (see Chapter 5), and from organizational forms that enable the fast spatial switching of productive capacity. In particular, this flexibility comes from the increased use of *external* subcontracting and strategic alliance relationships that allow firms to switch contracts between different firms and places without incurring the costs of moving production themselves (see more in Chapter 8). Third, and relatedly, understanding the geography of commodity chains is not as simple as locating each stage in a particular place or country. Global commodity chains also reveal the dynamics of *inter-place competition*. Firms in different localities may be vying for market share at different points along the chain. The global catfish commodity chain is revealing here (Figure 4.7). Catfish farmers in Vietnam's Mekong delta are not just connected to the US as a key export market, but also through relations of competition with catfish producers in the Mississippi delta region who are also seeking to sell in the US market. In other words, different localities involved in a global commodity chain may engage in competitive upgrading strategies to protect market shares and profitability (Box 4.2).

Box 4.2 Upgrading strategies in global commodity chains

It is important to view the value structures of commodity chains in dynamic terms. Upgrading refers to the potential for firms, or groups of firms, to improve their relative position within the system as a whole. It is useful to distinguish between four different types of upgrading (Humphrey and Schmitz, 2004):

- *Process upgrading*: Improving the efficiency of the production system by either reorganizing the production process or introducing superior technologies. For example, a car manufacturer might introduce robot technology to speed up its assembly lines.
- *Product upgrading*: Moving into making more sophisticated products or services. For example, a basic food processing firm might start making prepared frozen meals, or a financial firm might offer new kinds of insurance products.
- *Functional upgrading*: Acquiring new roles in the chain (and/or abandoning existing functions) in order to increase the overall skill content and level of 'value-added' of the activities undertaken. An electronics manufacturer, for example, might move from simple assembly to original equipment manufacturing (OEM) to own-design manufacture (ODM) to own-brand manufacturing (OBM) (see more in Chapter 8).
- *Inter-sectoral upgrading*: Using the knowledge derived from a particular chain to move into different sectors. For example, a firm might use its experience of making televisions to enable it to move into computer monitor making.

At the level of the individual firm, successful upgrading strategies can transform fortunes. The Turkish clothing firm Erak, for example, has developed from a clothing manufacturer supplying Germany's Hugo Boss into a global clothing brand and retailer through its Mavi jeans products. More broadly, successful upgrading lay behind the emergence of the newly industrializing economies of Asia and Latin America from the 1960s onwards (see also Box 7.2). The Taiwanese electronics industry, for example, has benefited from all four kinds of upgrading processes to develop from a base for foreign-owned electronics assembly into one of the world's leading centres for designing and producing new computer technologies in the global economy. For many other developing countries, facilitating upgrading across a wide range of sectors remains a key policy concern, as they seek to gain a greater share of the spoils of global commodity chains for clusters of local firms.

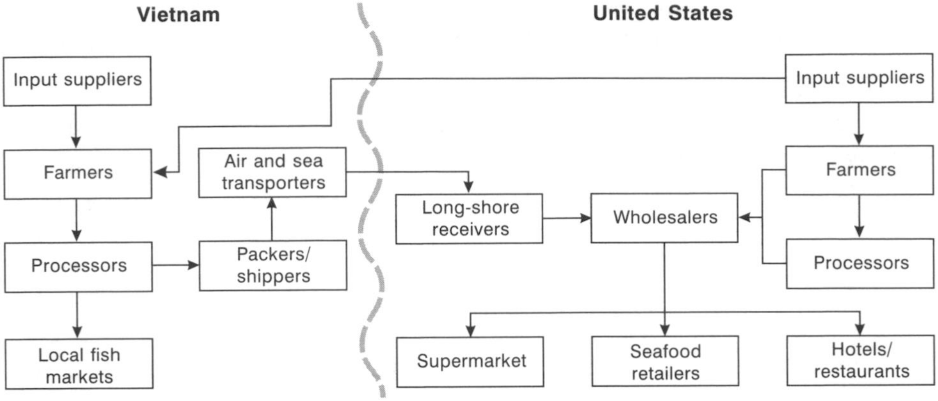

Figure 4.7 The catfish commodity chain
Source: Adapted from Duval-Diop and Grimes (2005), Figure 5.

Fourth, it is important to re-emphasize that global commodity chains are not just a feature of agricultural and manufacturing sectors, but are also apparent in many *service sectors*. For example, many service firms now find it advantageous to conduct routine data processing and software programming functions in overseas sites – India, Mauritius, Jamaica and Trinidad and Tobago are prime examples – where there is relatively low cost labour. Fifth, and finally, we need to connect these ideas about the geographical extensiveness and complexity of global commodity chains with the arguments about the geographical *clustering* of economic activity (see Chapter 5 for more). Some kinds of interactions within commodity chains will take place within the same locality, due to, for example, the intensity of transactions or the importance of place-specific knowledge to the activity in question. For these 'nodes in global networks', commodity chains are the organizational forms that connect clusters and agglomerations together.

Understanding the input–output structure and territoriality of a commodity chain is undoubtedly important, but still leaves questions unanswered. Think of Wal-Mart. Who controls the organizational structure and nature of its global commodity chain? Who decides where inputs are purchased from, and where final goods and services are sold? Who shapes the restless geographies of commodity chains? This brings us on to the important issue of governance.

Management processes

Thus far we have seen how commodity chains are constituted by a mix of intra-firm and inter-firm linkages, and a combination of near and distant connections. In most cases, however, the chain will have a primary coordinator or a lead firm 'driving' the system as a whole. In general terms, it is helpful to distinguish

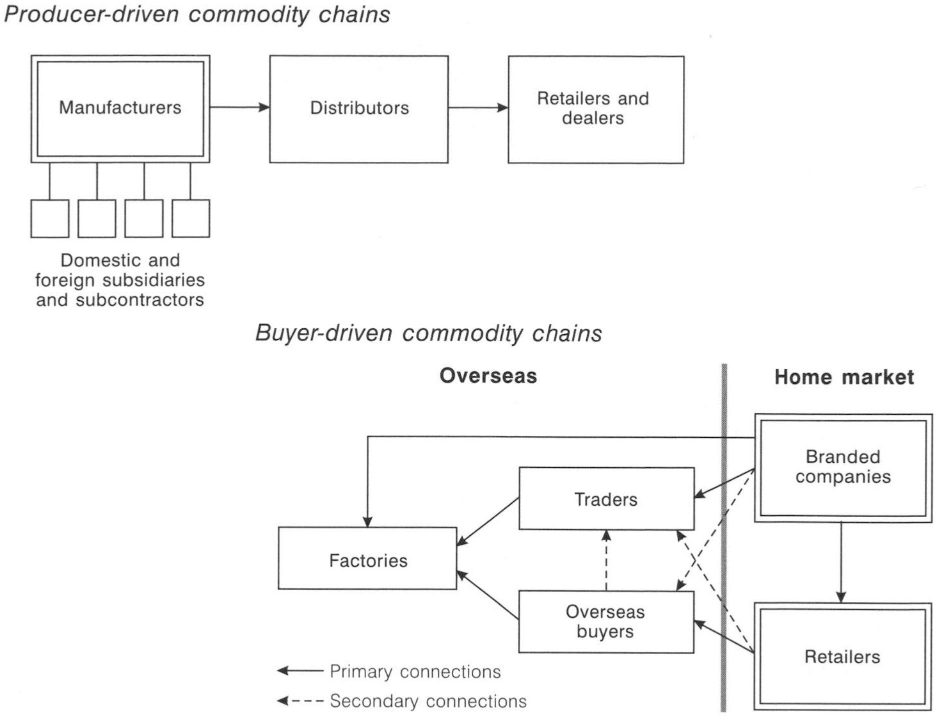

Figure 4.8 Producer-driven and buyer-driven commodity chains
Source: Adapted from Gereffi (1994), Figure 5.1.

between chains that are *producer-driven* and those that are *buyer-driven* (Gereffi, 1994: see Figure 4.8).

Producer-driven chains are commonly found in industries where large industrial transnational corporations (TNCs) play the central role in controlling the production system (see also Chapter 8). This situation is found in many capital- and technology-intensive industries such as aircraft, automobile, computer, semiconductor, and machinery manufacturing. These chains are notable for the degree of control exercised by the administrative headquarters operations of TNCs. Producers dominate such chains not only in terms of their earnings and profitability, but also through their ability to exert control over backward linkages to raw material and component suppliers, and forward linkages with distributors and retailers. Profits are secured through the scale and volume of production in combination with the ability to lead technological developments. The automobile industry provides an excellent example of this kind of chain. Leading assemblers such as Toyota and Ford coordinate production systems involving literally thousands of subsidiaries and tiers of subcontractor firms dotted around the world, as well as extensive global networks of distributors and dealers.

Table 4.1 Characteristics of producer-driven and buyer-driven chains

	Form of economic governance	
	Producer-driven	*Buyer*-driven
Controlling type of capital	Industrial	Commercial
Capital/technology intensity	High	Low
Labour characteristics	Skilled/high wage	Unskilled/low wage
Controlling firm	Manufacturer	Retailer
Production integration	Vertical/bureaucratic	Horizontal/networked
Control	Internalised/hierarchical	Externalised/market
Contracting/outsourcing	Moderate and increasing	High
Suppliers provide	Components	Finished goods
Examples	Automobiles, computers, aircraft, electrical machinery	Clothing, footwear, toys, consumer electronics

Source: Adapted from Kessler and Applebaum (1998).

Buyer-driven chains, on the other hand, tend to be found in industries where large retailers (e.g. Wal-Mart, Carrefour, or Ikea) and brand-name merchandisers (e.g. Adidas, Nike, and The Gap) play the central role in establishing and controlling production systems, usually located in export-oriented developing world countries (see Figure 4.8). This form of commodity chain is common in labour-intensive consumer goods sectors, such as clothing, footwear, toys, and handicrafts. Production is generally carried out through tiered networks of subcontractors that make finished goods subject to the specifications of powerful buyers. Profits in these chains are derived from the bringing together of design, sales, marketing, and financial expertise, allowing the retailers and merchandisers to connect overseas factories with the main consumer markets. Hence control is enacted through the ability of firms to shape mass consumption patterns through strong brand names and meet this demand through global sourcing strategies. The basic characteristics of producer- and buyer-driven commodity chains are summarized and contrasted in Table 4.1.

The distinction between producer- and buyer-driven commodity chains is a useful first step towards understanding chain governance. In reality, however, governance is far more complex and variable both within, and between, different economic sectors. Equally, a chain need not necessarily be coordinated by either a large manufacturer or a retailer. Box 4.3 describes a particular kind of intermediary in coordinating global commodity chains – the Japanese *sogo shosha*. As neither manufacturers nor retailers – but well positioned in networks of manufacturers and retailers, these very large trading companies possess strong organizational capabilities to coordinate even the most complex global commodity chains.

Box 4.3 Trading giants – the Japanese sogo shosha

The Japanese *sogo shosha* provide a fascinating example of the importance of logistics and distribution companies in the global economy. *Sogo shosha* translates directly as 'general trading company', but this does not really do justice to the range of functions they perform. The *sogo shosha* have long played an important role in the Japanese economy. During the 1960s, they were the first Japanese firms to venture overseas, acting as global marketing and intelligence gathering networks and thereby facilitating subsequent trade and Japanese foreign direct investment. The seven leading *sogo shosha* – Marubeni, Mitsui, Mitsubishi, Itochu, Nissho-Iwai Nichimen, Sumitomo, and Tomen – are now massive commercial, financial, and industrial conglomerates. They each operate huge networks of subsidiaries and affiliates across the global economy. In 2003, for example, Marubeni employed some 25,000 staff and registered total sales of almost US$70 billion across 263 affiliates in 38 countries, and Mitsubishi had 47,400 staff and total sales of US$134 billion from 309 affiliates in 31 countries (http://www.unctad.org, accessed 6 October 2005). Each of these firms handles tens of thousands of different products. More specifically, they perform four particular functions (Dicken and Miyamachi, 1998):

1 *Trading intermediation*: matching buyers and sellers in long-term contractual relationships.
2 *Financial intermediation*: serving as a buffer between buyers and suppliers.
3 *Information gathering*: gathering and synthesizing information on market conditions around the world.
4 *Organization and coordination of complex business systems*: as seen in the case of large infrastructure projects.

The role of the *sogo shosha* has changed in two important ways over the past two decades. First, their share of Japanese trade has fallen substantially: in 1991, they accounted for almost 60 per cent of Japan's imports and 50 per cent of exports, while by 2002 these figures had dropped to 22 and 12 per cent respectively. In part, these changes reflect how Japan's manufacturing companies have established their own marketing and sales networks overseas. Second, the nature of the business undertaken by the *soga shosha* has changed, with 79 per cent of foreign affiliates being involved in service activities – including finance, insurance, transportation, and project management – in contrast to the historic focus on manufacturing.

In this regard, it is useful to think of *relational* forms of governance that fall in between the producer- and buyer-driven models (Gereffi et al., 2005). These can be thought of as close inter-firm relationships that develop on a relatively even footing. The example of the trade in fresh vegetables such as peas and beans between Kenya and the UK is useful here (Dolan and Humphrey, 2004). Up until the mid-1980s, trade was handled through a series of market relationships between Kenyan farmers and exporters, and UK-based importers and traders, wholesalers and retailers. However, as UK supermarkets such as Tesco, Sainsburys and Asda have grown in size and expanded their market share, they have started to control and coordinate more directly the commodity chain. In order to try and attract customers, the supermarkets have introduced new varieties of fresh goods, placed heavy emphasis on quality, provided year-round supply, and increased the processing of products so they require little or no preparation before cooking. As a result, instead of buying vegetables through wholesale markets, they have developed closer, non-market-based relationships with UK importers. By the mid-1990s, a relational governance system had emerged in which the supermarkets work directly with a limited number of UK importers with whom they have established long-term relationships. The importers have moved beyond a trading role, offering a range of services including processing and handling, monitoring quality, finding new sources of supply, and supporting African producers, and assessing their performance. The governance of a commodity chain should always be seen in dynamic terms, however. Over time, the previously buyer-driven relationships between importers and Kenyan exporters are also becoming more relational as the capabilities of the exporters grow and they are able to take on more processing of the produce.

Institutional contexts

Commodity chain governance, then, is both a complex and a highly dynamic affair. Its nature does not just depend on the sector or industry in question, but also on the precise array of places that are connected together by the chain. This is because every point on the chain is connected to, and shaped by, the *institutional context* in which it is situated or embedded. In reality, the intersections between complex commodity chains (e.g. Figure 4.6) and their institutional contexts are many and varied.

We need to discern between different institutional contexts in two ways to make sense of this complexity. First, we can distinguish between *formal* and *informal* institutional frameworks. The former relates to the rules and regulations that determine how economic activity is undertaken in particular places (e.g. trade policy, tax policy, incentive schemes, health and safety/environmental regulations, etc.), while the latter describes less tangible, place-specific *ways-of-doing business* that relate to the entrepreneurial and political 'cultures' of particular places (see also Chapter 11). Second, it is useful to think about how institutional context is important at different *spatial scales*. At the sub-national scale, local and regional governments may implement a range of policies to try and promote certain kinds of economic development in the locality (e.g. tax holidays for firms that undertake more research and development). At the national scale, nation-states still wield a huge range of policy measures to try and promote, and steer, economic growth within their boundaries (see Chapter 7 for more). At the macro-regional scale, a variety of regional blocs have considerable influence on trade and investment flows within their jurisdiction. And at the global scale, institutions such as the World Trade Organization (WTO) and the International Monetary Fund (IMF) shape the rules-of-the-game for global financial and trade relationships. Even a relatively simple global commodity chain, then, will cross-cut and connect a wide range of multi-scalar institutional contexts. A banana commodity chain linking Ecuador and France, for example, may be affected by corporate strategies of French and American banana firms, Ecuadorian and French economic policies, by the rules and regulations of the Andean Common Market and the European Union, by WTO rules and regulations, in addition to any more localized policy initiatives within the two countries. In St Lucia, a small Caribbean island state with a population of about 150,000, two-thirds of its 10,000 banana farmers lost their revenue when its traditional export market, the UK, had to surrender its preferential trade arrangement with St Lucia in August 2005. This is because US-controlled banana companies such as Chiquita filed complaints with the WTO against the British trade preference for its former colony and the WTO ruled in Chiquita's favour.

We can further illustrate the profound impacts that changing institutional contexts can have on commodity chains by returning to the example of the coffee industry. Over 90 per cent of the world's coffee is grown in the developing world (Central and Latin America, Africa, and East and Southeast Asia), while the vast majority is consumed in developed countries. The coffee commodity chain that straddles these two groups of countries is depicted in Figure 4.9 (Ponte, 2002). The changing institutional context for the coffee chain can be considered at both the global and national scales. On the *global* scale, in the period 1962–89, the international trading of coffee was governed by a series of International Coffee Agreements (ICAs) managed by the International Coffee Organization (ICO), a supra-national organization constituted by representatives of a wide range of coffee importing and exporting countries (see http://www.ico.org). The ICAs combined price bands and export quotas to provide a coffee trading system that was widely credited with raising and stabilizing coffee prices. At the *national* scale, many exporting countries established coffee marketing boards. These were government institutions that controlled markets and monitored quality within producing countries and acted as a link to exporters and international traders (Figure 4.9). For individual farmers and growers, they provided an important buffer between themselves and international markets.

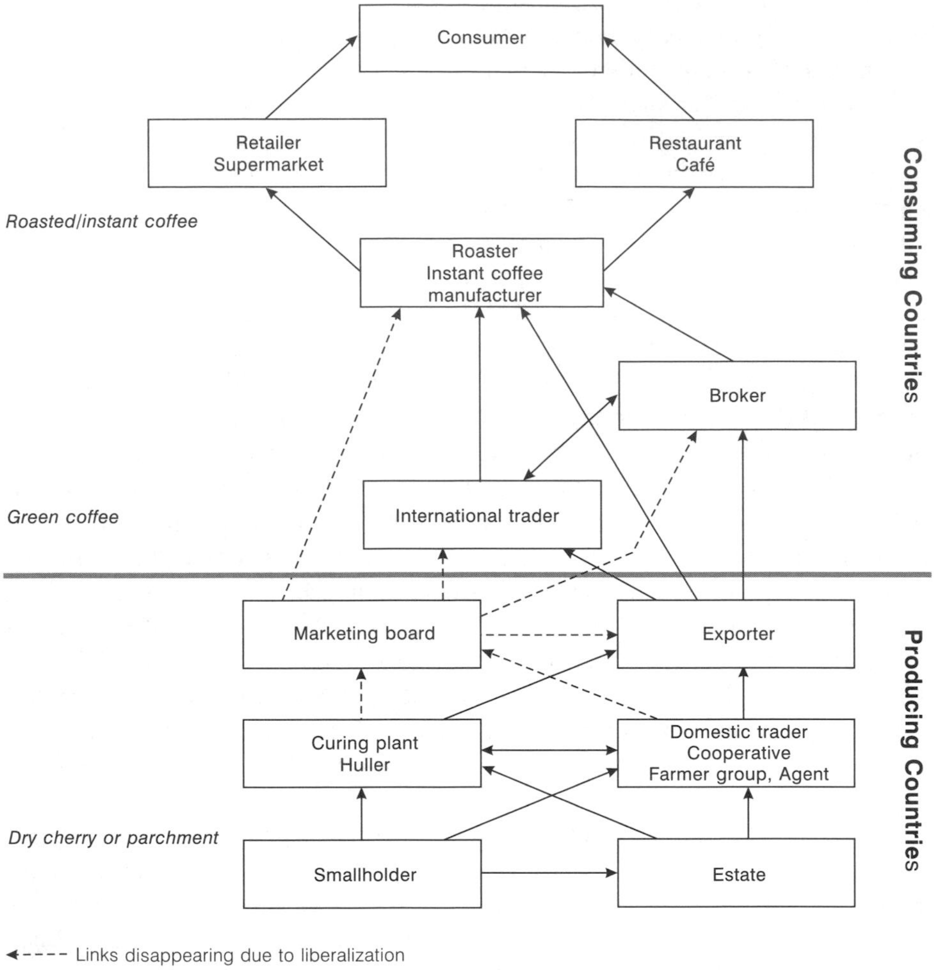

Figure 4.9 The coffee commodity chain: the changing institutional framework
Source: Adapted from Ponte (2002), Figure 1.

In 1989, however, the ICA was not renewed in the face of rising production levels and low-cost competition from non-member exporting countries. The ending of the ICA regime has dramatically altered the balance of power in the coffee chain, as the now liberalized, market-based coffee trade regime has led to lower and more volatile coffee prices. Power has been concentrated in the hands of consuming country firms, and in particular a small group of roasters and instant coffee manufacturers (Box 4.1). These firms are also increasingly applying stringent quality standards that have implications all the way down the chain to the farmers. It has been estimated that the percentage of income from the coffee chain that is retained in the developed markets has gone up from 55 to 78 per cent in the post-ICA era, while the proportion of income accrued by growers has fallen from 20 to 13 per cent (see also Figure 4.5). At the same time, the national coffee marketing boards in the exporting countries have either been eliminated, or have retreated into a restricted overseeing role that has left them marginalized within the commodity chain (Figure 4.9). As a result of these changes to the inter-linked international and national institutional contexts, millions of coffee smallholders and farmers worldwide are left open to the full force of the global coffee market and its fluctuating prices. At its worst, this can lead to a situation where farmers receive less for their crop than it actually costs them to produce it. The example of coffee clearly show how changing institutional frameworks can significantly affect all three of the other basic dimensions of a commodity chain, namely the input–output structure (e.g. the bypassing of coffee marketing boards), territoriality (e.g. the rapid growth of production in Vietnam), and governance (e.g. the accumulation of power with roasting/processing firms).

Overall, this section of the chapter has shown how commodity chains are organizational platforms that link producers and consumers together, within certain institutional contexts, across the global economy. The precise form taken by individual commodity chains varies greatly – both within and between different sectors of the economy – in terms of their structure, geography, governance and institutional context. Understanding this variability and complexity is a vital step towards trying to *change* how commodity chains function. We now move on to consider the potential for different ways of trying to change, or *re-regulate*, commodity chains in ways that negate some of the less desirable aspects of commodity production, circulation and consumption.

4.4 Re-regulating Commodity Chains: The World of Standards

There are many different ways in which various players may seek to alter the prevailing ways in which a commodity chain operates. In Chapter 6, we will look at how efforts can be made to mitigate the environmental impacts of certain commodity chains. In Chapter 9, we will see how certain groups of workers have the necessary *agency* to challenge established ways of working in their industry through engaging in different forms of *production politics*. Here, however, we want to explore the potential for different forms of *consumption politics*, i.e. interventions initiated at the consumption end of commodity chains through a desire to improve conditions at various points 'up' the chain. This is often also termed *ethical consumption* in that it seeks to be actively aware of social and geographical connections inherent to commodity chains and their implications.

Perhaps the simplest form of consumer campaign is to *boycott* – i.e. not purchase – the products of a particular company. For example, a boycott launched in 1977 against the giant Swiss foodstuffs company Nestlé, prompted by concerns over the company's promotion of baby milk formula in the developing world, is still in operation across 20 countries (see http://www.ibfan.org for more).

We can also identify *corporate campaigns* that seek to target highly visible corporations by mobilizing information regarding violations against workers or the environment. The United Students Against Sweatshops (USAS) campaign in the US, for example, used a series of university campus protests and sit-ins in 1999 to force university administrators and firms such as Nike to be far more transparent about their licensing arrangements and production systems in the US$2.5 billion college apparel sector (for more, see *Antipode*, 2004). However, the development of various kinds of *benchmarks* or *standards* – against which various end products and their production processes can be measured – are an increasingly important part of commodity chain regulation within the global economy. Indeed, while the kinds of consumer campaigns just described were particularly popular in the late 1980s and the 1990s as some of the downsides of globalization became evident to consumers and consumer groups (see also Section 8.6), most such campaigns are now tied to pushing for corporate adherence to standards and responsibilities of various kinds.

As Table 4.2 illustrates, however, the world of standards is a broad and varied one. First, they can be applied to different aspects of the commodity chain, for example, health or labour conditions. Second, the precise form they take may be a code of conduct, a label on a finished product, a tightly specified technical standard, a set of voluntary initiatives or some combination of all of these. Third, the standard may apply to a particular chain (e.g. fresh tomatoes), a sector (e.g. fresh fruit and vegetables), or be generic (e.g. a safety standard to sell an electrical good in a particular national market). Fourth, while consumer campaigns are clearly the domain of end consumers (and sometimes labour unions), standards may be developed by firms, NGOs, trade unions, or international organizations, and usually involve a combination of some, or all, of these institutions. Fifth, the certification or accreditation of standards (i.e. the assessing of whether they have been met) may be undertaken by a variety of different parties (both public and private, for profit and not for profit). Sixth, the regulatory impacts will vary from the voluntary (e.g. seeking 'Fairtrade' status for a product) to the mandatory (e.g. safety standards for plastic toys and organic certifications for food). Finally, standards are innately *geographical* in two different ways: in terms of the territory within which they apply to the consumption of particular products, and in the way in which they have implications further down a chain that may connect many disparate territories.

Here, we will use the first of these many lines of variation to explore some concrete examples of standards in practice, looking in turn at initiatives aimed primarily at the labour, economic, and quality assurance aspects of commodity chains. The UK's Ethical Trading Initiative (ETI) is a multi-stakeholder organization established in 1997, and funded jointly by its membership and the UK

Table 4.2 The world of standards

Attribute of standard	Variability
Field of application	• Quality assurance • Environmental • Health and safety • Labour • Social/economic • Ethical
Form	• Codes of conduct • Label • Standard
Coverage	• Firm/commodity chain specific • Sector specific • Generic
Key drivers	• International business • International NGOs • International trade unions • International organizations
Certification process	• First, second or third party • Private sector auditors • NGOs • Government
Regulatory implications	• Legally mandatory • Market competition requirement • Voluntary
Geographical scale	• Regional (e.g. a US-state) • National • Macro-regional (e.g. the EU) • Global

Source: Adapted from Nadvi and Waltring (2004).

government's Department for International Development (DFID) (the Fair Labor Association is a similar initiative in the US context). By 2005, the ETI was constituted by 35 corporate members (leading retailers and suppliers selling into the UK market, e.g. Asda, The Gap, Tesco), 17 NGOs (including many charities such as Christian Aid and Oxfam), and representatives from four international trade unions. The ETI has established a labour code that members can apply to their supply chain activities. The Base Code consists of the following nine provisions (http://www.ethicaltrade.org; accessed 29 September 2005):

- Employment is freely chosen.
- Freedom of association and the right to collective bargaining are respected.
- Working conditions are safe and hygienic.
- Child labour should not be used.
- Living wages are paid.
- Working hours are not excessive.
- No discrimination is practised.
- Regular employment is provided.
- No harsh or inhumane treatment is allowed.

While most of the retailers use large international, independent social auditors such as Bureau Veritas, some pass on the auditing role to key suppliers, and others seek the involvement of NGOs with an on-the-ground presence in key source areas (Hughes, 2005). Interestingly, the ETI explicitly chooses not to pursue a label, arguing that it is simply not possible for a retailer to know about labour conditions at all points on every commodity chain, and also that several of the Base Code provisions rely partly on government action in source countries (e.g. the right to form a labour union).

In contrast to this focus on labour conditions, the fair trade movement is more centrally concerned with the *economic* returns that primary producers secure for their commodities. One of the leading examples is the UK Fairtrade Foundation, which was established in 1992 by a group of charities and NGOs and is part of the Fairtrade Labelling Organizations International, which coordinates 20 national initiatives across Europe, North America, Japan, and Australia/New Zealand. The Fairtrade mark is a certification label awarded to products sourced from the developing world that meet international standards of fair trade, carrying the strap-line 'Guarantees a better deal for Third World producers'. Under Fairtrade labelling, there are two sets of standards, one that applies to small farmers, and one for workers in plantations and processing factories. These standards (http://www.fairtrade.org.uk; accessed 29 September 2005) stipulate that traders must do the following:

- pay a price to producers that covers the costs of sustainable production and living;
- pay a 'premium' that producers can invest in social, environmental and business improvements;
- make partial advance payments when requested by producers;
- sign contracts that allow for long-term planning and sustainable production practices.

By mid-2005, the Foundation was working with over 420 producer organizations in 49 countries, with benefits extending to approximately 5 million farmers, workers and family members. Sales of Fairtrade mark goods have expanded

Figure 4.10 Look for the label: Fairtrade coffee and bananas consumed in the UK
Source: The authors.

rapidly in the UK, up from £16.7 million in 1998 to £140 million in 2004. The mark can be found on 15 categories of foodstuffs, of which coffee, bananas, chocolate/cocoa, and tea are the most important, and four non-food products including cotton and cut flowers (Figure 4.10). Leading Fairtrade brands in the UK include Cafédirect (coffee and tea) and Traidcraft (food, drinks, crafts and gifts), and some supermarkets such as Tesco are now developing their own Fairtrade brands. While Fairtrade is, like the ETI, based on a central code/ standard, it also signifies to the consumer that a product meets that standard through its labelling system.

ISO9000 provides an excellent example of a *quality assurance* standard. The Geneva-based International Organization for Standardization (ISO) is perhaps the best-known global standards organization, and brings together national standards institutes from 130 countries. Its central objective is to facilitate international trade and investment by harmonizing national standards with international ones. While many ISO standards are technical, the ISO9000 standards (of which ISO9001:2000 is the current version) are procedural, outlining a comprehensive set of quality-management practices. Their purpose is to provide external quality assurance to customers by demonstrating that a supplier is conforming to a formalized quality management system. The requirements of ISO9000 – which are essentially a set of written rules – cover a wide range of activities such as a firm's customer focus, quality planning, product design, review of incoming orders, and monitoring of customer perceptions about the quality of the goods and services it provides (for more details, see www.iso.org).

The system can be certified in different ways: by a firm itself, by a customer firm, or by a third party accreditor. Adoption of the standard is growing rapidly.

Table 4.3 Regional share of ISO9001:2000 certificates (December 2004)

Region	Number of countries	Number of certifications	Share of world total (%)
Africa/West Asia	49	31,309	4.7
Central and South America	29	17,016	2.5
North America	3	49,962	7.5
Europe	50	326,895	48.8
Far East	21	225,220	33.6
Australia and New Zealand	2	19,997	3.0
World	154	670,300	100.0

Source: www.iso.org, accessed 29 September 2005.

ISO9001:2000, which was established in December 2000, expanded from 44,388 certifications in December 2001 to 670,399 by December 2004. Interestingly, the global geography of the uptake is highly uneven, with Europe and East Asia together accounting for over 80 per cent of total certifications (Table 4.3). At the country level, China, Italy, the UK, Japan, Spain, and the US are the top six adopters (in that order), with China alone accounting for almost 133,000 certifications in late 2004 (20 per cent of the global total). Mapping ISO9000 certification in this way is an interesting indicator of rates of growth and international connections with the global economy. Most importantly here, however, ISO9000 is illustrative of the way in which a global standard can transmit expectations about appropriate forms of undertaking business along a global commodity chain (there is also a parallel system – ISO 14000 – which applies to environmental management).

4.5 The Limits to Ethical Intervention?

We have now seen that there is a wide range of consumption-based initiatives that seek to improve the living and working conditions along the commodity chain. However, it is worth injecting a note of caution at this point, as there is always the danger that the various initiatives can simply become enrolled as part of the capitalist system, i.e. as another way for firms to make money.

Hence, it is important to think carefully about the motives of firms for participating in the kinds of schemes described above. While consumer campaigns may genuinely bring end-consumer pressure to bear on firms and effect change, many of the standards schemes are based on the *voluntary* participation of corporations. A more critical reading of a scheme such as the ETI, for example, would suggest that firms participate because it helps maintain their profitability and also serves to protect them from negative and critical media exposure. Any impact from consumers in such a process is indirect, and for the firms involved, the simple fact of participation may be just as, if not more, significant than any real improvements in working conditions for suppliers. Moreover, it is hard to judge the success of many such schemes. Again, in the context of the ETI, critics suggest that while only a few products may comply with the base code, retailers are happy for their association with the ETI to shed a favourable light on all their sourcing patterns. Others may resist the use of any kind of external auditor, while still publicly claiming they meet certain ethical standards.

Another potential pitfall may lie with the simple issue of *who* pays for ethical consumption? The rise of standards-based schemes has fuelled the emergence of a new category of commodity chain participant, the independent auditor, many of whom themselves are profit-seeking firms. The costs of auditing long and complicated commodity chains can become substantial, and whether firms are prepared to meet these costs will depend on both their financial resources and their corporate culture (Hughes, 2005). In some instances, they will expect suppliers to meet the extra production costs, but they do not pay a higher price for the products. Farmers involved in certified organic coffee production in Oaxaca State, Mexico, for example, pay between 10–30 per cent of their gross receipts to certify their produce. The result is that farmers are abandoning certified organic production in increasing numbers. In this instance, the global standard serves as a significant barrier to entry that increases production and certification costs to the extent that only the best organized and most well-funded farmers can make organic production into a profitable exercise (see Mutersbaugh, 2005). Label-based schemes often rely on the consumer to pay a premium voluntarily – and which may place them out of reach of certain consumer groups – but non-label schemes such as the ETI cannot raise funds in that way. Overall, it is perhaps hard to escape the conclusion that consumption-based schemes are driven at least as much by economic rationales as ethical ones.

Other critiques of ethical consumption initiatives are more explicitly *geographical*. Many of the codes of conduct or general standards are by necessity quite crude abstractions and simplifications. A simple statement such as 'living wages are paid' sounds promising on a first read, but in reality a living wage will vary dramatically from place to place, and even between different social groups within the same locality. Calculating meaningful wage rates would be hugely expensive and time-consuming. Equally, by focusing primarily on formal employment relations, these initiatives may make less of an impression in sectors and places where there is a significant informal (and often female) component to the workforce. This speaks to the wider issue that ethics – and understandings of what is wrong or right, acceptable or unacceptable – are geographically specific. As a result, the implementation of ethical standards may have ambiguous impacts in the areas they are meant to help.

Again, we can turn to the UK's ETI to illustrate this point. The strict implementation of the 'no child labour' provision in Zambia's horticultural export industry means that teenagers under the age of 16 can no longer help their mothers pick vegetables and thereby pay school fees. Equally, farms cannot offer jobs to Zambia's huge numbers of teenage AIDS orphans (Friedberg, 2004). For a huge retailer, the key issue is that they are not associated with child labour *anywhere*. In more dynamic terms, there is also an increasing tendency, driven by institutions such as the ISO and the European Union, to 'upscale' standards from a particular commodity chain and harmonize them across whole sectors (e.g. all organic foods) and larger geographic territories (e.g. the whole EU). Again, the result may be standards that are detached from sectoral and geographical contexts and hence serve to worsen the position of certain groups of people in particular places.

Overall, while there are many positive aspects to ethical consumption schemes, we must interrogate them with the same critical economic-geographical perspective that we would any other branch of the economy. In turn, this will necessitate approaches that appreciate the full complexity and interconnectedness of contemporary commodity systems. As Box 4.4 explores, in part, this may necessitate a shift to a language of multi-dimensional commodity *networks* rather than chains – a theme we develop in greater detail in Chapter 8.

Box 4.4 From chains to networks?

This chapter has talked almost exclusively in terms of commodity *chains*, understood as a sequential progression from sourcing and production through to distribution and consumption. But is this really the best metaphor for representing the complexity of contemporary production systems? What might be the advantages of a more open-ended *network* approach to economic systems? A central problem with representing these systems as chains is the inherent assumption of *linearity* – i.e. the step-wise progression from a grower/producer to an end consumer through a series of intermediaries. This conceptualization of production and consumption processes has important limitations. In reality, the basic linear structures of a commodity chain are enmeshed within complex and diffuse *networks* of relationships involving a wide range of corporate (e.g. financiers, management consultants, logistics providers) and institutional (e.g. standard setting bodies, NGOs, labour unions, etc.) actors. These networks of actor-specific relationships clearly influence the nature of value-adding processes and, yet, may be underplayed in 'chain' analyses. Moreover, a focus on the linear addition of value to commodities may obscure important connections between non-sequential stages of the system. In particular, it may downplay the complex exchanges of information within a production system that may play a vital role in its nature and operation. Information about quality standards may be passed directly from a retailer to a manufacturer, for example, or through an intermediary such as a logistics firm or importer/exporter. These insights suggest that it might be productive to move from talking about commodity chains to *commodity* or *production networks* (see also Chapter 8).

4.6 Summary

As an economic system, capitalism serves to conceal the intense connections between distant producers and consumers that underpin its operation. The result is that answering even such a deceptively simple question such as 'Where does your breakfast come from?' requires substantial 'geographical detective work' (Hartwick, 2000: 1178). While labels may reveal the country of origin of a particular product, they tell us little about the working conditions under which it was produced. In this chapter, we have seen how the notion of the commodity chain allows us to chart the complex geographical journeys taken by commodities, as they are transformed from initial raw materials and ideas into finished products and services, thereby serving to re-connect producers and consumers. The precise nature of the journey taken will vary tremendously from product to product. Each and every commodity chain is delineated by a particular sequence of value-adding activities, a distinct geographical configuration, different combinations of modes of governance, and crosses various institutional contexts. Commodity chains are thus important – and inherently geographical – organizational features of the contemporary global economy; they are the 'hidden' social relations that enable capitalism to extend its global reach. Conceptually, the commodity chain is an extremely important integrative idea that allows us to reveal the interconnections between the many actors – states, firms, workers and consumers – that are considered in more depth in Part III.

The careful mapping of commodity chains opens up the potential for intervening in the nature and operation of those chains. We have reviewed examples of initiatives through which consumers, governments, labour unions, and businesses, either working individually, or together, can seek to alter the conditions under which certain commodities are produced. In particular, the rise of different forms of standards is highly significant. While some of these schemes have had noticeable impacts, it is also important to recognize their potential limitations. On the one hand, they are susceptible to being incorporated into the profit-seeking capitalist mainstream. On the other, unless attempts to re-regulate commodity chains are alert to their inherent geographic variation and complexity, they in turn will have uneven, and perhaps unintended, outcomes. In the next chapter we move on to consider how technological developments are altering the geography and operation of contemporary production systems.

Further reading

- Gereffi (1994) and Gereffi et al. (2005) are two seminal statements on the global commodity/value chain perspective. See Dicken et al. (2001) for a geographical critique.
- Hughes and Reimer (2004) provide a wide range of geographical studies of commodity chains.
- For more details on the hard disk drive, catfish, fresh vegetable and coffee commodity chains described in this chapter, see Gourevitch et al. (2000), Duval-Diop and Grimes (2005), Dolan and Humphrey (2004), and Ponte (2002), respectively.
- Hartwick (2000) discusses a wide range of potential strategies of consumer politics, while Friedberg (2004) and Hughes (2005) offer more critical takes on consumption-based strategies.
- Henderson et al. (2002) and Barrett et al. (2004) discuss the potential for developing a production/commodity *network* approach to global economic formations.

Sample essay questions

- How does a commodity chain approach enable us to reconnect distant producers and consumers in the global economy?
- How and why are commodity chains governed in different ways?
- In what ways might the institutional contexts of a commodity chain affect its structure and operation?
- What are the pros and cons of adopting consumption-based approaches to commodity chain regulation?
- What are the potential advantages of a network-informed understanding of global economic connections?

Resources for further learning

- http://www.globalvaluechains.org: this site contains a wealth of conceptual and empirical material on global value chains.
- http://www.unido.org: the United Nations Industrial Development Organization (UNIDO) website also offers a wide range of data and reports on different commodity chains and the potential they offer for economic development in different localities.
- http://www.ico.org/index.asp: the website of the International Coffee Organization provides a range of information on the coffee industry, and in particular, its evolving regulatory structures.

- http://www.fairtrade.org.uk: the UK's Fairtrade Foundation is one of the best-known attempts to improve the economic returns offered to commodity producers.
- http://www.iso.org/iso/en/ISOOnline.frontpage: The International Standard's Organization's website is rich in information on a variety of international standard schemes.

References

Antipode (2004) Intervention symposium: geographies of anti-sweatshop activism, 36: 191–226.

Barratt, H.R., Browne, A.W. and Ilbery, B.W. (2004) From farm to supermarket, in A. Hughes and S. Reimer (eds) *Geographies of Commodity Chains*, London: Routledge, pp. 19–38.

Dicken, P. (2003) *Global Shift*, 4th edn, London: Sage.

Dicken, P., Kelly, P.F., Olds, K. and Yeung, H.W.C. (2001) Chains and networks, territories and scales: towards an analytical framework for the global economy, *Global Networks*, 1: 89–112.

Dicken, P. and Miyamachi, Y. (1998) 'From noodles to satellites': the changing geography of the Japanese sogo shosha, *Transactions of the Institute of British Geographers*, 23: 55–78.

Dolan, C. and Humphrey, J. (2004) Changing governance patterns in the trade of fresh vegetables between Africa and the United Kingdom, *Environment and Planning A*, 36: 491–509.

Duval-Diop, D.M. and Grimes, J.R. (2005) Tales from two deltas: catfish fillets, high-value foods, and globalization, *Economic Geography*, 81: 177–200.

Friedberg, S. (2004) The ethical complex of corporate food power, *Environment and Planning A*, 22: 513–31.

Gereffi, G. (1994) The organization of buyer-driven global commodity chains: how US retailers shape overseas production networks, in G. Gereffi and M. Korzeniewicz (eds) *Commodity Chains and Global Development,* Westport, CT: Praeger, pp. 95–122.

Gereffi, G., Humphrey, J. and Sturgeon, T. (2005) The governance of global value chains, *Review of International Political Economy*, 12: 78–104.

Gourevitch, P., Bohn, R. and McKendrick, D. (2000) Globalization of production, *World Development,* 28: 301–17.

Guardian, The (2003) Miles and miles and miles: how far has your basket of food travelled?, Food section, 10 May, p. 18.

Hartwick, E.R. (1998) Geographies of consumption: a commodity chain approach, *Environment and Planning D*, 16: 423–37.

Hartwick, E.R. (2000) Towards a geographical politics of consumption, *Environment and Planning A*, 32: 1177–92.

Harvey, D. (1989) Editorial: a breakfast vision, *Geography Review*, 3: 1.

Henderson, J., Dicken, P., Hess, M. Coe, N.M. and Yeung, H.W-C. (2002) Global production networks and the analysis of economic development, *Review of International Political Economy*, 9(3): 436–64.

Hughes, A. (2005) Corporate strategy and the management of ethical trade: the case of the UK food and clothing retailers, *Environment and Planning A*, 37: 1145–63.

Hughes, A. and Reimer, S. (eds) (2004) *Geographies of Commodity Chains*, London: Routledge.

Humphrey, J. and Schmitz, H. (2004) Chain governance and upgrading: taking stock, in H. Schmitz (ed.) *Local Enterprises in the Global Economy*, Cheltenham: Edward Elgar, pp. 349–77.

Kessler, J. and Appelbaum, R. (1998) The growing power of retailers in producer-driven commodity chains: a 'retail revolution' in the US automobile industry?, unpublished manuscript, Department of Sociology, University of California at Santa Barbara, USA.

Mutersbaugh, T. (2005) Fighting standards with standards: harmonization, rents, and social accountability in certified agrofood networks, *Environment and Planning A*, 37: 2033–51.

Nadvi, K. and Waltring, F. (2004) Making sense of global standards, in H. Schmitz (ed.) *Local Enterprises in the Global Economy: Issues of Governance and Upgrading*, Edward Elgar, Cheltenham, pp. 53–94.

Oxfam (2003) *Mugged: Poverty in your Coffee Cup*, Oxford: Oxfam International.

Ponte, S. (2002) The 'latte revolution'? Regulation, markets and consumption in the global coffee chain, *World Development*, 30: 1099–122.

Smith, M.D. (1996) The empire filters back: consumption, production, and the politics of Starbucks coffee, *Urban Geography*, 17: 502–24.

Watts, M. (1999) Commodities, in P. Cloke, P. Crang and M. Goodwin (eds) *Introducing Human Geographies*, London: Arnold, pp. 305–15.

CHAPTER 5

TECHNOLOGY AND AGGLOMERATION

Does technology eradicate distance?

Aims

- To demonstrate how certain kinds of technologies can be used to transcend time and space
- To appreciate the limits of the spatial impacts of technology on economic systems
- To understand why proximity still matters for many different kinds of economic activity
- To reflect on the importance of *relational* proximity in shaping contemporary economic geographies.

5.1 Introduction

eBay has been one of the few undisputed economic successes of the Internet age. The online auction site was established in the US in October 1995. By 2005, 150 million users worldwide were buying and selling goods worth over US$40 billion, producing a profit of over US$1 billion for the company. The basic principles underlying eBay are simple. While anyone can browse its website, both buyers and sellers have to register. Buyers – after searching through over 40,000 different types and subcategories of goods – enter the most they are prepared to pay for an item. eBay then bids on their behalf in pre-determined increments, or the shopper can buy instantly in some instances. The company makes its money from the 'registration fee' paid by the seller and a final-value fee on goods that are sold. eBay facilitates the processing of credit card payments through its PayPal system, but buyers may incur other costs associated with shipping, taxes, and import duties. The basic honesty trading system is protected by feedback profiling that records the online reputation of buyers and sellers, and is also policed by eBay's own category managers and security advisers. What started out as a niche market for trading specialists and highly collectable goods has become a huge online economy, trading products as diverse as used automobiles and capital goods such as catering equipment and hospital scanners. For some, eBay trading has become a way of life. For example, estimates suggested that by 2005 some 500,000 Americans were deriving some part of their living from the website, and that around 150,000 of these were full-time eBay traders (*The Economist*, 2005)!

As first glance, the huge success of eBay seems to illustrate clearly the incredible transformative potential of new information and communications technologies such as the Internet. The kind of person-to-person bidding historically associated with local trading places such as bazaars, markets, and auction houses has been up-scaled into a global exchange system facilitated by the ubiquity, speed, and reliability of the Internet. This is undoubtedly a hugely important economic-geographical phenomenon, as are other successful online brands, such as Google, Amazon, and Yahoo. To some, it signals the end of geography and the eradication of distance as a barrier to economic transactions. And yet, we should be cautious about over-emphasizing the ability of technology to re-shape or erase the geography of the economy. This can be seen on two levels.

First, even a global phenomenon such as eBay is nowhere near as global and omnipresent as one might think. As with all cyberspace activities, eBay is underpinned by a series of very real and uneven geographies. For example, eBay is largely a rich country phenomenon, relying as it does upon both the availability and affordability (in local terms) of Internet access, which is still very patchy at a global scale. Equally, the goods exchanged still have to be moved through the real world of postal, freight, and logistics services, and customs and import/export taxes and duties. As a result, most buyers and sellers on eBay actually transact *within* their 'home' national territory. Second, the eBay example only illustrates the potential impact of one kind of technology (the Internet) on one particular segment of the economy (online retailing of certain commodity types). Across the economy as a whole, the geographical impacts of a wide range of new technologies have been varied, uneven and complex. Some technologies have allowed the reorganization of economic activities at a global scale. Others, however, have reinforced pre-existing geographic inequities, or indeed, have *increased* the propensity for the agglomeration or clustering of economic activity in particular places. Moreover, in some areas of the economy, the seemingly transformative potential of technology has been heavily mitigated by the continued need for inter-personal social interaction.

We saw in Chapter 4 how the geography of different activities within commodity chains may be concentrated or dispersed, or indeed a combination of both: here, then, we look at the role technology plays in shaping these geographies. We unravel the complexity of geography-technology inter-relations in four substantive sections. First, we critique simplistic *end of geography* notions through which new technologies are suggested to have eradicated distance, and therefore location, as economic concerns (Section 5.2). Second, we introduce some key ideas that form the basis of a more nuanced view of technological change and its impacts (Section 5.3). We also review the main varieties of technological change to show that while some are space-transcending, by no means all are. Third, we then move on to look at the various ways in which proximity continues to matter for many economic activities, ensuring that cities and agglomerations remain vital components of the contemporary economic-geographical landscape (Section 5.4). Finally, we introduce the notion of *relational proximity* to capture the way in which much economic activity is actually undertaken through complex combinations of face-to-face contact *and* technologically mediated interaction (Section 5.5). Overall, we suggest the metaphor of *nodes in global networks* is perhaps the most apposite way of thinking about the complex configurations of near and distant connections that constitute the world economy.

5.2 The Rise of 'Placeless' Production?

The economist Richard O'Brien opened his 1992 book, *Global Financial Integration: The End of Geography*, with the following words:

> The end of geography, as a concept applied to international financial relationships, refers to a state of economic development where geographical location no longer matters in finance, or matters much less than hitherto. In this state, *financial market regulators* no longer hold full sway over their regulatory territory . . . For *financial firms*, this means that the choice of geographical location can be greatly widened provided that an appropriate investment in information and computer systems is made . . . *Stock exchanges* can no longer expect to monopolize trading in the shares of companies in their country or region, nor can trading be confined to specific cities of exchanges. Stock markets are now increasingly based on computer and telephone networks, not on trading floors. Indeed, markets almost have no fixed abode. For *the consumer of financial services*, the end of geography means that a wider range of services will be offered, outside the traditional services offered by local banks . . . Money, being fungible, will continue to try to avoid, and will largely succeed in escaping, the confines of the existing geography.
>
> (1992: 1–2; original italics)

This passage is symptomatic of a common approach to technological change – and the impacts of new information technologies in particular – that sees the constraints of geography being inevitably overcome. O'Brien's vision suggests that financial firms only have to invest in computer technology to open up a wide variety of potential locations. Equally, stock exchanges will become far less tied to particular cities, and consumers will be presented with a proliferation of products and services.

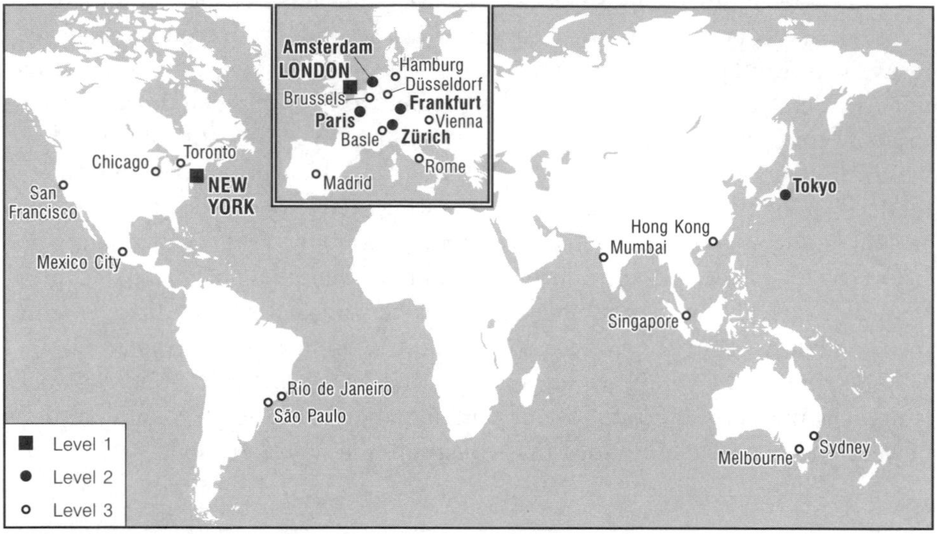

Figure 5.1 The global network of financial centres
Source: Dicken (2003), Figure 13.8. Reprinted by permission of Sage Publications.

Thinking about today's global financial system, we can see evidence for some elements of O'Brien's account. The workings of the financial services industry have indeed been profoundly transformed by the impacts of technologies such as computers, microchips, networks and satellites. Money has become largely electronic, replacing tangible forms such as coins, notes, cheques and credit notes. As a result, it can now be transferred almost instantaneously at a global scale. Overlapping financial markets in different time zones allow financial firms to engage in 24-hours-a-day trading. These connections between different markets at the international scale have enabled financial institutions to increase their international lending activities and to respond extremely rapidly to exchange rate changes in international currency markets. Finally, new technology has undoubtedly altered how financial firms interact with their clients. For example, as individuals we increasingly interact with our banks through the Internet rather than 'bricks-and-mortar' branches, and can use the search capability of the Internet to compare and contrast different financial products and services.

And yet, the core geographical element of O'Brien's argument – the end of geography – has clearly not come to pass. As illustrated in Figure 5.1, the overall geography of the global financial industry is curiously resistant to changes. A small number of cities – most notably London, New York, Tokyo, Paris, and Frankfurt – continue to control the lion's share of the international financial transactions in the global economy (see also Chapter 1). In 2004, London, for example, accounted for 31 per cent of the world's US$1.9 trillion daily foreign exchange trading and 50 per cent of Europe's investment banking, and was home to almost 500 foreign banks. Indeed by some measures, the concentration of control in these leading centres has actually *increased* at the same time as the use of information technologies has been expanding rapidly. How can we explain this apparent paradox? Why has the end of geography forecast by O'Brien not become more apparent in reality?

5.3 Understanding Technological Changes and Their Geographical Impacts

To answer these questions, we need to be more discerning in our conceptualization of technology, technological change, and its spatial ramifications. First, technology must be seen not as a technical process with a life of its own, but rather as a *social process* through which individuals and organizations deploy technologies to achieve certain ends. In the case of firms, technologies are usually used as part of the drive to make profits in a competitive capitalist environment. Viewing technology in this way allows us to avoid the trap of technological determinism, i.e. stating that technologies themselves *cause* various changes. Instead, we should think of how technologies *facilitate* and *enable* different kinds of change through their use by different types of economic actors.

Second, we need to distinguish between different kinds of technology. A key difference here is between what we might call space-shrinking transport and communications technologies that help overcome the frictions of space and time, and technological changes in production processes that may, or may not, as we shall see, have geographical ramifications (Dicken, 2003). Both types of technology are integral parts of the 'machinery' of capitalism outlined in Chapter 3. Space-shrinking technologies bring about time–space compression – and what has been termed the 'annihilation of space through time' (Harvey, 1989) – as capitalists seek to use new technologies to speed up the spatial circulation of capital in their pursuit of profits. Equally, production process technologies are crucial for increasing productivity and facilitating innovation in that same search for maintaining, and increasing, profitability.

Third, we need to think about different *levels* of technological change (Freeman and Perez, 1988). Incremental innovations are ongoing small-scale changes to existing products and production processes, such as adding more features to a mobile phone, or using computer software to replace paperwork in an office environment. Radical innovations are discontinuous events that significantly change an existing product or process. Examples here could include the invention of a new kind of vacuum cleaner (e.g. the Dyson) or the implementation of robot technology to replace human workers on automobile production lines. Changes in technology system are more extensive changes in technology that impact on several parts of the economy, and may have the potential to create new industries. These changes are linked to the emergence of generic technologies

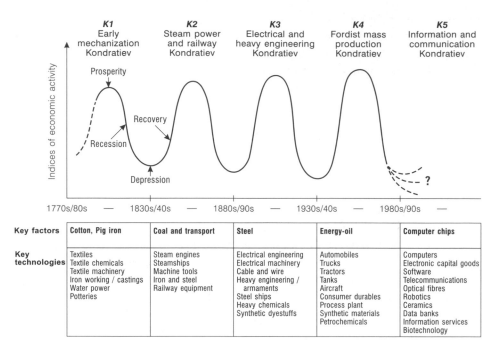

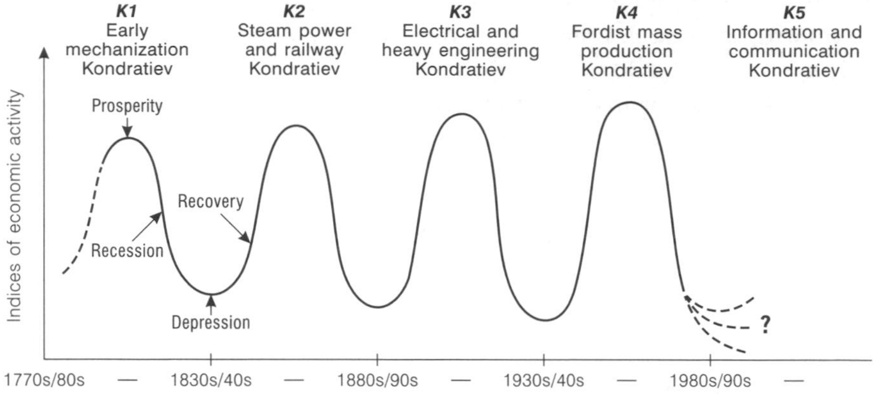

Key factors	**Cotton, Pig iron**	**Coal and transport**	**Steel**	**Energy-oil**	**Computer chips**
Key technologies	Textiles Textile chemicals Textile machinery Iron working / castings Water power Potteries	Steam engines Steamships Machine tools Iron and steel Railway equipment	Electrical engineering Electrical machinery Cable and wire Heavy engineering / armaments Steel ships Heavy chemicals Synthetic dyestuffs	Automobiles Trucks Tractors Tanks Aircraft Consumer durables Process plant Synthetic materials Petrochemicals	Computers Electronic capital goods Software Telecommunications Optical fibres Robotics Ceramics Data banks Information services Biotechnology

Figure 5.2 Kondratriev long waves and their characteristics
Source: Adapted from Dicken (2003), Figure 4.1. Reprinted by permission of Sage Publications.

with wide applications, such as biotechnology or a new energy technology. Changes in the techno-economic paradigm are revolutionary upheavals that have economy-wide effects. Clearly the advent of information technology in the contemporary era represents a new techno-economic paradigm: it is hard to think of an economic activity that has not been affected in some way by computers and the Internet. The power of this latest paradigm is derived from the way in which it enables users to both transmit and process information, activities that were previously conducted using separate technologies (e.g. telephones and free-standing computers).

Fourth, technological change needs to be placed in a long-term or *evolutionary* perspective. The information technology paradigm has come after a series of other earlier technological revolutions linked to periods of economic growth and prosperity, often known as long or Kondratiev waves, after the Russian economist who studied them in the 1920s (Figure 5.2). Similar to the idea of spatial divisions of labour in Chapter 3, these different transitions have been overlaid on one another sequentially over time. The different spatial outcomes of each wave of new technology have interacted with patterns of activity left over from the preceding wave. This observation starts to hint at a potential challenge to the alleged end of geography. London, for example, was established as a site of global financial power through the activities of the British Empire long before the information technology revolution. It was therefore extremely well placed to benefit from the advantages conferred by new technologies. This is known as *path-dependency*, with the fortunes of certain places under conditions of technological change being partly determined by pre-existing conditions.

Already then, we can start to see the simplifications inherent in O'Brien's argument. There are different kinds of technologies, different levels of technological change, and even the same technology may be used differently by different people in different places. We now move on to look at the two key categories of technology – space-shrinking and production process (a distinction made by Dicken, 2003) – and their geographic impacts in more detail.

Space-shrinking technologies

The functioning of the contemporary world economy depends on a wide range of space-shrinking technologies that connect firms, workers, governments, and consumers (i.e. commodity chains) together in different ways. What these technologies have in common, though, is the ability to help their users partially overcome the constraints of space and time. There are two important kinds of space-shrinking technologies that we need to introduce here (Dicken, 2003): (1) *transportation systems*, the means by which material goods (including people) are transferred between places; and (2) *communications systems*, which enable the transmission of various kinds of information (e.g. text, numbers, images, videoclips, music files, computer programmes) between places. In the past century, dramatic transformations of both systems have occurred, with profound implications for the geography of the world economy. We will now briefly consider the most important of these innovations (drawing on the typology of Dicken, 2003).

In terms of *transportation technologies*, it barely needs to be stated that the world has 'shrunk' in highly significant ways, particularly in the past two centuries. During the second Kondratiev wave in the mid-nineteenth century (see Figure 5.2), the advent of steam power and the use of iron and steel in railways and shipbuilding were vital elements in supporting the emergence of truly global trading empires. Two key transportation developments in the 1950s gave the development of a global economy fresh impetus, however:

1 *the advent of commercial jet aircraft*: while it seems obvious today that the extensive air travel network is an integral component of the global economy, it is worth remembering that it was only in the late 1950s that jet travel took off as an affordable and widely available mode of transport. The air travel network is not only important for transporting different groups of economically significant people (e.g. tourists, managers in transnational corporations, highly skilled migrants, contract workers, students), both within and between countries, but also in terms of circulating a wide range of high value (e.g.

documents, electronic components) and fresh products (e.g. fresh fruit, vegetables, cut flowers). The leading parcel companies have extensive air fleets of their own. For example, in 2005, FedEx – a company with 138,000 employees stationed across 235 countries – was using a fleet of 638 planes operating out of 235 airports, in combination with 43,000 other vehicles, to transport over 3.1 million packages daily (http://www.fedex.com).

2 *the advent of containerization*: another highly significant development since the mid-1950s has been the adoption of standardized 20-ft-long metal containers for land and sea freight, thereby vastly simplifying the transport and transhipment (i.e. transfer from ship to ship, or ship to train) of a huge range of commodities. Every day, over 15 million containers are moving around at sea or on land, or standing in a warehouse somewhere waiting to be delivered. It is estimated that they account for some 90 per cent of the world's traded goods by value, and in 2004, the total number of containers moved was 72 million. The global movement of containers is dominated by three flows: transpacific, transatlantic, and Europe–Far East, providing an interesting window on the nature of global trade patterns. The top 20 container ports – a list that includes Hong Kong, Singapore, Pusan, Shenzhen, Shanghai, Rotterdam, Hamburg and Los Angeles – alone handle about 50 per cent of global transhipments (Figure 5.3).

Figure 5.3 A key node in the global container system: the Port of Singapore
Source: The authors.

While innovations such as these have undoubtedly helped to 'shrink' the world in relative terms, it is important to remember that the shrinkage has been extremely uneven. Transportation developments have served to pull the world's leading cities and economies closer together, while other less industrialized and rural areas continue to be far less well connected, if not pulled further apart. This uneven 'shrinking' of the world further exacerbates global uneven development. For example, because of poor inter-island transport development, the movement of goods among the Caribbean countries can cost as much, if not more, than transatlantic routes. As a result, many rum, beer, and soft drink manufacturers in the Caribbean find it cheaper to import glass bottles from France and plastic bottles from China than source them locally in the Caribbean.

Meanwhile, huge advances in *communications systems* parallel these developments in transportation technologies. Key developments include:

1 *satellite and optical fibre technology*: since the mid-1960s, when the first geo-stationary communications satellite was launched, global communications have been transformed. Networks of satellites can transmit voices, images, and other forms of data across the globe almost instantly. Optical fibre systems have increasingly challenged satellite technology since the 1970s due to their ability to transmit huge quantities of information at very high speeds. An extensive network of ocean-bed optical fibres now criss-crosses the globe. Taken together, satellites and fibre optics offer affordable and high volume data transfer capability, and the capacity of fibre optics in particular continues to increase exponentially.

2 *the Internet*: clearly the Internet has revolutionized communications through the development of an interactive, mass user, computer network system. The Internet facilitates cheap and reliable inter-personal and inter-organizational email and information sharing on an unprecedented scale. The expansion of Internet use in the past decade has been astounding, rising from 16 million in 1995 to 939 million in 2005. The number of Internet host computers rose in equally impressive fashion from 6.6 million in July 1995 to 353.3 million in July 2005. Despite these growth rates, access to the Internet is still hugely uneven globally. In July 2005, Asia and North America accounted for 35 per cent and 24 per cent of global users respectively, while Africa and Latin America accounted for a meagre 2 per cent and 7 per cent. The geography of Internet domain names – a good indicator of the location of the Internet *industry* – is even more skewed, with the US, Germany, and the UK accounting respectively for 33, 16, and 13 per cent of the global total in July 2003. As noted in the introduction to this chapter, despite much hype to the contrary, the Internet still remains a largely urban, rich country phenomenon (data from http://www.nua.ie and Zook, 2005).

3 *mobile telecommunications*: the emergence of the Internet has been paralleled by significant developments in mobile telephony. Most importantly, in

the past two decades, the mobile phone has gone from being an expensive novelty to a seemingly ubiquitous presence in everyday life. Newer mobile phones and related devices allow wireless connection to the Internet, enabling users to access information and be in communication while on the move and in a greater variety of locations. In general, mobile telephones are more widely used in wealthy countries. In 2003, for example, the average number per 1,000 people was 708 in high-income countries, and just 24 in low-income countries. Interestingly, however, the level of mobile telephone use now exceeds that of fixed lines in many developing countries (in particular in the Asia Pacific, Central Asia, and Sub-Saharan Africa) with the latter technology effectively being bypassed by the pace of expansion of mobile phone usage (http://www.worldbank.org/data).

4 *the electronic mass media*: often overlooked in these debates, the electronic mass media play a crucial role in helping products and services to transcend space and time through their role in constructing markets and expectations in countries around the world. Radio and television are both extremely powerful media in this regard, carrying a wide range of advertising for branded products, for example. Indeed, leading television stations with global networks such as CNN, MTV, and BBC World have become significant brands in their own right. Since the 1980s, a combination of deregulation and new outlets such as cable and satellite television has led to the proliferation of television channels available in many countries. Again, however, it is vital to bear mind inequality in access to the media across the globe. In 2003, the highest level of television ownership was in Sweden, where there were 965 sets per 1,000 people, and, for all high-income countries, the average was 735. At the other end of the scale, just two sets per 1,000 people were recorded in Chad and the Democratic Republic of Congo, with the average of all low-income countries being 84 (http://www.worldbank.org/data).

Taken together, the space-shrinking attributes of transport and communications technologies have had significant impacts on the geography of the world economy. On a simple level, they allow us to buy goods from around the world in our local supermarkets. Not only do they facilitate the spatial separation of producers and consumers of good and services, they also allow the *disintegration* of different elements of the production process (or commodity chain) both within and between countries. As we saw in Chapter 4, the production of a computer hard disk drive in Thailand, for example, may involve assembling components and modules from Singapore, Malaysia, and Indonesia. Modern transportation and communication technologies allow the quick and efficient flow of components and information between different firms and facilities that such a process requires.

It is not just manufacturing processes that are being affected in this way, however. Space-shrinking technologies are also allowing the spatial disintegration

Figure 5.4 The offshoring of services: recruiting call centre workers in the Philippines
Source: The authors.

of service activities. The massive growth in *call centres* in recent years is a clear indication of these trends. Call centres represent a form of firm–customer interface whereby a wide range of support services (e.g. sales and marketing, technical support, claims enquiries, market research, reservations, information provision) are provided via telephone from dedicated centres to an often widely dispersed customer base. While the establishment of call centre operations in low-cost locations has long been a popular business strategy, the process has intensified in recent years, particularly in terms of the international relocation of these activities (Figure 5.4). This process has become known as the *offshoring* of services. In addition to call centres, two other important kinds of offshore services are back-offices (covering activities such as claims and accounts processing, payroll processing, data processing, invoice administration, etc.) and information technology services (e.g. software development, application testing, content development, etc.). Leading sectors involved in offshoring include financial services, telecommunications, transport services, health services and the public sector. India has been one of the key beneficiaries of these trends (Box 5.1).

Box 5.1 Offshore services in India

The Indian offshore services industry started to take off in the early 1990s, when companies such as American Express, British Airways, and Swissair set up back-office operations in India. By 2004, Indian software and service exports had expanded to US$12 billion per year, up from just US$0.6 billion in 1994. The country attracts a highly significant share of such investment at the global level: in 2002 and 2003, it attracted 12 per cent of the global total of international call centres (60 projects), 31 per cent of international back office investments (31 projects), and 19 per cent of international IT service centres (118 projects). Investors are attracted by the availability of a large, cheap, well-qualified English-speaking workforce who can easily be connected to home markets through international telecommunications. The salary differentials are startling. In 2003, an IT professional with three to five years experience was earning around US$96,000 per year in the UK, US$75,000 in the US, and just US$26,000 in India. A low-grade call centre worker in the UK was earning US$20,000, while the salary in India would be less than US$2,000. As a result, by 2003 India already had some 1,500 call centre firms employing over 102,000 young people, and the industry continues to grow rapidly. These offshore services can either be provided by directly owned subsidiaries (e.g. Bank of America's back-office operation in Hyderabad, India, which employed 1,000 people in mid-2005), or they can be *outsourced* to Indian firms (e.g. Delta Air Lines' contract with a subsidiary of the Indian firm Wipro to handle some of its call centre reservations). The trend has been to shift towards the outsourcing model, and the leading Indian providers such as Tata Consultancy, Wipro, and Infosys are huge transnational corporations in their own right, with extensive international operations. The process looks sets to continue – it is estimated that 200,000 jobs will be offshored to India from the UK alone by 2010. Other countries currently benefiting substantially from these trends are China, the Philippines, and the countries of Central and Eastern Europe.

Data from *The Economist* (2003) and UNCTAD (2004).

Production process technologies

The competitive capitalist economy dictates that not only must firms connect the different parts of the production process together efficiently (ranging from accessing resources and inputs right through to a variety of after-sales support services – see Chapters 4 and 8), but they must also strive to constantly improve the rate at which they develop new products and services, and the efficiency of their production processes. Firms need to take three important, and interrelated, decisions in this regard (Dicken, 2003). First, an appropriate *technique* of production needs to be selected, and in particular the precise combination of labour (i.e. workers) and capital (i.e. buildings, machinery, computers, telephones etc.) that will be used. Second, the *scale* of production must be determined. While economies of scale may be achievable by increasing production volumes, this may not always be the case and can vary hugely from industry to industry (e.g. television production compared to handmade furniture, or fast food retailing compared to corporate banking). Third, the *location* of production needs to be decided. This means making a relative assessment of the necessity of being close to labour (e.g. cheap or highly skilled), sources of inputs (e.g. raw materials, components or advice from management consultants) or the market (e.g. final consumers, other businesses or government departments).

As noted earlier, we are currently in the midst of an evolving techno-economic paradigm being driven by new information technologies. In the same way as these technologies are revolutionizing communication patterns, they are also driving profound transformations in production processes. There is much debate, however, about the nature of these production processes, which can be thought of as after-Fordist (Dicken, 2003) in relation to the preceding mass production wave often known as *Fordism* (Box 5.2). Over the past three decades, Fordism has been augmented by new modes of production, the chief characteristic of which is production *flexibility*. At the heart of this enhanced flexibility is the use of information technologies in machines and their operation to allow more sophisticated control over the production process. Most importantly, computerization enables the rapid and efficient switching from one part of a manufacturing process to another, which in turn allows firms to tailor products to the requirements of individual customers. Rather than engaging in mass production, firms are now able to make a wide variety of products for different market niches without compromising the cost savings traditionally associated with large volume production.

That being said, it is not possible to identify one kind of after-Fordist production regime (drawing on Dicken, 2003). Instead, it is more effective to identify three different kinds of industrial system that co-exist in the contemporary economy (Table 5.1). First, Fordism is alive and well in certain sectors where scale economies remain crucial, e.g. agriculture and food processing, routinized services activities, and the electronic component industries. Second, in other instances, flexible production technologies have led to the resurgence of craft-based production driven by innovative, small independent firms. At the heart of this so-called *flexibly specialized* production system are skilled craft workers using flexible machinery to produce small volumes of customized goods. These developments can be observed in craft industries producing specialized shoes, jewellery, clothing, furniture, ceramics, and the like. Third, there are forms of

Box 5.2 Fordism

Fordism signifies a range of industrial practices first associated with the workplace innovations of the American automobile manufacturer Henry Ford in Detroit, Michigan, during the second decade of the twentieth century. The Fordist production system has four important elements. First, it is characterized by a distinctive division of labour – i.e. the separation of different work tasks between different groups of workers – in which unskilled workers execute simple, repetitive tasks and skilled technical and managerial workers undertake functions related to research, design, quality control, finance, coordination, and marketing. Second, it is a system in which the manufacture of parts and components is highly standardized. Third, it is organized not around groups of similar machinery, but machines arranged in the correct sequence required to manufacture a product. Finally, the various parts of the production process are linked together by a moving conveyor belt – the assembly line – to facilitate the quick and efficient fulfilment of tasks. Together, these four attributes can provide *economies of scale* whereby the large volume of production reduces the cost of producing a single product. Prices can then be reduced, leading to increased sales and the potential development of *mass markets* for the commodity in question.

However, the term Fordism has come to connote much more than just the production line technologies of Henry Ford. For many commentators, it signifies a broader, society-wide model of economic growth that was prevalent in North America and Western Europe in the period stretching from the end of the Second World War until the early 1970s (sometimes known as Atlantic Fordism). This was a period of strong economic growth in those regions associated not only with mass production and the emergence of large, vertically integrated corporations (see Chapter 7), but also a particular set of national institutional and political conditions. In particular, it was based on agreements between governments, firms, and labour unions on wages and the provision of social welfare. Collectively, they sought to link annual wage increases to the productivity increases achieved through mass production techniques, with the result that workers had sufficient income to sustain mass consumption. These national agreements were supported by new post-war supranational organizations such as the International Monetary Fund (IMF) and World Bank that helped create a relatively stable global economy (see Section 7.5). The limits to the system became apparent in the early 1970s, when falling levels of profits, increased global competition, and rising oil prices initiated a period of intense economic restructuring, broadly labelled as the crisis of Fordism, during which the previously stable relations between firms, governments, and labour unions unravelled. This heralded the arrival of the *after-Fordist* era.

Table 5.1 Alternative production systems in after-Fordism

Characteristic	Flexible specialization	Fordist mass production	Japanese flexible production
Technology	Simple, flexible tools/machinery; non-standardized components	Complex, rigid single-purpose machinery; standardized components; difficult to switch products	Highly flexible (modular) methods of production; relatively easy to switch products
Labour force	Mostly highly skilled workers	Narrowly skilled professionals 'conceptualize' the product; semi/ unskilled workers 'execute' production in simple, repetitive, highly controlled sequences	Multi-skilled, flexible workers, with some responsibilities, operate in teams and switch between tasks
Supplier relationships	Very close contact between customer and supplier; suppliers in physical proximity	Arms-length supplier relationships; stocks held at assembly plant as buffer against disruption of supply	Very close supplier relationships in a tiered system; 'just-in-time' delivery of stocks requires 'close' supplier network
Production volume and variety	Low volume and wide (customized) variety	Very high volume of standardized products with minor 'tweaks'	Very high volume; total partially attained through the production of range of differentiated products

Source: Bryson and Henry (2005), Table 14.1 (adapted from Dicken, 2003, Table 4.2). Reprinted by permission of Sage Publications.

flexible production emanating originally from Japan that are distinctive from both the Fordist and flexibly specialized systems. Key here is the combination of information technologies with flexible ways of organizing workers (see Chapter 9 for more, particularly Box 9.3). Another important attribute of this system is the application of information technologies to supply chains, allowing the replacement of the expensive and inefficient 'just-in-case' systems inherent to Fordism with 'just-in-time' systems (Table 5.2). The overall result is perhaps

Table 5.2 The characteristics of 'just-in-case' and 'just-in-time' systems

'Just-in-case' system	'Just-in-time' system
Characteristics	
Components delivered in large, but infrequent, batches	Components delivered in small, very frequent, batches
Very large 'buffer' stocks held to protect against disruption in supply or discovery of faulty batches	Minimal stocks held – only sufficient to meet the immediate need
Quality control based on sample check after supplies received	Quality control 'built in' at all stages
Large warehousing spaces and staff required to hold and administer the stocks	Minimal staff and warehousing required
Use of large number of suppliers selected primarily on the basis of price	Use of small number of preferred suppliers within a tiered supply system
Remote relationships between customers and suppliers	Very close relationships between customers and suppliers
No incentive for suppliers to locate close to customers	Strong incentive for suppliers to locate close to customers
Disadvantages	
Lack of flexibility – difficult to balance flows and usage of different components	Must be applied throughout the entire supply chain
Very high costs of holding large stocks	Reliance on small number of preferred suppliers increases risk of interruption in supply
Remote relationships with suppliers prevents sharing of developmental tasks	
Requires a deep vertical hierarchy of control to coordinate tasks	

Source: Dicken (2003), Table 4.3. Reprinted by permission of Sage Publications.

best thought of as *mass customization*, with firms able to produce substantial volumes of a large range of products by combining mass produced components in different ways. While developed in the automobile industry, it has subsequently spread into other consumer goods industries (such as personal computers and mobile phones) and sectors such as chemicals and steel. Dell Computer provides an interesting case study of these technological and organizational dynamics (Box 5.3).

Box 5.3 Production process innovation: the case of Dell Computer

Dell, the American computer company, was formed in 1984. In 1994, it was the world's tenth largest computer firm, with a global market share of 2.4 per cent. By 2004, it was the world's largest personal computer manufacturer, with a global market share of 17.9 per cent. This remarkable growth was achieved not through product innovation – the personal computer was not invented by Dell and it has long been a largely standardized product – but by production and organizational innovations concerning the way in which it buys components, assembles computers, and sells them. There are three distinctive elements to Dell's business model. First, the company only sells direct to its customers, bypassing almost entirely distributors, resellers, and retailers. Second, Dell customizes all its products to meet the specific needs of its customers (i.e. mass customization). Third, it has used the Internet to establish not only its system of direct sales, but also the procurement and assembly operations underlying the mass customization. In short, 'by grafting its system of custom direct sales and procurement on to the Internet infrastructure, Dell has transformed these activities, creating what is arguably the most innovative procurement, production, and distribution organization ever built' (Fields, 2004: 166). The impacts of the Internet-based just-in-time procurement system have been particularly profound. In 1994, as Dell launched its Internet strategy, it held an average of 32 days of inventory in its production system. By 1997, the figure had shrunk to 13 days, and by mid-2002, it was down to just four days, compared to six weeks worth for its main competitor Compaq. On the ground, Dell's operations are organized around six assembly operations and their associated logistics centres that bring together components from suppliers. Figure 5.5 nicely illustrates the variable geographical impacts of technology being described in this chapter. On the one hand, Dell is able to maintain an extensive global business operation, meeting the needs of a global customer base through the intensive use of Internet technology. On the other hand, the assembly operations themselves are at the heart of six regional production clusters, combining local and global suppliers and logistic centres, and exhibiting the need for proximity inherent in just-in-time production systems. For more, see Fields (2004: Chapter 6).

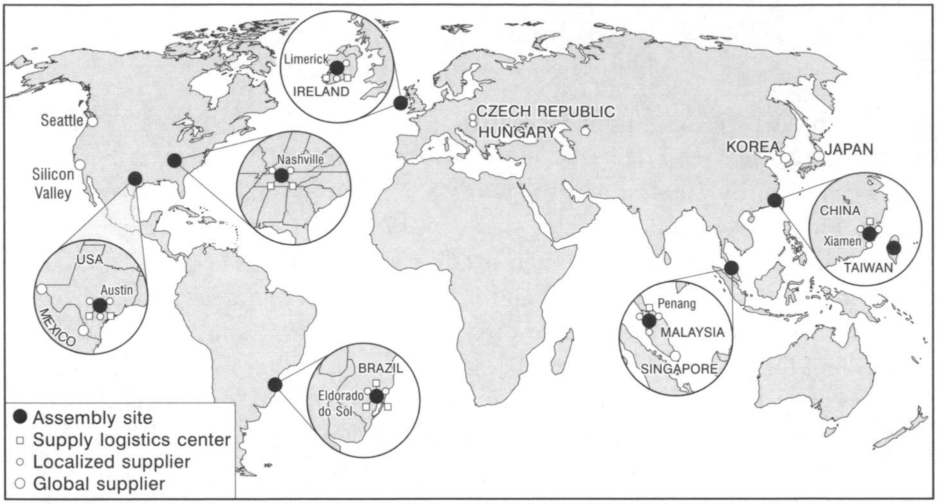

Figure 5.5 Dell's global operations, 2001
Source: Adapted from Fields (2004), Map 6.3.

But what are the geographical configurations of these three competing and overlapping production systems? Fordist production is associated with the formation of *branch plants* for undertaking the production process itself – as opposed to research and development, or sales and marketing, for instance – which can be relocated either nationally or internationally to benefit from variations in labour wages, skills, and levels of unionization. Intriguingly for economic geographers, the flexible specialization and Japanese flexible production systems both seem to exhibit a new-found tendency for firms to agglomerate – or group together – in particular places in what are commonly known now as *clusters*. It is to this apparently renewed importance of clusters/agglomerations (we use the terms interchangeably here) in the after-Fordist era that we now turn.

5.4 Proximity Matters: Traded and Untraded Interdependencies within Clusters

At first glance, the renewed agglomeration of economic activity in an era of advanced transportation and communications strategies seems rather counterintuitive. In this section, we will explore the different kinds of attractions or *agglomeration economies* (i.e. savings) that might cause firms to co-locate in particular places. Or, to put it another way, why do 'sticky places' hosting these clusters continue to exist in the 'slippery spaces' created by new technologies (Markusen, 1996)? At the outset, it is important to distinguish between two different types of agglomeration economies:

- *urbanization economies*: throughout history people have come together to create urban areas. The general clustering together of activities in towns and cities creates the potential for sharing the costs of a wide range of infrastructure (e.g. airports, road and rail networks) and services (e.g. universities, hospitals, libraries) among diverse firms, and for accessing large urban markets.
- *localization economies*: these refer to the cost savings that accrue to firms within the same or related industry through locating in the same place. These may derive, for example, from the presence of specialized pools of skilled labour, from access to industry-specific services and institutions, or from the development of a local knowledge base.

It is the latter category that is of particular interest to economic geographers, as it is central to after-Fordist agglomeration dynamics. We will now look at two kinds of localization economies that help to bind firms together in specialized clusters (or making places 'sticky') in contemporary capitalism.

The economic bases of agglomeration: traded interdependencies

The first set of localization economies we will consider has been termed *traded interdependencies* (Storper, 1997). These are created by firms co-locating in a cluster alongside suppliers, partners, and customers with which they have formal trading relationships. Proximity to these firms reduces the costs of transportation, communication, information exchange, and searching and scanning for potential customers. The greater the spatial dispersion of firms, the more onerous these costs will be. The consequence is that groups of firms with high interaction costs will choose to co-locate, thus creating a tendency to agglomeration. Hence, we can see that tendencies to agglomerate are inherent in both of the after-Fordist variants we presented in Table 5.1. On the one hand, the small firm complexes of the flexible specialization model require very close contact with nearby suppliers to facilitate the fine-grained division of labour between firms. On the other hand, under the Japanese flexible production system, suppliers often choose to locate close to customers in order to meet the needs of just-in-time supply systems. It is thus common to find clusters of suppliers surrounding assemblers and manufacturers in sectors where this production system is found (e.g. automobiles, personal computers). In both cases, a key process underlying the agglomeration of firms is *vertical disintegration*, whereby firms shed many activities and purchase them instead from their suppliers in order to focus on *core competencies* (Scott, 1988). This is particularly true in times of market volatility and uncertainty such as those that have prevailed since the mid-1970s. This disintegrative process creates more inter-firm relationships within a particular sector, and hence a strong propensity for spatial agglomeration to minimize the costs associated with sustaining those relationships. In this way, it can be argued

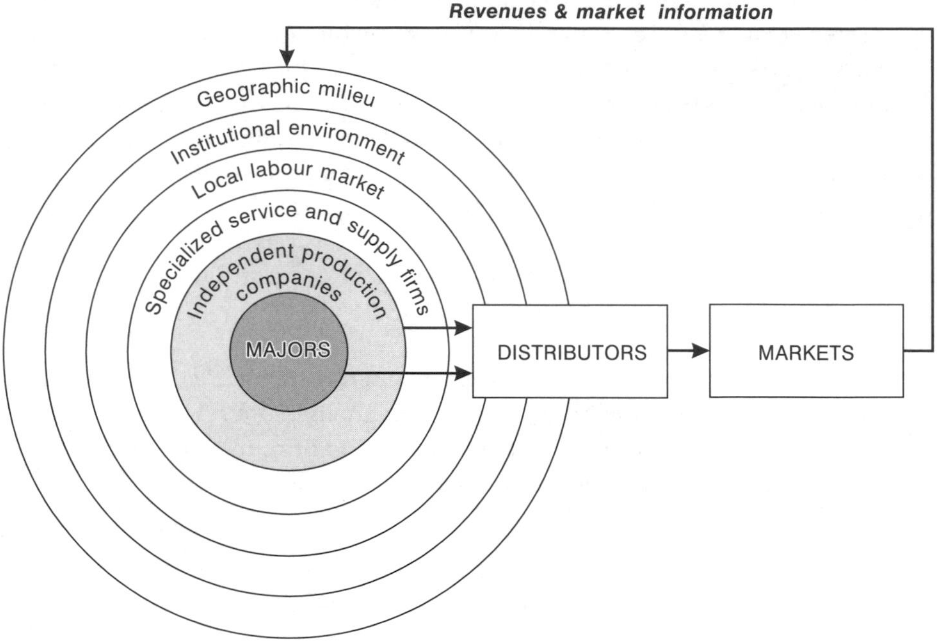

Figure 5.6 Schematic representation of the Hollywood film production agglomeration in Los Angeles
Source: Adapted from Scott (2002), Figure 3.

that agglomerative forces have become stronger in the after-Fordist period compared to the Fordist period.

We can illustrate these arguments with a brief exploration of the film and television production agglomeration centred in, and around, Hollywood, Los Angeles (Scott, 2002). Hollywood has long been the centre of the global film industry. The industry employs approximately 150,000 people in Los Angeles County (over 60 per cent of the national total for the sector) and is constituted by over 5,500 firms. Figure 5.6 characterizes the nature of this agglomeration in schematic terms. At the core of the system are the immensely powerful Hollywood studios (known as the 'majors') – MGM, Sony/Columbia, 20th Century Fox, Paramount, Universal, Warner Bros, and Disney – all of which are now part of global media conglomerates. While in the 1950s the studios were vertically integrated and used to undertake the whole production process 'in-house', regulatory and technological change meant that by the 1980s, a 'new' Hollywood had emerged and still prevails today. The new production system is characterized by vertically disintegrated production networks providing powerful traded dependencies between three groups of firms: the studios, a large base of independent production firms, and an even larger range of specialized service firms (e.g. script writing, lighting, costumes, catering, and cameras, etc.).

The studios, however, retain control over the entire production networks through their domination of finance and distribution activities. As Figure 5.6 shows, there are three other layers to the agglomeration. First, there is a local labour market containing a large number of skilled individuals. This labour market is constantly being renewed by the in-migration of talents from all over the world. Second, there is a rich institutional environment comprised of many organizations and associations representing firms (e.g. the Alliance of Motion Picture and Television Producers), workers (e.g. the Screen Actors Guild), and government departments (e.g. the Entertainment Industry Development Corporation). These institutions help oil the wheels of the industry and represent its common interests. Third, the whole agglomeration is embedded in a particular place and landscape that, through path-dependent processes, has developed a reputation and image as *the* key place to make films (see also Box 3.5 on California's historical development). Taken together, these various dimensions create a cluster that is rich in traded interdependencies and therefore offers strong localization economies to its participants. These economies have sustained Hollywood's leadership in the industry and its strong exports to countries around the globe.

The social and cultural determinants of agglomeration: untraded interdependencies

In some instances, then, agglomerations or clusters may be formed by localized transactional networks between firms. However, in many cases, revealing the formal economic relationships within an agglomeration may not be enough to explain its scale, formation and significance. We then need to bring into view the social and cultural bases of economic clusters, or what we might call *untraded interdependencies* (Storper, 1997). These are the less tangible benefits of being located in the same place, for example, the emergence of a particular pool of specialized workers, or the development of local institutions such as business associations, chambers of commerce, government offices, and universities with particular attributes and capabilities. In particular, clusters can facilitate patterns of intense and ongoing face-to-face communication between people working in the same or closely related industries. These interactions may in turn have a very real impact on the success of firms in the cluster by facilitating innovation and knowledge exchange (Box 5.4). Localized interaction is meant to be particularly important for the transfer of what is known as *tacit knowledge*. Tacit knowledge is best thought of as 'know-how' that can only be effectively created and shared through people actually doing things together in real life (see an example in Section 5.5 on the production of this book). It is often contrasted with *codified knowledge* – ideas and know-how that can be made tangible through, for example, writing it down or creating a diagram (an example being the book you are reading right now). While codified knowledge is able to travel across

Box 5.4 The creative class and spatial clustering

In his influential book, *The Rise of the Creative Class*, Richard Florida (2002) charts the rapid emergence of a so-called creative class as the key contemporary employment dynamic affecting the US economy. By 1999, the grouping accounted for some 38 million workers or roughly 30 per cent of the entire US workforce. In Florida's conceptualization, the creative class has two main components. First, there is a super-creative core of scientists, engineers, professors, entertainers, actors, writers, and the like (12 per cent of total workforce). Second, there is a wide range of creative professionals working in high-tech sectors, financial services, the legal and health care professions, and business management (18 per cent of total workforce). The class is defined in economic terms as workers who add economic value through their creativity, and covers many of the most highly paid professions in the economy. What the disparate activities supposedly have in common are creative class values based on notions of individuality, meritocracy, and diversity/openness to difference.

Most importantly for our analysis here, Florida argues that the creative class tends to cluster in particular places, which he terms *creative centres*. These centres are thriving economically as they offer an attractive quality of place to creative class workers combining elements of what's there (e.g. the built and natural environment), who's there (e.g. social diversity and high levels of interaction), and what's going on there (e.g. vibrancy of street and outdoor life). Florida's overall prognosis is that for places to succeed economically, they need to offer all three Ts of Technology, Talent, and Tolerance. Constructing a *creativity index* from various measures of these dimensions, he found that San Francisco, Austin, San Diego, Boston, and Seattle were the leading creative centres, emerging from a new geographic sorting of classes that is bypassing many places in the US. Intriguingly, in contrast to accounts of regional economic development that look purely at industrial structure and innovation levels to explain growth rates, Florida argues that a vibrant, diverse and bohemian social scene is a necessary condition for creativity-led growth. He found the concentration of gay people living in a particular locality to be an excellent measure in this regard. It is worth noting, however, that Florida's vision is a vigorously contested one. Concerns have been raised, for example, as to whether his dimensions of creativity are the cause of economic growth, or simply side-effects, and about what happens to the remaining 70 per cent of the workforce if the creative class are given priority in urban policy initiatives.

space, tacit knowledge is much stickier geographically, and may only be accessible by being immersed in the untraded social and cultural interactions of a locality.

But what specific form do these untraded interdependencies take in reality? The case of Britain's Motorsport Valley provides an interesting window into these dynamics (drawing on Pinch and Henry, 1999). The term refers to a dense agglomeration of motorsport activity in Southern England, and more specifically, an areas with a radius of 50 miles around the counties of Oxfordshire and Northamptonshire (north-west of London) (Figure 5.7). The majority of the British motorsport industry – which has some 40,000 employees, 25,000 of which are highly skilled engineers, and has an annual turnover of £5 billion (or US$9 billion) – is located in this area. Over 75 per cent of the single-seat racing cars used around the world are UK-built, and six of the ten Formula 1 teams are UK-based. Motorsport Valley acts as the global hub for the industry, with drivers, expertise, engineers, components, money, and sponsorship coming from all over the world (for more, see http://www.motorsportvalley.com). What is of interest to us here is the way in which specialist knowledge circulates between the hundreds of small firms that constitute the Motorsport Valley *outside* of formal business transactions, i.e. as untraded interdependencies. It is possible to identify six interacting ways that knowledge is disseminated:

1 *staff turnover*: the ongoing circulation of key personnel – engineers, drivers, designers, and mechanics – means that crucial information about how things are done is transferred between firms and teams. Analysis of career trajectories has revealed that an average designer or engineer will move firm every 3.7 years, and eight times during their career.
2 *shared suppliers*: another mechanism for knowledge transfer is through the links that motorsport teams have with numerous component and services suppliers. While confidentiality agreements are in place to limit the transfer of successful ideas from one team to another through a common supplier, in reality there is a tendency for technical know-how to leak, as suppliers want to provide their clients with the best and most up-to-date advice.
3 *firm births and deaths*: motorsport is a very expensive and risky economic enterprise. As a result, the industry is characterized by high levels of firm births and deaths. Each of these individual events represents an opportunity for staff mixing and hence the diffusion of knowledge.
4 *informal collaboration*: racing car companies are on the one hand extremely competitive and secretive, yet on the other hand are involved in a tightly regulated collective endeavour that necessitates interaction and involvement in working groups. Collective discussion of the regulations, and how to respond to proposed changes in them, provide another mode of inter-firm knowledge transfer.

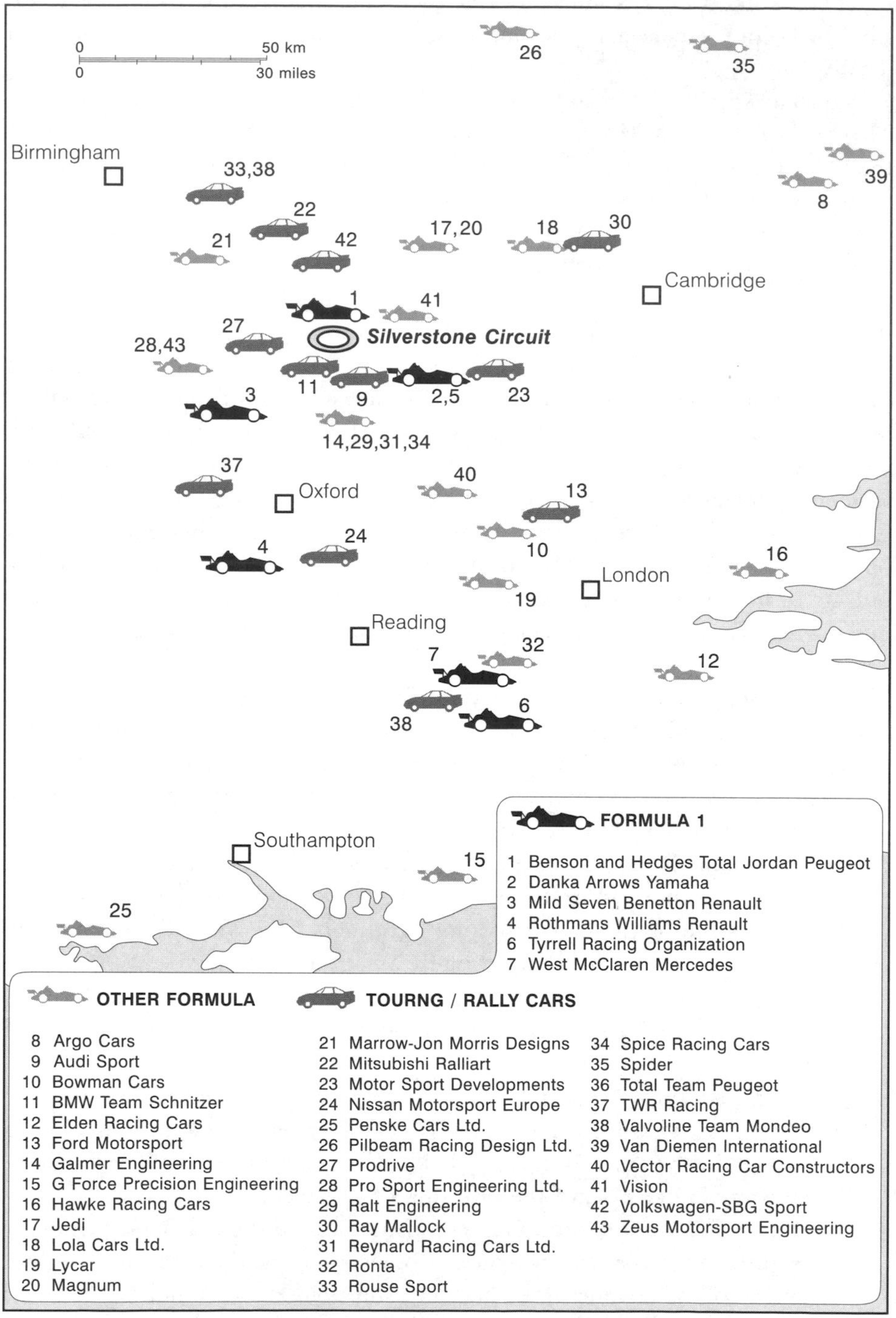

Figure 5.7 The Motorsport Valley in the UK
Source: Adapted from Henry and Pinch (1999), Figure 1.

5 *industry gossip*: inter-personal networking in the Valley is extremely strong, in large part because of the dynamics described above. These networks criss-cross firms and teams, and are activated for a number of reasons, including facilitating staff recruitment and for advice on technical issues.
6 *trackside observation*: racing car technology is revealed on the track during testing and racing. As soon as teams observe something different on another car, they will endeavour to imitate and test the changes themselves. It is not uncommon for new innovations to appear almost simultaneously on a number of competing cars.

As this brief case study has shown, there are numerous mechanisms for transferring tacit knowledge that do not depend on contractual relationships between two firms. In the Motorsport Valley, strong rates of new enterprise generation, highs level of inter-firm interaction, and rapid flows of skilled personnel between firms are at the heart of knowledge dynamics. The precise combination of mechanisms used will vary from sector to sector, and from place to place, however. In sectors such as finance, for example, non-workplace spaces of informal interaction (e.g. bars, clubs, and cafés) are very important for information-sharing. In certain areas of scientific research (e.g. computer science), presentations and interaction at conferences, publications in journals, and online forums are crucial avenues for knowledge dissemination and transfer. What is most important here, though, is that many different kinds of agglomeration generate economies from the nature of the localized social-cultural interaction that takes place within them. In an era of advanced information and communications technologies, it may well be that these patterns of social interaction are a more significant form of glue, than traded economic relationships, in holding agglomerations together.

A typology of agglomerations

There is a wide variety of different kinds of agglomerations in the contemporary economy, held together by varying combinations of urbanization and localization economies, and of traded and untraded interdependencies. A large city such as New York, for example, offers huge urbanization economies at the metropolitan scale to a wide variety of different economic activities. Look within the city, however, and you will find many more focused clusters of activity offering additional localization economies of both the traded and untraded varieties to firms in the financial sector (Wall Street district), advertising (Madison Avenue), and the fashion industry (midtown Manhattan), to name but a few. Recognizing this variability in terms of geographical scale, sectoral composition and causal dynamics is vital if economic geographers are to effectively understand the contemporary space economy. Hence, we conclude this section by offering a sevenfold typology of significant cluster types:

- *labour-intensive craft production clusters*: these are often found in industries such as clothing where work is characterized by sweatshop conditions and often very high levels of immigrant labour. Firms are involved in tight subcontracting networks, and may also use home-workers. Examples include the garment production districts of Los Angeles, New York, and Paris.
- *design-intensive craft production clusters*: these refer to dense agglomerations of small and medium-sized firms specializing in the high-quality production of a particular good or service. They are characterized by a highly disintegrated production system in which individual firms perform specialized, narrowly-defined roles. Examples include the renowned towns and districts of the Third Italy – e.g. Arezzo (gold jewellery), Carpi (knitwear), Sassuolo (ceramics), and Ancona (shoes) – and Jutland, Denmark (furniture) and Baden Württemberg, Germany (machine tools).
- *high-technology innovative clusters*: these are characteristic of 'new' after-Fordist sectors such as computers and biotechnology. These clusters tend to have a large base of innovative small and medium-sized firms and flexible, highly-skilled labour markets. They have often grown up in areas with little history of industrialization and unionization. Examples include the Silicon Valley and Route 128, Boston, districts in the US; Grenoble, France; and Cambridge, UK.
- *flexible production hub-and-spoke clusters*: in these clusters, a single large firm, or small group of large firms, buys components from an extensive range to local suppliers to make products for markets external to the cluster. These clusters represent the spatial logic of just-in-time production systems. Examples include Boeing in Seattle, US, and Toyota in Toyota City, Japan.
- *production satellite clusters*: these clusters represent congregations of externally owned production facilities. These range from relatively 'low tech' assembly activity, through to more advanced plants with research capacity, but all are relatively 'stand alone'. Firms have generally co-located to access the same labour market conditions or financial incentives. Examples are to be found across the export processing zones (EPZs) of the developing world. The electronics industry on the Malaysian island of Penang is a specific instance of this kind of cluster.
- *business service clusters*: business services activities such as financial services, advertising, law, accountancy, and so on are often concentrated in the central districts of leading cities (e.g. New York, London, and Tokyo), and in some cases, their hinterland regions (e.g. software and computer services in the Western Arc of counties outside London).
- *state-anchored clusters*: a disparate category of clusters has developed due to the location decisions of government facilities, such as universities, defence industry research establishments, prisons, or government offices. Examples include agglomerations that have developed due to government research investment (Colorado Springs, US; Taejon, South Korea; M4 Corridor, UK) and universities (Madison, Wisconsin, US, and Oxford/Cambridge, UK).

Figure 5.8 A multi-faceted cluster? High-tech business in Silicon Valley, California
Source: The authors.

In reality, of course, applying any such typology is not an easy task. Some places are constituted by clusters that are *hybrid* forms merging the characteristics of two or more of these types. For example, Silicon Valley, a cluster in California employing some 400,000 high technology workers, can be seen to exhibit the traits of a high technology innovative cluster, a flexible production hub-and-spoke cluster (due to the presence of large manufacturers such as Intel), and a state-anchored district (due to the significance of huge levels of government defence spending) (see Figure 5.8 and Chapter 11). In addition to this complexity, Silicon Valley is not just defined by local interactions. Its firms are also very well plugged into a variety of important national and global connections with partners, suppliers, and customers. In other words, clusters are not necessarily shaped by local forces only. We now move on to consider how we might conceptualize clusters as being defined by patterns of *both* local and non-local interaction.

5.5 Neither Here Nor There: Thinking Relationally

As the preceding section has made clear, economic geographers continue to be fascinated by the importance of spatial proximity – i.e. people being in the same

Box 5.5 Global cities

There have been many attempts over the past 20 years to try and capture the essence of the global urban hierarchy and its interactions with the changing geography of the world economy. A leading example is Sassen's (2001) detailed research on the so-called *global cities* of New York, London, and Tokyo. Her thesis is that globalization has created a new strategic role for major cities with control and organization functions. There are four stages to her basic argument: (1) cities are key command points in the world economy; (2) this creates a demand for financial and business services that are leading sectors in advanced economies; (3) cities become key sites for production and innovation in these sectors; and (4) cities in turn constitute the main markets for business and financial services. In this view, global cities – of which New York, London, and Tokyo are leading examples – are not just the site of corporate headquarters, but full-blown *global service centres* with a wide array of innovative support activities such as legal, banking, insurance, accountancy, management consultancy, advertising, and software firms.

Sassen's work also highlights some of the downsides of this concentration of knowledge and expertise in a few cities, most notably the polarization of the labour market into an extremely well-paid service class and an underclass of (often immigrant) workers filling a number of jobs to meet their work and leisure needs (e.g. cleaners, security guards, caterers, waiters, kitchen hands, shop assistants, etc.). While other scholars have used different measures to construct global urban hierarchies (e.g. number of *Fortune 500* firm headquarters, scale of the financial sector, etc.), most studies agree that there is a leading group of cities – most notably New York, London, Tokyo, Paris, Zurich, and Frankfurt – that play a disproportionate role in controlling and coordinating the global economy. Three caveats about this approach should be noted. First, beyond this very top tier, the ranking can alter significantly, depending on what kind of measure is being used. Second, any ranking of cities must be seen in dynamic terms: the recent rise of cities such as Beijing and Shanghai, for example, will profoundly re-shape existing patterns. Third, the next challenge for economic geographers is to reveal the nature of the connections *between* key cities, rather than their static attributes. It may well be that the notion of a hierarchy is out-dated in today's interconnected world (Taylor, 2004).

place at the same time – in enabling socio-economic interaction. Yet how do we reconcile the apparent importance of spatial proximity with what we know about globalization and the power of various information and communication technologies to transcend space and time? In order to make sense of this apparent paradox, we need to advance our understanding in two further ways.

First, we need to recognize that the concentration of some economic activities in particular places is part of the very same processes that are seeing other kinds of activities dispersed across the global economy. As we will see in more detail in Chapter 8, within the production system for a particular product or service, there are some operations such as corporate decision-making which need to be located in agglomerations, while other activities such as production, sales and marketing, and customer support, can be more spatially disparate. Put another way, the spatial decentralization of economic activity facilitated by space-shrinking technologies *requires* the very kinds of control and coordination that can only be achieved through locating certain important corporate functions in key *global cities* (Box 5.5). More specifically, this dispersion creates different challenges of integration and coordination for firms (Amin and Thrift, 1992). Information has to be acquired about what is going on across the production system as a whole. Moreover, social interaction is required to allow important interpersonal relations such as trust to develop. Finally, firms somehow have to keep up with innovations in both products and processes within their industry.

Clusters offer a solution to these problems by acting as centres of tacit knowledge and control within wider global networks, for example, through the interaction of corporate headquarters, government departments and industry-specific media. Clusters act as centres of sociability, places where inter-personal relationships are created and maintained. And they act to provide a critical mass to help generate innovations through the intense ongoing interaction of numerous users and suppliers of technologies. Clusters, then, should not be considered on their own, and purely in terms of the local relationships they contain, but as *nodes in global networks*, where certain kinds of local relationships are used to facilitate and support other kinds of non-local (and possibly global) relationships.

Second, we can open up our economic-geographical interpretation of clusters still further by thinking of spatial proximity as just one of several different kinds of proximity (or a sense of nearness) that enable the global economy to function. Additionally, we can think of the following:

- *institutional proximity*: a nearness created through operating within the same legal and institutional frameworks as others (e.g. within the German economy);
- *cultural proximity*: a nearness derived from a shared cultural heritage and linguistic background (e.g. the Irish, Jewish or Chinese diasporas; see more in Chapter 13);
- *organizational proximity*: a nearness engendered through both written rules and codes, and unwritten ways of doing things within a particular firm or

institution (e.g. the corporate culture of a large transnational firm, see Chapter 11);
- *relational proximity*: a closeness derived from informal inter-personal relations (e.g. long-standing friends in distant places).

In some situations, these other forms of proximity may be as, if not more, important than physical proximity. Relational proximity, for example, may facilitate the same transfer of tacit knowledge that is argued by many to depend on physical proximity. While relational proximity may depend to a certain extent upon face-to-face interaction, it can also be achieved through modern communications technology, and through the mobility of knowledgeable individuals. In this way, nearness is achieved through combinations of inter-personal and electronic communication, and travel.

If all this sounds a little abstract, a brief consideration of the production process of this book is revealing. Although we knew one another beforehand, the relationships underpinning the collaboration were really forged in 1998–2000, when all three of us were working as lecturers at the National University of Singapore. Here, a combination of *spatial* and *organizational* proximity helped to engender strong inter-personal relations. Since 2000, when two of us left Singapore for Manchester and Toronto, this strong *relational* proximity has sustained a number of collaborations including this book. The book itself has been produced through a mixture of intermittent travel and face-to-face contact, and intense ongoing 'electronic' communication. Four periods of face-to-face contact (or spatial proximity) were vital at different stages of the process to augment the ongoing circulation of draft chapters, comments, and ideas via email. A sequence of meetings in Manchester (crystallizing the initial idea), Singapore (detailed book planning), Manchester (discussion of early chapter drafts and book format), and Bellagio, Italy (discussion of final chapter drafts) provided an intensity and quality of interaction that electronic communication could not. Through this combination of strong relational proximity, and intermittent physical proximity, we were able to produce this knowledge-intensive product despite being physically separated by thousands of miles for most of the time. The nature of this production process also points to the increasing significance of project-based forms of working as a way of engendering spatial and temporal flexibility in many sectors of the economy (Box 5.6).

Box 5.6 Project working

While projects – the bringing together of a set of people to fulfil a complex task – are a long established way of working in industries organized around one-off schemes such as architecture, advertising, construction, ship-building, and film making, there is evidence that they are spreading into new sectors (e.g. automobiles, chemicals), while many new industries (e.g. software, new media, or business consulting) are dominated by project working. Projects can be thought of as temporary social systems in which people of varied professional and organizational backgrounds work together in an environment that is insulated from the other ongoing activities of the firm or organization. Projects can vary greatly in their constitution, duration, and location. In terms of their constitution, they made be made up a range of different skilled professionals from within a single corporation or from a range of different firms. In the latter case, the project team can be thought of as a virtual firm. In terms of their duration, a project may range from a few weeks (e.g. the filming of a TV movie-of-the-week on location) to many months and even years (e.g. a large scale software project for a major multinational client).

In terms of the location, a project team may be based with a firm's office network, on a client's site, at a neutral site, or even in cyberspace (e.g. a team of far-flung specialists working simultaneously or in sequence on a particular piece of software or an architectural blueprint). Many will combine both face-to-face and electronically mediated forms of communication. While project members may be drawn from within a single office building or a national subsidiary of a company, firms are increasingly assembling *transnational* project teams to meet the needs of their clients. Whatever the precise configuration of different project teams may be, two things are certain. First, projects represent an increasingly pervasive form of working that is extremely flexible in both time and space, and one that facilitates the transfer of various forms of tacit 'know-how' within and across organizations. Second, projects pose an important challenge to economic geographers by blurring the boundaries of key analytical categories such as 'firms' and 'clusters'. Indeed, even just positioning 'where' economic activity is taking place can be problematic in the world of project working. See Grabher (2002) for more on project working in the advertising industry.

5.6 Summary

This chapter has explored the interactions between new technologies and economic geography. New transportation and communication technologies have transformed the workings of the global economy by allowing the fragmentation and separation of production processes over space. However, they have done so in ways that are hugely uneven in social and geographical terms. Even a seemingly placeless technology such as the Internet is in reality full of spatial inequalities and contradictions of different kinds. In many cases, new space-shrinking technologies have served to reinforce existing patterns of inequality, rather than allowing more people and places to interact with the global capitalist system as some accounts of technological change might suggest. New information technologies have also transformed how products and services are made, engendering greater production flexibility in different forms of after-Fordist production. These changes have created a propensity for related firms to agglomerate in clusters in order to reduce the costs of transacting with one another. In addition, there is a wide range of other untraded relationships that may compel firms to co-locate. The result is a complex mosaic of different kinds of cluster operating as nodes in global economic networks.

So, can technology eradicate distance? In some instances, the answer is that technology can help to at least overcome geographic separation. The ability to move people, money, products, and technologies quickly, efficiently, and cheaply around the world has been highly significant in enabling the globalization of economic activity. Yet, in other instances, technologically-mediated communication and interaction are not enough. In industries where tacit knowledges are important, people still need to locate together to immerse themselves in the buzz of day-to-day interaction, sociability, and innovation. Intriguingly, both of these dynamics occur within the same economic activity. Returning to financial services, for example, large financial deals often rely on the development of interpersonal trust and the richness of knowledge flows within the financial districts of a few global cities. These financial districts, however, are now almost seamlessly integrated into a global trading system through information and communication technologies. In sum, the contemporary global economy operates through complex combinations of near and distant relations, and technology plays an important, but not determining, role in its operation.

Further reading

- Dicken (2003: Chapter 4) provides a clear account of different kinds of technology and their geographical impacts.
- See Dodge and Kitchin (2000) and Zook (2005) for more on the emerging geographies of the Internet world.
- Bryson and Henry (2005) offer a cogent account of the supposed shift from Fordism to after-Fordism and its spatial implications.
- UNCTAD (2004), Scott (2002), and Pinch and Henry (1999) provide more detail on the case studies of offshore services, Hollywood, and the Motorsport Valley respectively.
- Storper (1997) explores the various economic, social, and cultural factors behind regional agglomerations.

Sample essay questions

- Why is it important to take an evolutionary perspective to technological change?

- Why do space-shrinking technologies have varied and uneven geographical impacts?
- Does the Internet have geographies?
- What are the geographical implications of the supposed shift from Fordism to after-Fordism?
- What are the different factors that may cause agglomerations to develop?

Resources for further learning

- http://www.worldbank.org: the World Bank offers a wide range of data on the worldwide usage rates of information technologies such as fixed and mobile telephones in its World Development Reports and World Development Indicators.
- http://www.unctad.org/en/docs/wir2004_en.pdf: UNCTAD's *2004 World Investment Report,* 'The shift towards services', provides a wealth of information on the offshoring of service activity and can be downloaded for free.
- http://www.dell.com: personal computer manufacturer Dell's homepage contains a range of information on the company, its strategies and its history (click on 'About Dell').
- http://www.creativeclass.org: see this website for a wide range of materials on Richard Florida's controversial ideas about the creative class.
- http://www.lboro.ac.uk/gawc: check out the *Globalization and World Cities* (GaWC) website hosted by Loughborough University for a huge range of resources on world cities and the networks that connect them.
- http://www.jointventure.org: the Joint Venture Silicon Valley Network provides a range of perspectives on this world-renowned high-tech cluster.

References

Amin, A. and Thrift, N. (1992) Neo-Marshallian nodes in global networks, *International Journal of Urban and Regional Research*, 16: 571–87.

Bryson, J. and Henry, N. (2005) The global production system: from Fordism to post-Fordism, in P. Daniels, M. Bradshaw, D. Shaw, and J. Sidaway (eds) *An Introduction to Human Geography: Issues for the 21st Century*, Harlow: Pearson, pp. 313–35.

Dicken, P. (2003) *Global Shift*, 4th edn, London: Sage.

Dodge, M. and Kitchin, R. (2000) *Mapping Cyberspace*, Harlow: Addison-Wesley.

Economist, The (2002) Special report – Container Trade, 6 April, pp. 73–5.

Economist, The (2003) Special report – Offshoring, 13 December, pp. 79–82.

Economist, The (2005) Special report – eBay: Meg and the power of many, 11 June, pp. 71–4.

Fields, G. (2004) *Territories of Profit*, Stanford, CA: Stanford Business Books.

Florida, R. (2002) *The Rise of the Creative Class*, New York: Basic Books.

Freeman, C. and Perez, C. (1988) Structural crises of adjustment, business cycles and investment behaviour, in G. Dosi, C. Freeman, R. Nelson, G. Silverberg, and L. Soete (eds) *Technical Change and Economic Theory*, London: Pinter, pp. 38–66.

Grabher, G. (2002) The project ecology of advertising: tasks, talents and teams, *Regional Studies*, 36: 245–62.

Harvey, D. (1989) *The Condition of Postmodernity*, Oxford: Blackwell.

Markusen, A. (1996) Sticky places in slippery spaces: a typology of industrial districts, *Economic Geography*, 72: 293–313.

O'Brien, R. (1992) *Global Financial Integration: The End of Geography*, London: Pinter.

Pinch, S. and Henry, N. (1999) Paul Krugman's geographical economics, industrial clustering and the British motor sport industry, *Regional Studies*, 33: 815–27.

Rimmer, P.J. (2000) Neighbours: Australian and Indonesian telecommunications connections, in M.I. Wilson and K.E. Corey (eds) *Information Tectonics*, Chichester: John Wiley & Sons, Ltd, pp. 165–98.

Sassen, S. (2001) *The Global City*, 2nd edn, Princeton, NJ: Princeton University Press.

Scott, A.J. (1988) Flexible production systems and regional development: the rise of new industrial spaces in North America and Western Europe, *International Journal of Urban and Regional Research*, 12: 171–86.

Scott, A.J. (2002) A new map of Hollywood: the production and distribution of American motion pictures, *Regional Studies*, 36: 957–75.

Storper, M. (1997) *The Regional World*, New York: Guilford Press.

Taylor, P. (2004) *World City Network: A Global Analysis*, London: Routledge.

UNCTAD (2004) *World Investment Report 2004: The Shift Towards Services*, New York: UNCTAD.

Zook, M. (2005) *The Geography of the Internet Industry*, Oxford: Blackwell.

CHAPTER 6

ENVIRONMENT/ECONOMY

Can nature be a commodity?

Aims

- To examine conventional economic approaches viewing nature as a simple commodity
- To understand how nature is brought into economic systems through processes of commodification
- To explain how attempts to protect the environment can become commodified and seen as value-added activity
- To think more creatively about the role of 'nature' within economic geography.

6.1 Introduction

In June 2006, the Oscar-winning British actress Vanessa Redgrave was awarded a lifetime achievement award at the International Transylvanian Film Festival in Romania. She dedicated her award to a local community organization – Alburnus Maior – which was fighting against the plans of the Canadian mining company Gabriel Resources to develop the biggest open cast gold mine in Europe around the village of Rosia Montana in central Romania (see Figure 6.1). Concerned about the environmental impacts of the proposed development, she told the festival audience: 'our planet is dying and we have no right to destroy an ecosystem'. In return, Alburnus Maior gave the actress a square metre of land in the village in order to create a highly symbolic tangible and legal tie. Gabriel Resources, a sponsor of the film festival, were outraged. The company's response was to publish a full page 'open letter' to Redgrave in the British *Guardian* newspaper challenging her intervention on behalf of villagers backing the mining development (Box 6.1). Redgrave responded, 'I'm amazed that they should be so

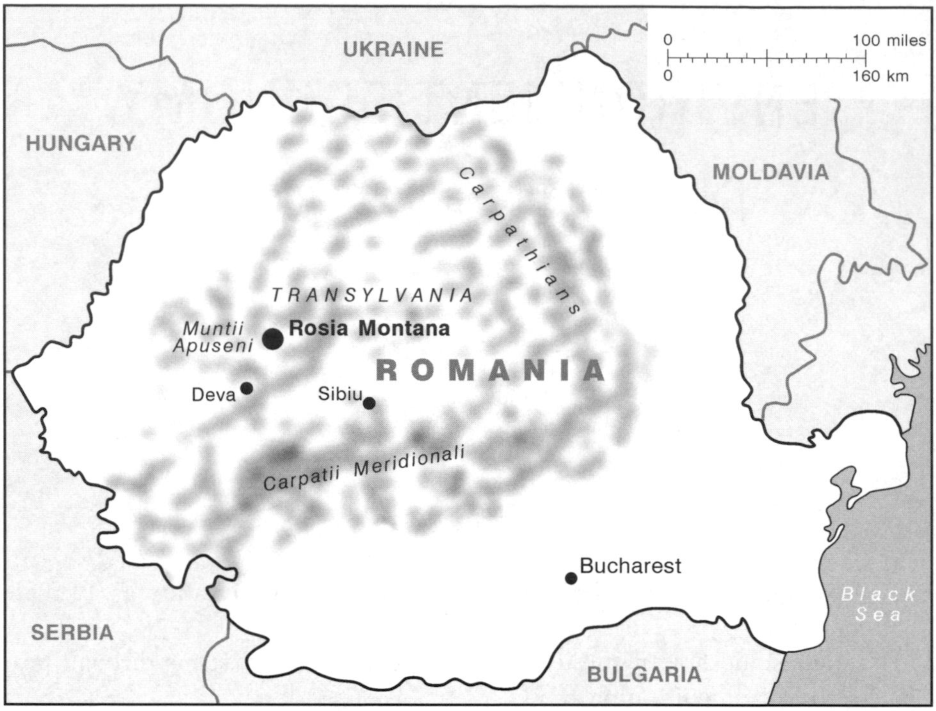

Figure 6.1 Location map of proposed mine project in Rosia Montana, Romania

desperate as to put in an advert criticising me, but it's for their shareholders back in Canada' (*Guardian*, 2006a).

Gabriel Resources first initiated the project back in 1997 in partnership with Minvest, a Romanian state-owned mining company, which owned a 20 per cent stake. They argued that this would be a 'model' mine in an area of huge unemployment (70 per cent), providing 1,200 jobs during construction, and 600 during operation, as well as a further 10 indirect jobs in the service sector for each direct one. If approved, the mine would start producing gold in 2009. It was designed to run for 16 years, and would produce some 500,000 ounces of gold during its lifetime, with silver being an important by-product of the operations. Campaigners against the mine argued, however, that the mine would destroy four mountains, cause pollution, drastically affect biodiversity, erase many Roman and prehistoric sites, and necessitate the demolition of the village of Rosia Montana and the forcible relocation of its 2,000 residents. Of particular concern were the plans for a 600-hectare unlined cyanide storage 'pond', held back by a 180-metre high dam, and the estimated 200 million tonnes of cyanide-laced waste that the mine would produce. The issue split the local community down the middle. Some – like the 77 who signed the open letter to Redgrave – were supportive of the scheme and willing to accept the US$40–

Box 6.1 Excerpts from an open letter to Vanessa Redgrave

As a new neighbour, we ask you:
How can you argue that you know better for us when you don't know the profound reality of our village?

When you condemn the mining profession, without any other viable alternatives for long- and average-term development, you condemn our future.

You say you did it to save our environment:
But our village is polluted now, from almost 2,000 years of poor mining practices. A new and modern mine will actually help make our streams and rivers cleaner. A powerful economy means a healthy environment; poverty means misery!

You argue that you want to save our village:
But we have 70% unemployment. No full-time doctor. No regular running water. Only wood to burn for heat in the winter. If the mining project doesn't start as soon as possible, the inhabitants have to start wandering, looking for a new chance of survival. We don't think this is the salvation for Rosia Montana!

We have always been a mining community. We want the chance to decide what will keep our village alive, and create a future for our village and children.

We support the investment for the mine rehabilitation for Rosia Montana and we will fight, so that mass-media and the protesters from the European capitals will understand the reality.

Miss Redgrave, please hear our community's voice!

The Guardian, 23 June 2006, p. 3 (emphasis in original).

50,000 per home offered by Gabriel Resources to relocate to a nearby purpose-built village. Others, including more than 350 families represented by Alburnus Maior, felt they were being forced out of their home village against their wishes, and that if they refused to sell, they would have their land and property expropriated (for more, see http://www.rosiamontana.org/, and for other such examples, visit http://www.nodirtygold.org/).

This example nicely illustrates the key themes of this chapter. Most fundamentally, it demonstrates how a natural substance – in this case gold embedded in Romanian rocks – becomes a commodity, i.e. something that can be bought and sold. Once gold is extracted, Gabriel Resources would be able to sell it profitably on to the jewellery industry, which uses 80 per cent of the gold that is mined globally. It also raises questions about the ownership of natural resources, and who benefits from the commodification process. How is it that distant Canadian shareholders would make money from the gold mine, while the benefits accruing to the Romanian government and the people of Rosia Montana are far less clear? And the case also brings into view the environmental consequences of resource extraction and use, and the problems of accounting for them. It is estimated that the production of one gold ring generates 20 tonnes of mine waste. How do we put a *value* on the potential environmental damage of a gold mine, for example, the aesthetic impact of hillsides being destroyed? How do we weigh this against the direct jobs and investment that the project would supply? How can we balance short-term economic benefits with environmental consequences that may last for generations? As we shall see, conventional approaches to resource exploitation under capitalism continue to underplay these kinds of questions.

We explore these issues in this chapter in four stages. Next, in Section 6.2, we critique economists' approaches that either exclude or simplify the role of nature in economic processes. Instead, we propose an approach that sees material flows and transformations as an integral part of the global commodity and production chains first introduced in Chapter 4. In Section 6.3, we profile the various ways in which natural resources can be *commodified* and thus incorporated into the capitalist economy. In Section 6.4, we explore the ways in which the *outputs* or *impacts* of the capitalist system on the natural environment can also become entangled in processes of commodification. Here we emphasize how nature is being valued as an asset that can be translated into corporate profits. Finally, in Section 6.5, we consider some recent attempts to rethink and 'blur' the relationships between nature and economy/society.

6.2 How Is Nature Counted in Economic Thought?

In this section we present two ways of analysing the worth of natural resources in contemporary economics. In *conventional economics*, the valuation of nature is deceptively straightforward. It is established in the form of a price set for units of a given natural material such as wheat, gold or lumber. The actual mechanism for determining a price is in a market exchange where buyers and sellers come together. In some cases (such as a local cattle auction) the price reached (for a head of cattle) will be locally specific. But for many natural raw materials (such as metals or oil), a global price is established at large commodity exchanges. Such exchanges now exist in many countries around the world, but the largest and most important ones operate in just a few major centres: the Chicago Board of Trade and China's Dalian Commodity Exchange for agricultural products;

the London Metal Exchange for metals such as tin, aluminium and copper; and the New York Mercantile Exchange for energy products such as oil and natural gas. Some trading at these exchanges is for immediate purchases (or 'spot trades'), but most is for 'futures' – whereby future deliveries of a commodity are purchased at a guaranteed price.

Conventional economic analysis views these exchanges as highly efficient mechanisms for establishing the value of natural raw materials. They allow producers and traders of natural substances to find readily a buyer for their commodities even before they are extracted from the earth. Such exchanges also allow an equilibrium price to be reached so that supply and demand are matched. In establishing an equilibrium price for a commodity, it is the volume of this commodity entering the market, along with the volume of its demand that together determine the price. The volume of *supply* can, in some cases, respond relatively quickly to market prices. If, for example, the price of tuna fish increases, then more fishing boats will quickly seek to capitalize on the windfall (and the increasing supply will bring the price back down again). But if a cartel of producers restricts supply, as in the case of the Organization of Petroleum Exporting Countries (OPEC), then prices can be pushed higher. Alternatively, a change in *demand* for a product can move its price. The increasing demand for artificial sweeteners in the second half of the twentieth century, for example, led to the long-term decline in the world price of sugar, devastating many Caribbean economies. More recently, the industrial expansion and increasing wealth of China in the late 1990s and early 2000s have pushed up the global prices of many raw material commodities. Between 1999 and 2004, for example, China's annual consumption of aluminium doubled, reaching 5.9 million tonnes (20 per cent of global consumption). Over the same period, the global price for aluminium at the London Metal Exchange jumped from under US$1,200 per tonne to over US$1,600 (USITC, 2006).

While commodity exchanges 'discover' the price for a natural raw material, the price in question is the outcome of a range of factors. The ways in which these factors come together is the product of daily analysis and guesswork by economists and traders in the various commodity exchanges around the world. What is absent in this calculation, however, is any evaluation of broader environmental impacts (in the present or the future) of extracting and using a resource. The market is, in other words, entirely blind to the environmental effects of natural resource extraction. As we saw in the case of the Romanian gold mine in the introduction, the effects of stripping hillsides is seldom part of any cost calculation. Thus, the economic valuation of commodities as purely the outcome of a supply and demand equation resulting in an equilibrium price obscures many of the environmental costs of a product.

There are, however, techniques available to calculate some of these consequences – largely from the field of *environmental economics* that emerged in the 1970s. They involve the quantification of a wide range of costs associated with environmental degradation. To take one example, we can examine the effects of the rapid expansion of oil palm plantations in Southeast Asia since the 1990s. Palm oil is a type of vegetable oil that is now widely used as an industrial raw material in the food industry. It is also a source of *biofuel* and has therefore been promoted as an alternative to fossil fuels, with far fewer consequences for climate change. Indonesia and Malaysia are the two largest producers of palm oil in the world and, in the past 20 years, vast tracts of land in both countries have been converted into palm oil plantations.

Environmental groups have raised concerns about the loss of biodiversity as rainforests are converted into plantations, and the destruction of habitats for endangered species such as orang-utans. Of more immediate concern, however, have been the effects of land clearance in parts of Sumatra and Borneo, where vegetation is burned to make way for plantation crops. Since 1997, this burning has intensified and has spread a thick pall of smoke and haze across the region for several months each year. In Singapore and peninsular Malaysia, visibility and air quality are now sometimes reduced to dangerous levels. If a more complete accounting of the costs of palm oil production were calculated, what would it look like? Table 6.1 provides some initial clues in the form of an estimation of the costs of atmospheric haze in Singapore alone (here we draw on Quah, 2002). A variety of proxy measures are used to assess the costs of air pollution (e.g. the costs of treating haze-related illnesses, lost productivity and decreases in tourism receipts). The loss of scenic views, due to poor visibility, is

Table 6.1 Estimates of costs to Singapore resulting from the 1997 haze

Impacts of haze damage	Estimate Range (US$)
Direct Damages	
Direct cost of illness	1,186,900–1,535,668
Self-medication expenses	678,943
Loss in earnings/productivity	1,907,341–2,068,109
Preventive expenditures	3,524–234,909
Loss in tourism revenue	136,577,290–210,449,067
Loss to local businesses	*Not calculated*
Indirect damages	
Loss in visibility and views	23,057,133–71,137,941
Loss in recreational activities	94,170
Damages to biodiversity	*Not calculated*
Total costs	163,505,300–286,198,807
As percentage of 1996 GDP	0.18–0.32

Source: Adapted from Quah (2002).

evaluated using a technique that assesses the willingness of the general public to pay for the preservation of a view, as revealed in survey questionnaires.

There are, however, three problems with this approach. First, costs associated with environmental damage are seldom incorporated into the pricing of a commodity. In the case of the Indonesian forest fires, the costs of burning stubble in order to develop plantations will never be factored into the cost of the palm oil that they produce. Second, the techniques of environmental economics, while useful, are still limited. Like most economic analysis they are limited to the numerical measurement of environmental costs. Anything that falls outside of the quantifiable realm is neglected or at best severely underplayed (e.g. loss of orang-utan habitat, or impacts on the general quality of life in areas affected by haze). Finally, the costing of environmental degradation involves viewing nature as an *object* – something that economic processes *act upon* and something that has meaning only when it affects an economic process. The economy is thus seen as being fundamentally *separate* from the natural world. Geographical analysis has tended to take a rather different view – seeing nature and economy/society as two sides of the same coin, each providing the basis for the other.

Thus far, we have critiqued approaches to nature that see the natural realm as distinct, measurable and quantifiable. In the next two sections we will pursue this idea by examining the ways in which nature becomes part of the economy and, in turn, the economy becomes part of nature. In exploring these processes, our argument is less explicitly 'geographical' – in the sense of adopting an inherently *spatial* perspective – than in other chapters. Instead, here we tap into another long-standing element of geographical research – namely an exploration of the *people/environment interface* (e.g. Simmons, 1996) – to explore the real-world problems of valuing nature in capitalist societies.

6.3 Incorporating Nature: Commodification, Ownership and Marketization

In this section we move away from the economic approach to evaluating nature and start to explore the ways in which nature and economy/society are intimately integrated. We noted in the introduction to this chapter that economies are based upon the transformation of nature. At its most basic level, an economic activity involves the extraction, transformation and trading of substances that we, as human societies, have found on earth. In most cases, this means the geological deposits, life forms, waters and atmosphere of the planet. Whether in relatively basic forms such as fresh food, water and fuel, or transformed through production processes of varying complexity into clothes, vehicles, computers, machinery, and so on, all commodities are derived from the natural environment's resource base.

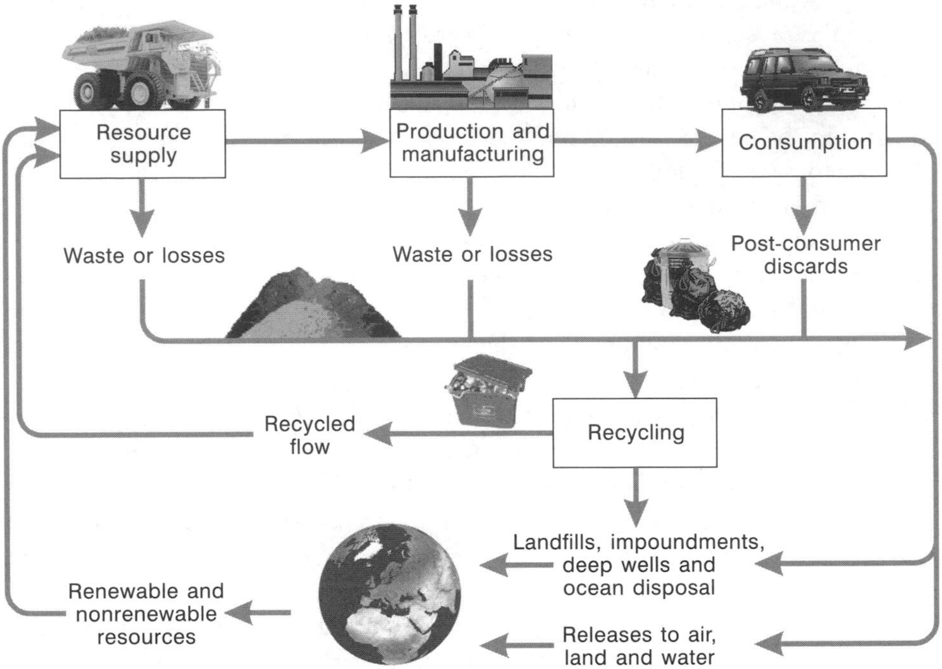

Figure 6.2 The economy as a system of material flows
Source: Adapted from United States Geological Survey (2002), Figure 2.

In this sense, it is useful to think of a commodity chain – a concept we introduced in Chapter 4 – not only as a series of value-adding activities, but as a system of material flows and balances (Figure 6.2). This notion of the material flows cycle allows us to trace the flows of physical and chemical inputs from extraction through production, manufacturing, and utilization to recycling or disposal. Such an analysis reiterates how economic activity places two sets of demands on the natural environment (Dicken, 2003). First, it acts as provider of inputs to the production process in the form of *natural resources*. For example, in 2004, the UK economy consumed some 693 million tonnes of natural resources and products, including 131 million tonnes of plant and animal biomass, 300 million tonnes of minerals (ores, sand and gravel, stone, etc.) and 246 million tonnes of fossil fuels. Second, it acts as a receiver of the outputs of production processes in the form of *waste disposal* (e.g. landfills and ocean disposal) and *pollution* (i.e. releases onto land or into the air/water). In 2002–03, the UK economy generated some 330 million tonnes of material waste, and 732 million tonnes of greenhouse gases (data from http://www.statistics.gov.uk, accessed 26 June 2006). Waste and pollution can be produced at any stage of the commodity chain. Figure 6.3, for example, shows the Karabash copper smelter, located

Figure 6.3 The Karabash copper smelter, Russia
Source: Gavin Bridge, with permission.

on the eastern flank of the southern Urals in Russia, which is a major polluter of the local area through both waste dumps and airborne contaminants.

In this section we will focus on the first of these two dimensions and explore the ways in which the 'natural world' becomes converted into useable, ownable and tradable commodities, i.e. how nature becomes commodified and thus a part of economic processes. In so doing, we are bringing together the commodity chain concept introduced in Chapter 4 with notions of the commodification of nature that are inherent to any commodity system. It is important to remember, however, that the dividing line between nature and society (or humanity) is a very blurred one (Box 6.2). Indeed, we can use the term *socio-nature* as a

Box 6.2 Nature/society

It is worth remembering that the notional division between humanity and nature is one that *we* have created, and is the product of several millennia of social development, leading to complex knowledge, divisions of labour, and a disconnection from the material bases of our survival (especially in urban industrial society). Indeed, it is this 'disconnect' that has provided us with the incentive to measure and value nature, and develop the tools to manipulate and use nature that we are discussing in this chapter.

However, we can problematize this simple separation of nature and society on several levels. First, human beings are themselves a part of nature: our bodies are products of, and participants in, the complex global ecosystem of life on earth. The separation of humans from nature does not necessarily exclude human bodies from the same kind of commodification that affects wood, water, minerals and other substances. While human bodies are seldom (anymore at least) sold as ownable commodities, parts of our bodies certainly are commodified. Consider, for example, the contemporary economic importance of genes, organs, blood, and so on.

Second, the idea that 'nature' comprises the *untouched* aspects of the earth's mineral, water, air and living things does not bear scrutiny very well. There are very few ecosystems or environments that are unaffected by human habitation of the earth in some way. A trip to the 'countryside' in most places is a journey through landscapes with multiple layers of human modification. A hike or a camping trip in a national park or conservation area is an adventure into a carefully defined and protected environment. And even when an ecosystem is in a supposedly pristine state, it may well be commodified by a tourism industry that explicitly trades on its beauty and wildness.

Third, and taking things one step further, we might want to argue that what we conventionally call nature – seeds, fields, forests, rivers – are actually the *outcomes* of economic processes, that is they are *produced* by the needs of our economic systems. Thinking back to the example of poverty in Niger in Chapter 1, we argued that what might at first glance appear to be a *natural* disaster could also be read as the outcome of economic, social and political processes operating at different spatial scales. To give another example, from this perspective, the wide range and huge productivity of crops achieved by Californian agriculture are not seen as a natural result of favourable soils and climate, but rather an outcome of intense capitalist farming practices (Walker, 2004). This *production of nature* thesis fundamentally challenges the notion of nature as external to the economy (Smith, 1984). It seeks to emphasize how nature is reworked over time, rather than there ever being a pristine domain into which the economy has intruded, transformed and damaged nature as it goes. The key questions it seeks to answer are '*how* we produce nature and *who* controls this production of nature' in capitalist societies (Smith, 1984: 63).

way of signalling that nature is socially created and changed, *and* that society is necessarily founded on the transformation of nature. Here we will consider the ways in which elements of nature *cross* that line and become used and valued by human societies. We will do this first by considering the ways in which commodities are *created*, then by examining how *ownership* over them is established.

Creating a commodity

In some instances, the harnessing and incorporation of nature for economic purposes, its commodification, is fairly obvious: a mine, a quarry, a farm, or a dam are clearly all sites where natural materials are converted into commodities. But it is important to note the ways in which nature becomes a resource. The key transformation that occurs in the process of commodification is the *valuing* of natural substances. This might be value based on the *uses* that they have (as food, shelter, clothing or inputs into making something else). It might also be based on the value that they have in *exchange* for other commodities. While the extraction of natural resources for their *use* value is still practised (for example, in subsistence agriculture), in most cases our engagement with the natural world is mediated by exchange relationships, for example, in supermarkets, building supply stores, clothing shops, or water, gas and electricity supply systems.

In order for the exchange of natural materials to occur, they must be assigned a monetary value. But the specific elements of the natural world that will be valued are variable over time and across space. There are two principal variables that have determined the history and geography of nature's commodification:

1 *Technological and scientific knowledge*: Knowledge can reveal an element of nature that was not previously valued as a commodity. This can happen in two ways. First, new technologies might create a demand for a substance that previously did not exist. Uranium, for example, was not a commodity one hundred years ago, but is now a prized economic resource for use in nuclear power generation, medical technologies, and weapons manufacturing. Only with demand for the substance does it become a commodity to which a value in assigned. Second, technology may 'create' a commodity through new forms of extraction, which make it possible to use something that was always there but never before useable. Sunlight, for example, has always been freely available, but only with the development of photo-electric cells has it been possible to harness the sun's heat and light to generate electricity that can then be stored and traded.
2 *Economic circumstances*: Changing economic conditions can also turn a natural substance into a commodity, sometimes very quickly. Knowledge about the usefulness of a substance might exist, and the technology to obtain

it might be available, but it might be just too expensive to extract it as a tradable commodity. But when circumstances change – for example, other supplies of the substance run dry, or demand increases – a part of nature can quickly become an important resource. On example is found in the 'oil sands' of Western Canada (located mostly in the province of Alberta). Scientists and mining engineers had been aware of these oil-bearing sand deposits for some time, but the high cost of extraction (double that of conventional oil wells) had effectively rendered them without any status as commodity. It has only been in recent years that the global price of crude oil has risen to heights that make the oil sands worth owning and exploiting. In this way, largely desolate landscapes have suddenly become booming sites of economic activity.

The correct technological and economic circumstances are therefore required before a piece of nature becomes an economically valued resource. It is worth noting, however, that once nature becomes a resource, it does still retain some *material* properties and these mean that nature is not entirely subsumed by human attempts to tame it. For this reason, different resources have very different characteristics as commodities. One obvious dimension of this point is that many natural resources occur in certain places and not others. There is some flexibility and human agency in this process, for example, agricultural crops or forestry can be cultivated wherever appropriate soil and climate conditions exist (and even further afield if artificial environments, such as greenhouses, are created). But in many cases, the location of a commodity (a coal seam underground, a fishing ground at sea, or a climate appropriate to tropical fruits) cannot easily be moved or modified. This means that extractive industries have a logic to their distribution across space that is rather different from the manufacturing or service sectors. In essence, resource extraction activities must generally locate where resources are to be found.

A second feature of natural resources is that, once extracted, each commodity has distinctive characteristics that determine the ways in which it can be used. Milk, for example, can be produced anywhere that a herd of cows can be fed – either through open grazing or in controlled feedlots – but its physical properties require that it be transported to a consumer as quickly as possible in refrigerated containers. Likewise, there is no getting around the fact that water is a bulky and heavy commodity that is prone to leaking and evaporating, and so a water supply is best drawn from sources relatively close to its consumers. On the other hand, although diamond deposits are concentrated in very few places around the world, once processed, diamonds are eminently transportable and virtually indestructible. Thus the concentration of diamond dealers in a few cities, such as Antwerp (Belgium) and Tel Aviv (Israel) is perfectly feasible. Each commodity, then, has its own post-extraction geography, as we have already argued in Chapter 4.

Figure 6.4 Government involvement in natural resource development
Source: The authors.

A third characteristic of natural resource extraction, especially on the scale now demanded by urban-industrial societies, is that it requires significant levels of investment, infrastructure and logistical support. As a result, governments have often been central to the emergence of resource-based economies. For mining, agricultural or forestry activities to develop, for example, it is usually governments that provide the infrastructure that enables them to happen in far-flung places. In developed countries, schools, hospitals, roads, railways, sewers, policing, etc. are all facilities usually laid on by agencies of the state. Figure 6.4, for example, shows grain wagons owned and operated by the Canadian and Province of Alberta governments.

When all these features of commodity creation, and commodity transformation, are put together, they can lead to very distinctive patterns of development and decline across space. Furthermore, each commodity, as we have noted, has its own peculiar history and geography. These are most evident in economies that have historically been highly dependent on resource extraction, as in the case of Canada. Box 6.3 describes one influential attempt to understand the distinctive geography of the Canadian resource economy.

Box 6.3 Canada, Harold Innis and the geography of resource economies

Canada has long been an economy rich in natural resources. The stereotype of Canadians as 'hewers of wood and drawers of water' is no longer very accurate, but wood, oil, water, wheat, fish and minerals together represent significant economic sectors in various parts of the country (forestry in British Columbia; oil in Alberta; farming in Saskatchewan and Manitoba; mining in northern Ontario, hydropower in Northern Quebec, and fisheries on the East coast). The twentieth-century Canadian political economist Harold Innis sought to devise a model of how a resource (or in his terminology 'staples') economy develops. Innis brought together three sets of issues:

- *technology*: Referring to the technological requirements for extracting a resource, and changes in technology leading to demand for different kinds of resources (for example, from water to coal to uranium in power generation).
- *geography/ecology*: Resources are unevenly available across space – oil is only present in certain places, as are fertile soils, forests, fish, etc. Only when the necessary quantities of a resource are present in a technologically-feasible location will it become economically viable.
- *institutions*: Technology and geography can only be brought together when an appropriate institutional structure exists. Most resource extraction in peripheral areas requires large amounts of capital, which means involvement by the state and/or large corporations.

These three factors mean that while resource extraction often occurs in remote areas, it is controlled from metropolitan cores where financial, technological and communications power is concentrated. 'Staples economies' thus tend to have a geography based on *core–periphery* divisions. These might draw resource wealth to regional or national centres (forestry wealth in Vancouver, oil in Calgary, mining in Toronto, for example). In other instances, the wealth generated by resource extraction may be siphoned overseas. In the colonial era, it was the resources of Asia, Africa and the Americas that built the fine urban landscapes of nineteenth-century London, Amsterdam and Paris. Today, mining, forestry and agro-food corporations draw wealth into a system of major world cities.

The geography of resource extraction is also dynamic. Innis argued that when the three factors come together – the right technology; the right geography; and the right institutions – the result is a 'cyclonic frenzy' of investment, extraction and wealth accumulation. Ultimately, however, it cannot last, and development based on extractive resources (oil fields, mines, logging concessions) moves on elsewhere. This leaves behind abandoned resource regions and communities. There is thus also a geography of cyclical development and decline across space. See Barnes (1996) and Hayter (2000) for further details on Harold Innis and the Canadian resource economy.

Establishing ownership

We have seen that natural resources can enter the economic sphere in various ways and can create a variety of economic geographies. We have also seen that to become commodities, natural substances need to be capable of being packaged into units that can be valued and sold. If, however, nature is to be sold, then a further requirement is that it has to *belong* to someone who will do the selling. This will, in most instances, be an individual, a group of people, a firm, or a state.

In some cases, ownership of nature has long been established in deeply rooted legal traditions. In Western Europe, for example, the ownership of agricultural land has been entrenched in legal structures of private property for several centuries. For other resources, or in other places, however, the definition of who owns a resource is more uncertain and contested. In countries with minority indigenous populations, their claims to territory – often based on very different legal traditions – have usually been neglected by majority populations. In Canada and Australia, for example, indigenous groups have outstanding claims to land and mineral resources across much of the territories now administered by those nation-states. Such land claims are only slowly being settled by contemporary governments and native groups. In many parts of the world, marine resources provide a particular problem, precisely because of the characteristics of the natural processes being harnessed. Who, for example, will own migratory species of fish, such as salmon, when they traverse the US-Canada border? And by whom will offshore oil reserves be extracted when they lie in the South China Sea in between Vietnam, Malaysia, the Philippines and Vietnam? Clearly, ownership and tenure rights are not always straightforward in the commodification of nature.

While ownership is often contested, we have seen that it is a necessity of commodification for ownership of resources to be assigned. In very simple terms, we can identify four possible models:

- *communal access*: such rights allow individuals to *use* resources but not to *own* them. The rights of shifting cultivators to access and use forest resources would be one example. In reality, though, there are now very few examples of land over which ownership is not claimed in some sense by a state, which then 'permits' communal use rights. One example, however, would be the fishery resources of the open seas. While some legal frameworks do apply to fishing in open waters (a limited international ban on whaling, for example), in general, marine resources can be appropriated by fishing boat operators as they cannot be said to *belong* to anyone while they remain in the sea.

- *state ownership and state exploitation*: in this model the benefits of extraction are shared (as government revenue) by all citizens. The development of hydroelectricity or water supply projects by public utilities would be one example. Just two or three decades ago, it was common to find public bodies engaged in such activities, but widespread privatization has reduced the role of the public sector, as we will discuss shortly.
- *state ownership and private exploitation*: here, corporations pay some kind of licensing fee to extract a publicly owned natural asset. For example, when governments grant timber licences to logging firms entitling them to extract wood from a defined territory, or when oil companies purchase exploration rights. In both cases, royalties are then usually paid to governments based on the amount of the resource that is extracted.
- *private ownership and private exploitation*: as, for example, when a farmer owns the land that he/she cultivates and is entitled to reap all of the profits that result from the operation.

In reality, these categories of ownership are overlapping and increasingly blurred. There has also been a widespread trend around the world away from the first two and towards the latter two: in other words, a process of *privatization* in which natural resources are increasingly owned and exploited by the private sector (we will discuss how privatization relates to broader processes of *neoliberalization* in Chapter 7).

One commodity where ongoing privatization has received particular attention in recent years is *water*. There are a variety of ways in which people might be supplied with fresh drinking water in urban areas, ranging from individual access to wells and water courses, to private corporate ownership of large-scale water supply systems. Figure 6.5 depicts the various options along two axes – of community versus corporate control, and artisanal (small-scale) versus industrial (large-scale) systems. Increasingly, water supply has been moving into the bottom-right portion of the diagram. In a number of poorer countries, development aid and loan packages from international financial institutions have been conditional upon the privatization of public utilities. At the same time, such private ownership has become increasingly concentrated in just a few hands. Two companies in particular, ONDEO/Suez Lyonnaise des Eaux and Vivendi/Générale des Eaux, both based in France, are together estimated to control around 70 per cent of the privatized water market worldwide (Bakker, 2003a).

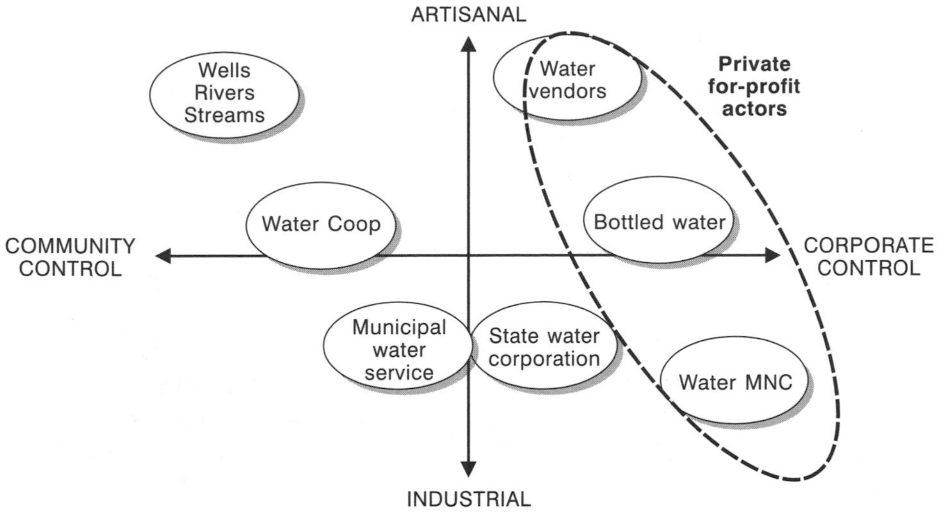

Figure 6.5 Modes of urban water supply provision
Source: Redrawn from Bakker (2003a), Figure 1. With permission of Blackwell Publishing.

In some countries, governments have voluntarily transferred ownership and management of water systems to the private sector. The UK provides a particularly intriguing example. Until relatively recently, water supply in the UK was provided through state-owned utility companies. The commodity was managed based on principles of satisfying human needs rather than profit generation. Not only was this a *politically* necessary approach, but it was also *practically* necessary, given the expense involved in the development of infrastructure for water collection, purification, storage, delivery and disposal. By the 1980s, however, economic crises and ideological shifts made state provision in this way less attractive. Having already privatized several British public utility industries, by 1989, Margaret Thatcher's Conservative government was selling shares in English and Welsh regional water companies.

While water companies entered private ownership, they remained closely regulated in some respects by the UK government. Prices were set by a national 'watchdog' agency, and environmental/health compliance was also closely monitored, although labour conditions in the industry fell outside of government regulation. Water companies were also spared any real competition, as each one enjoyed a monopoly within its territory for the 25 years of the licences they were granted by the national government. The end result was rising water bills through the 1990s, and increasing profits for the companies involved.

Water's particular characteristics as a commodity made it 'uncooperative' as a privatized resource (Bakker, 2003b). In reality, it was very difficult to extend complete private ownership over a substance that was so essential to human health, so ubiquitous, so bulky and heavy, so dependent on massive infrastructure, and so central to the country's ecosystems. The end result in the UK was a monopolistic private sector supplier in each region, responsive not to an equilibrium price established by supply and demand, but bound by negotiations with its central government regulator.

One last characteristic of water supply leaves it uncooperative to full commodification. A principle behind privatization was that water resources would be priced in such a way that they reflected the true cost of supplying the water. For this reason, attempts were made to ensure that all water would be metered at the point of use. People would no longer be able to use unlimited amounts for one fixed payment. Essentially, this was an effort to 'package' water into units (i.e. litres), which could be priced in relation to the costs of producing them. We have already noted that the idea of clean water as a 'right' made this politically impossible to implement, but it also failed in another sense. It was virtually impossible to set a price for water supply that genuinely represented its true cost. Extracting water from the environment – through dams or groundwater extraction, for example – leads to various and complex environmental consequences. The disposal of waste water and sewage also has indeterminate costs associated with it. In short, the centrality of water in the ecosystem means that the true cost of water supply can never truly be factored into the price charged.

This last point is fundamental to the commodification of nature. As soon as a commodity is extracted from the natural system of which it is a part, and is packaged, owned, transported, and used in some way, a cost is incurred by the earth's natural system – a cost that is never fully accounted or even understood in economic calculations. Nature, then, is a very particular kind of commodity. Some use the term *fictitious commodity* (Polanyi, 1944) to capture the way in which nature is treated as a commodity in economic systems, yet is not directly produced by humans – as, for instance, a toothbrush or a television are – but rather occurs naturally. While water provides a particularly compelling example of this argument, in the case of virtually any natural material, the integrated and interdependent characteristics of ecological relationships mean that the true cost of commodifying and privatizing nature can never be truly counted.

6.4 Valuing Nature: The Commodification of Environmental Degradation

In the last section we considered the ways in which nature becomes harnessed for economic purposes. We now turn to the side-effects and after-effects of economic processes, and consider how environmental *protection* can itself become commoditized. In a general sense, being *seen* to be environmentally responsible makes good business sense, thus we now see numerous companies promoting their own environmental ethics. But in more specific ways, the protection of natural environments from the waste, outputs and effects of economic production is itself becoming a commodity with a tradable value.

Such ideas tap into a wider school of thought about the economy–nature interface known as *ecological modernization* (here we draw on Gibbs, 2006). This is a relatively positive view on human–environment interactions that assumes there can be synergy between economic development and environmental protection as opposed to there always being conflict. Put another way, ecological modernization approaches seek to resolve the competing pressures of allowing businesses to generate profits and provide work while protecting the natural environment. It focuses, in part, on the potential for new technologies to offset the detrimental environmental impacts of economic activity at the firm-level. Firms, in turn, may benefit financially in several ways from such developments: through greater efficiencies due to reduced pollution and waste production; by avoiding future financial liabilities, such as the potential cost of clearing-up contaminated land; through potential sales of more 'environmentally-friendly' products and services; and, most directly, through the sale of pollution prevention and abatement technologies.

Equally, however, ecological modernization is concerned with society-wide dynamics, such as new market and regulatory mechanisms designed to reduce environmental impacts, and the emergence of new kinds of 'environmentally-friendly consumption'. Efforts to try and place a value on environmental degradation are often integral these dynamics. Seen from this wider perspective, ecological modernization relies on significant and cohesive political and institutional changes within society in order to be effective. While some countries – for example, Japan and the Netherlands – have integrated these ideas into their national economic planning and as a result now consume less natural resources and produce less waste, others remain far more sceptical about an approach which retains the capitalist profit mechanism at its heart.

We will illustrate these arguments with two case studies. The first relates to the rise of markets for trading emission credits, thereby turning pollution and degradation into commodities that can be traded. The second concerns the growth of certification programmes in different resource industries, designed to give certified products clear environmental credentials that allow producers to charge a premium for them in the marketplace.

Trading pollution

In the US, the Environmental Protection Agency's 'Acid Rain Program' was initiated in 1995 to try and reduce sulphur dioxide emissions from coal-burning electricity generation plants (http://www.epa.gov/airmarkets/arp/index.html, accessed 26 July 2006). In the first phase of the programme, from 1995–2000, emissions were reduced by almost 40 per cent below the target level. The scheme has now expanded to include all forms of fossil fuel-fired power stations and also encompasses nitrogen oxide emissions. Central to the programme is a trading mechanism which allows power producers that achieve cuts in emissions to sell their allowances (in units of tonnes of sulphur dioxide released) on an open market to companies that have exceeded their target and hence face the possibility of financial penalties or being forced to surrender future allowances to meet the shortfall.

Such moves to turn environmental protection directly into a commodity have recently taken on wider importance due to attempts to establish a global carbon credit trading system under the auspices of the *Kyoto Protocol*. In 1997, the signatories to the United Nations Framework Convention on Climate Change met in Kyoto, Japan, and developed an agreement that would be legally binding when ratified by national governments (http://www.unfccc.int). The Kyoto Protocol establishes a commitment to reducing greenhouse gas emissions by a set per centage from their 1990 levels by the year 2012 at the latest. The European Union's target reduction, for example, is 8 per cent by 2010. The Protocol came into force in February 2005, and by April 2006 a total of 163 countries had ratified it, collectively representing some 61.6 per cent of global greenhouse emissions (the most notable exception being the United States, which alone accounts for 24 per cent of global greenhouse gas emissions).

With each country establishing a limit on its carbon dioxide emissions, governments then allocate emission permits (known as Emission Reduction Units, or ERUs) to companies within their territories, allowing them to release a certain quantity of carbon dioxide over a given period of time. If companies reduce their emissions through investments in new technology or renewable energy, they can then sell their surplus ERUs to those who will exceed their allocation. The price of ERUs is variable, the intention being to encourage emissions by companies in order to avoid costly purchases of further emission permits. A global market for carbon dioxide emissions has sprung up rapidly, and by 2005 was worth more than US$10bn (http://www.ieta.org, accessed 16 June 2006).

The European Union (EU) Emissions Trading System (ETS) is by far the most developed market of its kind, contributing 75 per cent of the total global market in 2005. Its first phase, running from 2005 to 2007 covered 12,000 big industrial polluters across member countries. It aimed to deal with a thorny problem: cutting carbon dioxide emissions by the worst offenders without putting them out of business. In the scheme, individual countries were told how much carbon dioxide they could emit, which in turn was divided up among firms in the sectors covered by the scheme: power generation, steel, glass, cement, pulp and paper, and ceramics. In the first phase, for example, the UK was allocated 736 million tonnes of emissions, or 11.2 per cent of the EU total, across

1,078 installations in those sectors. Firms emitting more than their permitted level could buy other companies' unused allocation, or pay fines (40 Euros per tonne of excess), while those who were reducing their pollution could sell unused permits. If there was extensive over-pollution, the price of permits would rise, therefore supposedly inducing firms to cut pollution levels. Overall, some 230 million tonnes in permits, worth 4 billion Euros, were traded in 2005.

As with any other market, state regulation is crucial to the EU Emission Trading System. Critics have argued that the first phase targets were extremely generous, keeping prices for credits low and not forcing firms to invest in cleaner technologies. Some firms, in turn, were able to make money by simply trading their overly-generous allowances (*Guardian*, 2006b). Overall, though, the rise of carbon trading, of which the ETS is the best example, provides clear illustration of the possibilities and problems of attempts to put a value on environmental degradation.

Certifying environmental protection

The second area we will consider is the rise of certification programmes that aim to demonstrate the environmentally-friendly nature of certain products. Often, these goods are sold at a premium over regular products, reflecting an attempt to incorporate an element of environmental protection into the costing of the product. In this way, nature becomes commodified as a part of the overall value-adding process. Such initiatives are directly analogous to those involving 'fair' or 'ethical' trade practices discussed in relation to commodity chains in Section 4.4.

Forestry provides an interesting case study in this regard (here we draw on Stringer, 2006). Sustainable forestry initiatives emerged out of the Rio de Janeiro 'Earth Summit' in 1992, fusing together environmental concerns about forestry – and in particular deforestation in tropical countries – with concerns about the value of forestry to indigenous peoples. The best-known such initiative is the Forest Stewardship Council (FSC). The FSC was established in 1993 as the world's first independent scheme for identifying and labelling products from well-managed forests. It is a stakeholder-owned organization – made up of representatives from environmental and social groups, the wood and forestry industries, indigenous people's organizations, community forestry groups, and forest product certification organizations – that sets international standards for responsible forest management through consultative processes. The FSC accredits independent third party organizations that can certify forest managers and forest product producers to FSC standards, in accordance with 10 guiding principles that promote sustainability for both the forest environment and the communities who live and work there (Box 6.4).

The FSC offers both Forest Management Certificates to organizations that manage forests, and Chain of Custody Certificates to processing operations that can track certified forest products through to the finished product. The FSC's

Box 6.4 The Forest Stewardship Council's ten principles

Principle 1: Compliance with laws and FSC principles
Forest management shall respect all applicable laws of the country in which they occur, and international treaties and agreements to which the country is a signatory, and comply with all FSC Principles and Criteria.

Principle 2: Tenure and use rights and responsibilities
Long-term tenure and use rights to the land and forest resources shall be clearly defined, documented and legally established.

Principle 3: Indigenous people's rights
The legal and customary rights of indigenous peoples to own, use and manage their lands, territories, and resources shall be recognised and respected.

Principle 4: Community relations and worker's rights
Forest management operations shall maintain or enhance the long-term social and economic well-being of forest workers and local communities.

Principle 5: Benefits from the forest
Forest management operations shall encourage the efficient use of the forest's multiple products and services to ensure economic viability and a wide range of environmental and social benefits.

Principle 6: Environmental impacts
Forest management shall conserve biological diversity and its associated values, water resources, soils, and unique and fragile ecosystems and landscapes, and, by so doing, maintain the ecological functions and the integrity of the forest.

Principle 7: Management plan
A management plan – appropriate to the scale and intensity of the operations – shall be written, implemented, and kept up to date. The long-term objectives of management, and the means of achieving them, shall be clearly stated.

Principle 8: Monitoring and assessment
Monitoring shall be conducted – appropriate to the scale and intensity of forest management – to assess the condition of the forest, yields of forest products, chain of custody, management activities and their social and environmental impacts.

Principle 9: Maintenance of high conservation value forests
Management activities in high conservation value forests shall maintain or enhance the attributes which define such forests. Decisions regarding high

conservation value forests shall always be considered in the context of a precautionary approach.

Principle 10: Plantations
Plantations shall be planned and managed in accordance with Principles 1–9. While plantations can provide an array of social and economic benefits, and can contribute to satisfying the world's needs for forest products, they should complement the management of, reduce pressures on, and promote the restoration and conservation of natural forests.

http://www.fsc-uk.org/about/principles/, accessed 24 July 2006.

product label then allows consumers to recognize products derived from approved forests. By late 2005, 67 million hectares in over 60 countries had been certified according to FSC standards, while tens of thousands of products were being produced using FSC certified wood and displaying the FSC label. Canada is currently the leading country in the world for FSC certified forests (with 14.3 million hectares, or 23 per cent of the global total, in 2005). In September 2005, the FSC announced the largest certification to date, a 5.5 million hectare area of Alberta, Canada. The certification followed a ten-month assessment according to FSC Canada's National Boreal Standard, which combines the ten FSC principles with 56 more specific conditions. The audit – undertaken by Smartwood, an arm of another environmental NGO, the Rainforest Alliance – involved over 100 local stakeholders (see http://www.fsc.org/en for more).

Since its inception in 1993, the FSC has now been joined by two competing certification schemes, albeit with more regional foci. The Programme for the Endorsement of Forest Certification (PEFC) was established in 1999 by several national forest interest groups in Europe in response to concerns about the costs of certifying small-scale forest-owners through the FSC. PEFC is an umbrella scheme that endorses national initiatives that meet its auditing and certification procedures. By late 2005, almost 130 million hectares of forest was covered by some 2300 PEFC certificates (see http://www.pefc.org/). The Sustainable Forests Initiative (SFI) was established by the American Forest and Paper Association in 1995 in response to ongoing public worries about the ability of the industry to protect wildlife and practise sustainable forestry. The SFI is only concerned with environmental issues and does not address the social and economic issues covered by the FSC. In mid-2005, some 53 million hectares of North American forests were covered by the SFI (see http://www.aboutsfi.org/). PEFC and SFI differ from the FSC, however, in that they are industry dominated and lack the same level of stakeholder participation (see Table 6.2 for a full comparison of the schemes).

Table 6.2 A comparison of forest certification schemes

	FSC (Forest Stewardship Council)	PEFC (Programme for Endorsement of Forest Certification)	SFI (Sustainable Forestry Initiative)
Standard	Performance-based Includes social and environmental issues	System-based Non-binding criteria; varies between countries	System-based on ISO 14001 Ignores social and economic issues
Standard development process	Comprises three chambers: social, economic and environmental	Private forest owners Industry representatives	Industry dominated
Logo	Yes	Yes	Yes
Labelling and chain of custody	Chain of custody number is included on the label Tracking of certified and uncertified wood	Three tier structure based upon the process of inclusion or separation of certified wood in the production process	No chain of custody certification
Strengths	Economic, social and environmental interests included	Increasing level of transparency within some countries Regular reviews of national standards	Recent inclusion of NGOs
Weaknesses	Non-certification of wholesalers and retailers Independent verification of chain of custody difficult	Governed by forest owners and forest industry Does not recognise land rights	No clear performance requirements Wood not originating from SFI-certified forests can be included under label

Source: Adapted from Stringer, 2006, Table 1. With permission of the author.

What exactly is going on in such schemes? Are they really concerned with more fully valuing the environment and forest resources? The emergence of more business-driven schemes such as PEFC and SFI is indicative here, as perhaps is the fact that the vast majority of certified forests are located in the temperate, industrialized and wealthy countries of Europe and North America, while only 5 per cent by area are in the tropical zones of Africa, Asia and Latin America. Critical observers worry that these initiatives effectively act as technical barriers to trade, impeding access to developed country markets for developing country producers unable to meet the stringent monitoring criteria built into these schemes. Certification, regardless of which body provides the accreditation, also provides security against consumer boycotts in environmentally sensitive markets, thereby potentially increasing market share, while Chain of Custody certification also allows firms to capture extra 'value added' in their products. In sum, certification can perhaps best be viewed 'as a marketing and market access tool which allows the company to gain higher prices and a competitive edge in the market place rather than a mechanism to encourage sustainable forest management in forests under most threat' (Stringer, 2006: 16).

6.5 Bringing Nature to Life?

The last two sections have illustrated some of the ways in which nature is an integral part of economic processes – first, as natural materials are incorporated as the basis for economic activity, and, second, as an economy has emerged around the protection of ecosystems. In both cases we have seen that nature is far from being 'out there' and external to economic processes – the nature–economy interface is far more complicated than that.

Clearly we need to bring natural resources more fully into our understanding of economic systems. Commodity chains are simultaneously both linked sequences of value adding activities *and* systems of material flow and transformation. But coming to terms with a blurred boundary between the 'natural' and the 'economic' is a difficult conceptual manoeuvre. One way in which some scholars have sought a way around this issue is to think of the natural world as a 'player' in commodity chains just as we would consider human participants as actors (playing the parts of miners, farmers, traders, shippers, etc.). This means that we have to take seriously the physical properties of natural materials when studying the configuration of their commodity chains.

These ideas have become increasingly prevalent in economic geography through the adoption of Actor-Network Theory (ANT). ANT seeks to argue that we should collapse the artificial binary distinctions between nature/culture and human/non-human that pervade our ways of thinking, and instead focus on the sprawling networks of both animate and non-animate actors through which social and economic life is played out (Box 6.5).

Box 6.5 Actor-network theory (ANT)

Actor-network theory (ANT) emerged in the 1980s from attempts to understand the sociology of science and technology. Its original concern was to show how, rather than being some kind of absolute truth, scientific knowledge was the product of a wide variety of things: papers in academic journals and presentations, laboratory equipment and experiments, skills embodied in scientists, reputations, research funding and so on. In short, ANT represents a way of thinking about and uncovering the many connections and relationships between various actors – human and non-human, tangible and intangible – through which social life is played out. Since the early-to-mid 1990s, ANT has become an increasingly popular theoretical-cum-methodological approach to economic geography as part of the so-called 'cultural turn' (see Box 1.2).

ANT rests on a series of underlying principles:

- The many binaries through which we understand the world – nature/culture, structure/agency, human/things, global/local – should be set aside.
- Instead, the world should be viewed as a series of network connections (or *heterogeneous associations*) between different actors.
- It is these connections between actors, rather than the actors themselves, that have agency and cause things to happen.
- The activities of non-humans within these networks may be as important as those of humans.
- The continuation of actor-networks depends on the circulation of devices (known in ANT as *immutable mobiles*) that can move from one location to another without changing form (e.g. money, animals, coffee beans, technologies, etc.).

Geographers have been particularly interested in the ways in which ANT challenges existing notions of space, place and scale. In particular, rather than seeing the world as organized into hierarchical scales, from the local through to the global, social systems might be thought of as operating through combinations of 'short' and 'long' network connections. For more, see Bosco (2006).

We can illustrate these ideas using the example of a Fairtrade coffee network connecting Peru and the UK (as introduced in Chapter 4). Figure 6.6 illustrates one way of thinking about this system, namely as an interlocking series of connections between people, documents, electronic information and coffee beans in North America, Latin America and the UK (here we draw on Whatmore and

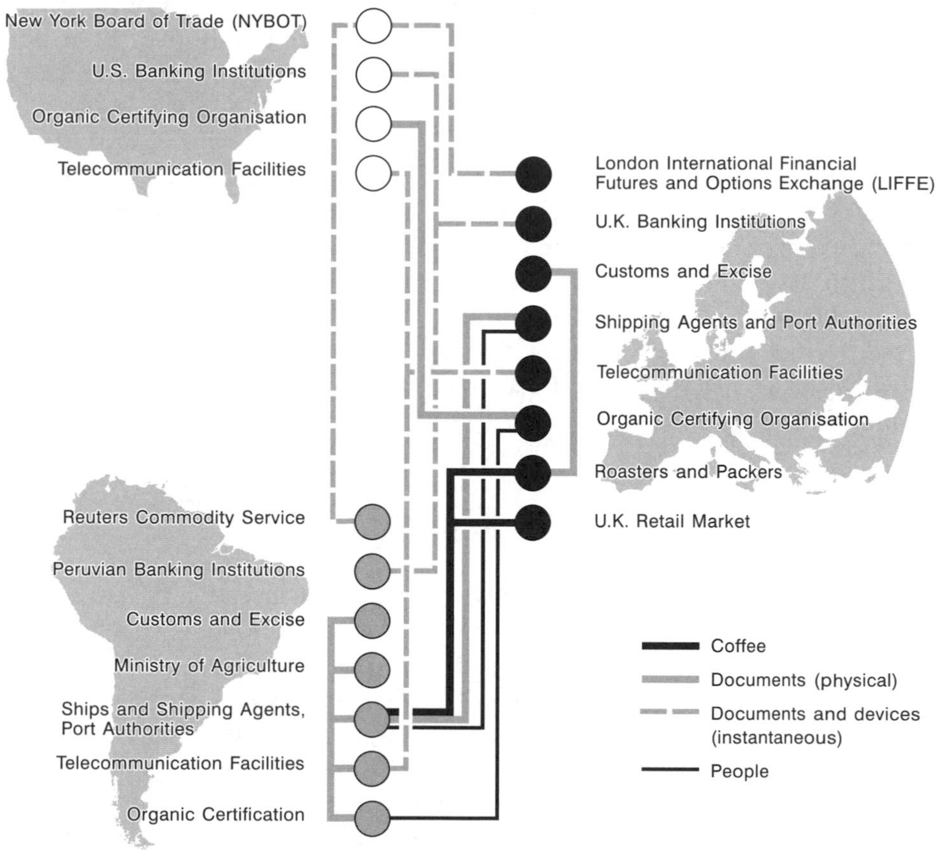

Figure 6.6 A Fairtrade coffee network connecting Peru and the UK
Source: Adapted and updated from Whatmore and Thorne (1997), Figure 11.1.

Thorne, 1997). In part, this characterization reflects ANT's suggestion that documents and technologies are as much a part of this system as the various firms and institutions involved. Most importantly from our perspective here, however, is the status of the coffee beans themselves as an integral actor in the Fairtrade network: 'the coffee is fraught with variabilities that reverberate through the network as a whole' (Whatmore and Thorne, 1997: 299). In particular, high quality beans are required for selling to the Fairtrade buyers. If the beans are of low quality – for example, due to rainfall variations, pests, or fermentation – then it will not be suitable and will be sold to commercial buyers for the market price as set by New York's Board of Trade. Equally, if the crop in a competitor country such as Brazil is ravaged by frost, then the open market price available might be high enough that there is no need to sell to the Fairtrade buyers. These 'natural' contributory factors to the quality of coffee beans are beyond direct human control. Hence, the very characteristics, operation and sustainability of the network as a whole depend on the biophysical characteristics of the coffee beans and the environment in which they are produced. In this way, the quality of the coffee beans that can be commodified from the natural environment at a given time and place affects what happens in the network as a whole.

Clearly, human and non-human actors do not possess the *same kind* of agency within such networks: coffee beans are not sentient beings capable of thought and strategic actions. They do, however, possess important and changeable attributes that are an inherent and potentially crucial element in the networks of which they are part. Recognizing these qualities and how they shape the commodification process is an important step in understanding the particular characteristics of individual commodity chains. The 'nature of nature' is thus an integral and inseparable part of the complex economic geographies that surround us.

6.6 Summary

In this chapter we have worked our way towards a conception of nature that blurs the line between the natural world on the one hand and economic processes on the other. We started out with approaches rooted in economics and identified some of their shortcomings. The analysis of commodity pricing, for example, is purely an exercise (albeit often a very sophisticated one) in establishing the relationship between supply and demand. In this case, nature is treated as nothing more than the source of commodities to be traded – it is the trade itself that is seen as the more interesting focus of attention. While environmental economists do have tools to examine more carefully the economic consequences of environmental change, we also saw that these analyses tend to be limited to those effects that can be quantified in monetary terms.

What both economic approaches had in common was a very clear separation of the economic and the natural worlds. Natural resources were given prices, and environmental degradation was given a cost, but neither the integration of natural ecosystems, nor the interdependence of human society with natural environments, were acknowledged. A geographical approach to the interface between nature and the economy places much greater emphasis on these features.

We demonstrated the complexity of nature–economy relations first by exploring the ways in which natural materials become commodities – a process that has both a history and a geography associated with it. But once commodified, natural materials are often awkward and elusive when attempts are made to assert ownership over them. The case of water privatization, in particular, illustrated how 'uncooperative' a commodity can be when treated as a privately-owned asset. The second nature–economy interface that we explored was in the arena of environmental protection. Here, the global trade in carbon credits illustrates how a market in environmental protection might be established, along with some of the pitfalls that arise. The environmental certification of resources such as timber represents another way in which the environment can be commodified.

In both these sections, we saw how the line between nature and economy is blurred in various ways. In the final section, we addressed this 'blurring' more explicitly by examining Actor Network Theory as a conceptual framework that attempts to dissolve the nature–society distinction. By viewing non-human participants (coffee beans, animals, tractors, documents, etc.) as actors in commodity networks, ANT has become an increasingly popular framework for thinking about the ways in which the geographies of commodities are constructed.

Further reading

- See Angel (2000) and Gibbs (2006) for discussions of the prospects for an environmental economic geography.
- Castree (2003) and Bakker and Bridge (2006) provide difficult but useful summaries of debates concerning the commodification of nature.
- While Bakker (2003b) provides a detailed discussion of the UK water privatization case, Bridge (2000), Mansfield (2004), McCarthy (2005) and Prudham (2003) provide fascinating discussions of the copper industry, ocean fisheries, community forestry and commercial forestry, respectively.
- For more on ecological modernization, see Gouldson and Murphy (1996), *Geoforum* (2000) and Gibbs (2006). Stringer (2006) provides a lucid discussion of certification schemes in the forestry sector.
- For a student-friendly introduction to actor-network theory (ANT), see Bosco (2006). See Dicken et al. (2001) for how ANT relates to understanding the global economy.

Sample essay questions

- Why, and how, is nature undervalued in economic systems?
- Why may different natural resources become commodified in different ways?
- In what ways can attempts to value nature also lead to its commodification?
- How might we more fully account for the active role played by nature in economic systems?

Resources for further learning

- http://www.env-econ.net: this website contains the best collection of news, opinions, and analysis by economists on nature and resources.

- http://www.unep.org: this website provides a broad view on the international efforts in caring for the environment.
- The websites of various commodity exchanges explain their history and operation in some detail. See, for example, http://www.nybot.com/ (New York Board of Trade), http://www.cbot.com/ (Chicago Board of Trade) and http://www.lme.co.uk/ (London Metals Exchange).
- http://unfccc.int: this website gives full information on the history and details of the Kyoto Protocol.
- http://ec.europa.eu/environment/climat/emission.htm: this EU website provides further information on the Emission Trading Scheme.

References

Angel, D. (2000) Environmental innovation and regulation, in G.L. Clark, M. Feldman and M.S. Gertler (eds) *The Oxford Handbook of Economic Geography*, Oxford: Oxford University Press, pp. 607–22.

Bakker, K. (2003a) Archipelagos and networks: urbanization and water privatization in the South, *Geographical Journal*, 169(4): 328–41.

Bakker, K. (2003b) *An Uncooperative Commodity: Privatising Water in England and Wales*, Oxford: Oxford University Press.

Bakker, K. and Bridge, G. (2006) Material worlds? Resource geographies and the 'matter of nature', *Progress in Human Geography*, 30(1): 5–27.

Barnes, T. (1996) *Logics of Dislocation*, New York: Guilford Press.

Bosco, F.J. (2006) Actor-network theory, networks and relational approaches in human geography, in S. Aitken and G. Valentine (eds) *Approaches to Human Geography*, London: Sage, pp. 136–46.

Bridge, G. (2000) The social regulation of resource access and environmental impact: cases from the U.S. copper industry, *Geoforum*, 31(2): 237–56.

Castree, N. (2003) Commodifying what nature? *Progress in Human Geography*, 27(3): 273–97.

Dicken, P. (2003) *Global Shift,* 4th edn, London: Sage.

Dicken, P., Kelly, P.F., Olds, K. and Yeung, H.W-C. (2001) Chains and networks, territories and scales: towards a relational framework for analysing the global economy, *Global Networks*, 1: 89–112.

Geoforum (2000) Special issue on 'Ecological Modernisation', 31(1).

Gibbs, D. (2006) Prospects for an environmental economic geography: linking ecological modernisation and regulationist approaches, *Economic Geography*, 82: 193–215.

Gouldson, A. and Murphy, J. (1996) Ecological modernisation and the European Union, *Geoforum*, 27(1): 11–21.

Guardian, The (2006a) Redgrave centre stage in campaign to halt Romanian gold mine that has split village, 23 June, p. 3.

Guardian, The (2006b) Governments accused of giving industries permission to pollute, 16 May, p. 4.

Hayter, R. (2000) Single industry resource towns, in E. Sheppard and T. Barnes (eds) *The Companion to Economic Geography*, Oxford: Blackwell, pp. 290–307.

Mansfield, B. (2004) Neoliberalism in the oceans: 'rationalisation', property rights and the commons question, *Geoforum*, 35: 313–26.

McCarthy, J. (2005) Devolution in the woods: community forestry as hybrid neoliberalism, *Environment and Planning A*, 37(6): 995–1014.

Polanyi, K. (1944) *The Great Transformation*, New York: Farrar and Rinehart.

Prudham, S. (2003) Taming trees: capital, science and nature in Pacific slope tree improvement, *Annals of the Association of American Geographers*, 93(3): 636–56.

Quah, E. (2002) Transboundary pollution in Southeast Asia: the Indonesian fires, *World Development*, 30(3): 429–41.

Simmons, I.G. (1996) *Changing the Face of the Earth: Culture, Environment, History*, 2nd edn, Oxford: Blackwell.

Smith, N. (1984) *Uneven Development: Nature, Capital and the Production of Space*, Oxford: Blackwell.

Stringer, C. (2006) Forest certification and changing global commodity chains, *Journal of Economic Geography*, 6(5), pp. 701–22.

United States Geological Survey (2002) *Circular 1221: Material Flows in the Economy*, Denver, CO: US Geological Service.

USITC (US International Trade Commission) (2006) *The Effects of Increasing Chinese Demand on Global Commodity Markets*. Staff Study 28, June: http://hotdocs.usitc.gov/docs/pubs/research_working_papers/pub3864-200606.pdf

Walker, R. (2004) *The Conquest of Bread: 150 Years of Agribusiness in California*, New York: New Press.

Whatmore, S. and Thorne, L. (1997) Nourishing networks: alternative geographies of food, in D. Goodman and M. Watts (eds) *Globalising Food: Agrarian Questions and Global Restructuring*, London: Routledge, pp. 287–304.

PART III

ACTORS IN ECONOMIC SPACE

CHAPTER 7

THE STATE

Who controls the economy: firms or governments?

Aims

- To understand how state and supra-national institutions shape economic processes
- To recognize the different kinds of states within the global economy
- To appreciate the changing role of the state in an era of globalization
- To demonstrate why geographical scales matter in the reconfiguration of the state.

7.1 Introduction

On 22 June 2005, China National Offshore Oil Corporation (CNOOC) made an unsolicited US$18.5 billion cash offer to woo Unocal (the Union Oil Company of California) from the latter's American suitor, Chevron. The bid apparently made a lot of sense to CNOOC, China's third largest state-owned oil company, because a successful takeover would allow it to tap into Unocal's huge oil and natural gas reserves in Asia. CNOOC's cash bid was also 15 per cent more than Chevron's offer of US$16.5 billion. Three days after CNOOC had announced its intention to discuss its takeover bid with Unocal and the US regulator, however, a group of 41 US Congressional representatives asked the US government to closely scrutinize CNOOC's bid on the grounds that China might gain control of precious US energy resources at a time of sky-high oil prices and growing tensions over China's huge trade surplus with the US. The representatives also asked the US Treasury Department to review the possible stripping of Unocal's core technology and assets by CNOOC. As the political pressure mounted, the House of Representatives voted in favour of forbidding the Treasury Department from using federal funds to process a recommendation for the approval of CNOOC's bid. On 3 August 2005, CNOOC decided to drop its bid for Unocal, blaming US political opposition to the deal.

This example of the proposed acquisition of a privately-owned US oil company by its state-owned Chinese counterpart illustrates the *inherent limits* to the global reach of such giant corporate entities: even though CNOOC's bid was financially more attractive than Chevron's offer, the final verdict effectively remained in the hands of the US state. Like other failed attempts to buy into 'corporate America', this example showcases the complex political-economic issues involved in the globalization of firms *and* the continued importance of nation-states in the political-economic governance of their domestic economies. It also reiterates the fact that the nation-state retains important regulatory controls of its national space-economy in an era of accelerated globalization of economic activities. With large firms becoming increasingly globalized, they are likely to confront different national regulatory regimes and to engage in tussles over the control of their economic interests *vis-à-vis* the nation-state.

As this suggests, as we move into Part III, we are switching the focus to the key *actors* in economic space, and the role they play in creating economic geographies: the four chapters focus on the role of the state, transnational corporations, workers and consumers, respectively. As we know from our understanding of the commodity chain developed in Chapter 4, in reality these actors are interconnected and interdependent. In this Part of the book, however, we focus on each actor individually in order to explore their importance and influence in more detail.

In this chapter we will explore the extent to which the state remains a powerful shaper of economic geographies. We will profile the ongoing role of the nation-state in economic governance, and shed light on the changing abilities of nation-states to address political-economic processes such as uneven development, economic restructuring, and poverty. This integral role of the nation-state in the modern economy highlights an important point – the economy is intermeshed with the *politics* of the nation-state at a variety of *geographical scales*. As will be explained more explicitly later in this chapter, these state politics may be about *inter-national* geopolitical imperatives such as the failed CNOOC bid above, debates within a particular *macro-regional* grouping of states such as the European Union, the *intra-national* politics of resource transfers and investment within nation-states, or *local* politics. In all instances, we simply cannot underestimate the continued significance of the state for our understandings of the dynamics of economic development.

The chapter is divided into five major sections. The next section (7.2) introduces the contemporary debate on the (ir)relevance in economic governance. Here, we present some grossly oversimplified views on the nation-state championed by enthusiastic proponents of globalization known as *ultra-globalists*. In Section 7.3, we offer an explanation of the dependent relationships between states and firms and discuss different state functions that concern this state-firm relationship. This section explains the mechanisms through which the state manipulates the economy via its 'visible hand'. Section 7.4 further elaborates on the *geographical diversity* of nation-states in today's global economy. We show that state functions tend to work out differently in states dominated by contrasting political-economic ideologies in different places. In Section 7.5, we demonstrate how these state functions are changing in relation to the reconfiguration of state capacity at different geographical scales, while we then move to on problematize the state somewhat by considering how control over its geographical territory can in reality be limited and partial (Section 7.6). Throughout the analysis, we emphasize both state functions and their different manifestations in different *places* (i.e. their geographies).

7.2 The 'Globalization Excuse' and the End of the Nation-state?

We begin by reviewing how the nation-state has become increasingly viewed as irrelevant in today's global economy. This is commonly known as the *ultra-globalist* position, particularly promoted through the popular accounts offered by business gurus such as Kenichi Ohmae (1995) and popular journalists such as Thomas Friedman (1999). The following quotation sums up this position nicely:

> What is new today is the degree and intensity with which the world is being tied together into a single globalized marketplace and village. What is also new is the sheer number of people and countries able to partake of today's globalized economy and information networks, and to be affected by them . . . This new era of globalization . . . is turbocharged.
>
> (Friedman, 1999: xvii, xviii)

In this view, globalization is not just a phenomenon to be reckoned with. More importantly, it is a force that integrates capital, technology, and people in a 'borderless' world. Globalization, we are told, thus creates a single global market and a gigantic 'global village' that is neither stoppable nor controllable by individual nation-states. In outlining this *new* global system, these ultra-globalists have created a popular imagination for the end of the nation-state as institutions of political-economic governance (see also Section 2.5).

This explanation of the demise of the nation-state can be thought of as the 'globalization excuse', i.e. explaining phenomena such as the reconfiguration of state functions by invoking an external factor known as globalization. Globalization becomes the convenient explanation for whatever happens to the nation-state, not as something (e.g. a set of processes) that needs to be explained. Indeed, we can observe this line of explanation throughout the media, popular books, political speeches, and so on. Typically, this explanation is offered in the following manner: 'Because of globalization, we have to . . .'.

These ultra-globalist stories often start with technological innovations in transport and communications that facilitate global flows of capital, people, information, and culture – all different aspects of globalization (see also Chapter 5). In the economic sphere, giant corporations and international financial institutions are claimed to be orchestrating global flows of capital and technology. This rise of corporate power and global finance has been presented as one major explanation for the demise of the nation-state. The global reach of these corporations and institutions has been described as 'effortless' and 'limitless': the world is their oyster. In some extreme cases, the annual sales of these large global corporations (e.g. Wal-Mart, General Motors, and Toyota) are compared with the annual production of national economies to demonstrate their corporate power.

In this account, the nation-state is deemed powerless in its capacity to control its national economic affairs (e.g. inward investments) and its own corporations (e.g. outward investments). In deciding on whether to invest in a particular country, these global corporations often play one nation-state off against another in order to obtain maximum benefits such as political support, financial incentives, lax environmental regulation, and market access. The successful bargains achieved by these corporations in both advanced industrialized countries and poor developing countries has led to the universalistic claim by ultra-globalists that host nation-states are no longer able to effectively govern their economies. Meanwhile, the successful global reach of these corporations is seen to offer supporting evidence against the nation-state's relevance in domestic economic regulation. This occurs primarily when global corporations are capable of evading the regulatory and monitoring systems of particular countries of origin.

While we have already introduced the grounded nature of technological change in Chapter 5 and will offer a more nuanced alternative to this uncritical view of the global reach of transnational corporations in Chapter 8, it is useful to point out here that such a view has widespread purchase in the *political arena*. Many politicians jump on this globalization 'bandwagon' in order to rationalize what is occurring within their territories. In this way, globalization becomes a 'scapegoat' used to explain the failure of economic policies or to justify policy interventions. 'Because of globalization' and 'there is no alternative (TINA)' explanations are commonly invoked today in the report cards of most governments and authorities throughout the world. By willingly acknowledging the inexorability of globalization, some political leaders are seemingly surrendering the various economic policy toolkits that they used to deploy so effectively to manage national economies.

7.3 Functions of the State (in Relation to the Economy): Long Live the State!

Is the nation-state really withering away as these ultra-globalists allege? Not really. In this chapter, we argue for an alternative and more nuanced economic-geographical reading of the role of the nation-state in today's global economy. The state–firm nexus continues to produce all kinds of *political-economic geographies*, ranging from how certain regions and electorates receive favourable treatment because of intra-national politics, to how the national government intervenes to prevent 'encroachment' of foreign companies (e.g. the CNOOC example above). As shown in this and the next section, the nation-state performs a range of important functions in relation to the economy, though its capacity to execute these functions depends on the historical and geographical context of the state in question.

More specifically, our broader theoretical rationale for the intertwined role of nation-states and politics in governing the modern economy can be understood in three ways. First, we reject the ultra-globalist position that polarizes the nation-state and the global firm. In particular, we note that both firms and markets are engaged with the nation-state in a *mutually dependent* relationship. Firms need the state to function in the five areas described in this section, while the state's political legitimacy will be challenged if it fails to deliver enough economic development and opportunities through the activities of firms and markets within its borders.

Second, the ultra-globalist story depicting the nation-state as being the same everywhere is clearly a gross oversimplification of reality. Indeed, the nation-state comes in different shapes and sizes; there are many *different varieties* of nation-states (see next section) and they cannot all be described as 'powerless'. In arguing for this geographical diversity of nation-states, we bring into our account the highly uneven nature of the global political economy (see also Chapters 2 and 3). This recognition of geographical diversity is also important because it showcases different *power relations* within the global economy. While some nation-states may succumb to global forces and hence their stories can be seen to corroborate the claims of ultra-globalists, there are many other nation-states that consciously shape and fashion globalization processes to their own benefit. In fact, some nation-states have arguably been reshaping the global economy in their own image in recent decades (e.g. the US).

Third, and perhaps most importantly, we subscribe to a more measured view of the nation-state that understands it as always remaking itself and undertaking necessary adjustments to new global realities. In this view, the nation-state does not exist as a static body of institutions and norms. On the contrary, it is a dynamic entity capable of self-transformation for the sake of political renewal and, sometimes, survival. We should not be misled by the ultra-globalist position into seeing globalization as the dynamo reshaping the global economy and the nation-state as the passive victim. In the penultimate section (7.5), we shall consider in detail some of these reconfiguration processes of the nation-state.

To begin our unpacking of the intermeshing of state politics and the modern economy, we need to understand some of the most fundamental issues in relation to state functions – how do nation-states interact with firms and markets? Here, we introduce the full range of economic functions performed by modern nation-states today and show how they can work out very differently in *different places*. We will emphasize their geographical concreteness rather than abstract characteristics. While the state is also expected to take care of many non-economic functions (e.g. national defence, foreign policy, cultural development, law and order, public health, and so on), we have identified five important state functions in managing the national economy:

- ultimate guarantor and institution of last resort;
- regulator of economic activities;
- architect of the national economy;
- owner of public enterprises;
- provider of public goods and services.

Ultimate guarantor

Markets can fail miserably. It is in these situations that the state steps in, thus becoming an institution of *last resort* and the ultimate guarantor. While the market may appear to be self-organizing, it can fail in the wake of unforeseeable forces (e.g. financial crises and natural disasters). This is the most basic role for the state, and the one which is most recognized by neoclassical economics. We can think of four important aspects to this role:

1. *Dealing with financial crises*: market failure is particularly likely in the financial industry where bankruptcy of major financial institutions can result in severe financial crises in different national economies. Whereas some nation-states in advanced capitalist economies choose to leave these bankruptcies to the 'creative destruction' of the market mechanism, other states may decide to intervene in the market for political and social reasons. These latter states may seek to avert a banking crisis by offering financial backing to failed banks. For example, the Japanese government stepped in to assist the Long Term Credit Bank of Japan in October 1998 when the latter had collapsed under mounting non-performing loans. Such intervention can help to restore public confidence in the financial system, and stabilize the economy at large.
2. *Guaranteeing national economic instruments*: the international economic credibility of a nation-state in part depends on its ability to maintain the value

of its currency and government bonds. For example, US Treasury Bills, a form of government-issued IOU (I Owe You), have strong credibility in the international financial community. Many institutional investors such as pension funds and insurance companies have purchased these Treasury Bills because of their attractive interest payments and the security of repayment upon maturity. In comparison, the bonds issued by many developing countries (e.g. Brazil and Argentina) may be subject to a lot more uncertainty and are therefore less attractive to potential investors. In terms of currencies, states ensure that their currency is the universal standard, or legal tender, accepted in their territory and try to maintain the relative value of that currency. The experiences of some South American countries during the early-to-mid 1990s (e.g. Argentina and Mexico) and some Asian economies in the late 1990s (e.g. Indonesia, Thailand and South Korea) are telling here. In both instances, the nation-states in question were unable to guarantee their national currency with sufficient foreign reserves or gold, resulting in massive depreciation and subsequent financial crises as people sought to rapidly sell reserves of the currency.

3 *Securing international economic treaties*: the nation-state is also important in that no other institutions (e.g. firms and markets) have the political legitimacy to conclude and sign international trade and investment agreements. Nation-states often engage in bilateral free trade agreements (FTAs) with one another in order to advance their common economic interests. By the end of 2005, there were close to 300 such FTAs either in the planning and negotiation stages, or concluded (http://www.wto.org, accessed 1 December 2005). Smaller states such as Singapore have deployed these FTAs highly effectively to expand their economic space and connectivity in the global economy. Nation-states also engage in bilateral investment guarantee pacts in order to protect the commercial interests of their national firms in foreign territories.

4 *Property rights and the rule of the law*: nation-states are also crucial in establishing and maintaining private property rights and upholding the rule of the law through a well-defined legal system. Property rights refer to the right of an individual or a corporate entity to own properties (e.g. land, building, machineries, ideas, designs, and so on) and derive income from these properties. This aspect of state functions defines the quintessential character of *modern capitalism* (see Chapter 3) because without effective property rights, capitalists (individuals or firms) cannot capture profits derived from their properties and therefore do not have any economic incentives to invest in these properties. A nation-state thus has numerous state departments to deal with property rights issues such as land titles, patents and trademarks, and business registries, and it also enacts many laws to protect these property rights.

Regulator

States are also the primary regulators of the economic activities that take place within, and across, their borders. This aspect of state functions can be interpreted in different ways. For those of a pro-market persuasion, the nation-state should simply seek to enable and protect market-driven activities. And yet, the nation-state's primary source of legitimacy originates from its citizens. In turn, it is expected to protect its citizens from harmful or negative effects of the market and firm activities. The nation-state therefore often engages in a wide variety of forms of regulation of economic activities, ranging from economic and environmental to social and ethical considerations:

- *market regulation*: the nation-state ensures the fairness of the market mechanism. The extent to which the state actively regulates market behaviour varies in different national economies. The US, for example, is particularly known for its strong preference for open market competition and anti-monopoly stance. For example, in twentieth-century America, many large corporate empires were broken down into smaller business units in order to prevent excessive market power held by these corporate giants, e.g. the dismantling of Rockefeller's Standard Oil into a few oil giants in 1911, including today's ExxonMobil and Chevron, and the former AT&T into many local 'baby' Bell companies in 1984. Many other nation-states have established fair trade commissions and anti-monopoly units in their departments of commerce to ensure market competition in different industries. This anti-monopoly approach to regulating the market economy can be contrasted with the experience of other nation-states that direct own and operate monopolies, mostly in developing countries. In these latter cases (see the next subsection), the state becomes a direct owner and manager of business enterprises, replacing the market as a primary mechanism of economic governance.
- *regulating economic flows*: the contemporary nation-state plays a very important role in regulating its borders. This function is particularly important in an era of accelerated globalization associated with massive cross-border flows of capital, commodities, people, and information. In regulating capital flows, some states may be very restrictive and do not allow financial capital to enter and leave their countries easily without completing cumbersome regulatory procedures or paying hefty taxes. Border regulation is of particular importance in relation to labour flows. The border crossing point between China's Shenzhen and Hong Kong's Lo Wu shown in Figure 7.1 is the most heavily-used checkpoint between the two administrative systems in one country (China). Millions of people cross this border every year for work, leisure, and social visits. More importantly, however, only a fraction of these people from mainland China are allowed to migrate to Hong Kong on a permanent

Figure 7.1 The border crossing between China's Shenzhen and Hong Kong's Lo Wu
Source: Martin Hess, with permission.

basis. As globalization has intensified, so levels of international migration of various kinds (e.g. short-term/permanent, high skill/low skill) have increased. Increasingly, destination countries (e.g. Canada or Australia) are seeking to regulate the *kinds* of migrants that they allow in, using points systems to assess the skills, qualifications and financial resources that potential migrants can bring into the economy. As another example, the accession of ten Eastern and Central European countries to the European Union in 2005 saw different initial responses from existing member states. While many countries negotiated that they would not have to allow migrants from these states for a set period, by contrast the UK, Ireland and Sweden immediately opened their borders to documented migrants in sectors where there were labour shortages (e.g. construction).

Architect of the national economy

In managing their national economies, most nation-states pursue a wide range of economic policies in order to sustain, shape and promote economic development. These economic policies may be linked to trade, investment, industry, and

TRADE POLICIES

Imports

1. Tariffs
2. Non-tariff barriers
Import quotas
Import licenses
Import deposit schemes
Import surcharges
Rules of origin
Anti-dumping measures
Special labelling and packaging regulations
Health and safety regulations
Customs procedures and documentation requirements
Subsidies to domestic producers of import-competing goods
Countervailing duties on subsidized imports
Local content requirements
Government procurement preference for domestic producers
Exchange rate manipulation

Exports

Financial and fiscal incentives to exporters
Export credits and guarantees
Setting of export targets
State-sponsored overseas export promotion agencies
Establishment of export processing zones and/or free trade areas
Voluntary export restraints through export quotas
Embargo on strategic exports
Exchange rate manipulation

FOREIGN DIRECT INVESTMENT POLICIES

Inward FDI

Government screening of investment proposals
Sectoral restrictions or prohibitions
Ownership restrictions
Requirement for local personnel in managerial positions
Compliance with national codes of business conduct
Insistence on local content of production
Insistence on minimum level of exports
Technology transfer requirements
Locational restrictions or preferences
Restrictions on remittances of capital or profits abroad
Level and methods of taxing profits
Direct promotion of FDI through competitive bidding, overseas promotional agencies, and investment incentives

Outward FDI

Restrictions on capital export (e.g. exchange control regulation)
Necessity for government approval
Direct government ownership in outward FDI
State-level FDI negotiations on behalf of national firms
Direct promotion of outward FDI through overseas promotional agencies and investment incentives (e.g. taxation benefits)

INDUSTRIAL POLICIES

Promotional

Investment incentives: capital- and tax-related
Labour policies: subsidies and training
State procurement policies
Technology policies
Small firm policies
Policies to encourage industrial restructuring
Policies to promote investment

Regulatory

Merger and competition policies
Company legislation
Taxation policies
Labour market regulation: labour union and immigration policies
National technical and product standards
State ownership of production assets
Environmental regulations
Health and safety regulations

Criteria affecting the selectivity of these policies

1. Particular sectors of industry
To bolster declining industries
To stimulate new industries
To preserve key strategic industries
2. Particular types of firm
To encourage entrepreneurship and new firm formation
To attract foreign firms
To help domestic firms against foreign competition
To encourage firms in import-substituting or export activities
3. Particular geographical areas
Economically depressed areas
Areas of growth potential or new settlement

Figure 7.2 Major types of economic policies pursued by nation-states
Source: Adapted from Dicken (2003), Figures 5.4, 5.6 and 5.8. Reprinted by permission of Sage Publications.

technology (Figure 7.2). In many cases, economic policies will also involve labour and finance. The most critical issue here is whether a particular nation-state pursues these economic policies in a *strategic* manner in order to seek or advance its national competitive advantage. If so, the relevant economic policies can be termed *strategic economic policies*:

- *strategic industrial policies*: these policies stand out as the most significant dimension as the nation-state gets directly involved in nurturing and promoting certain industries (e.g. automobile and electronics) and firms. Sometimes, explicit trade and investment policies are combined to favour these industries and firms.
- *strategic trade policies*: a strategic state will *manage* trade actively in order to protect the interests of its domestic producers. In addition to managing imports, nation-states often pursue strategic trade policies (e.g. export subsidies) to ensure their most favoured national firms – known as 'national champions' – are able to export globally and earn significant levels of foreign currency.
- *attracting foreign investment*: the role of the state becomes even more visible in these policy interventions. As the competition for global investment steps up significantly during the 1990s and the twenty-first century, many nation-states are now pursuing economic policies that are highly favourable to foreign investors. For example, general tax incentives, the availability of prime land, training of the domestic workforce, and so on are often combined to form an attractive package to attract foreign investors.
- *regional development policies*: in most countries, the nation-state actively pursues these policies to balance inherent inequality in the development of different regions (see also Chapter 3). These regional development policies may range from direct grants and financial incentives offered to poorly developed regions to central policy interventions (e.g. training the labour force and favourable industrial policies to target specific regions).

The above strategic economic policies often work out very differently in contrasting economies. For example, the US, widely known as *the* champion of free trade, does not necessarily apply the doctrine to its *imports* from around the world. As recently as 2003 and 2005, the US imposed punitive tariffs on imports of steel from Europe, Japan and many developing countries, and imports of textile and garments from China. The US justified these punitive tariffs on the basis of the 'strategic' importance of domestic producers in these sectors and the unfair 'cheap dumping' by exporting countries. By contrast, unlike the US' focus on regulating imports, the Asian newly industrializing economies (NIEs) explicitly engage in the strategic promotion of *exports*. For example, generous export subsidies and tax incentives have been offered to national champions such as Hyundai and Samsung from South Korea, and Acer and Tatung from Taiwan. In Taiwan (and to a lesser extent South Korea), the state has been highly active in growing its IT-related industries since the 1990s. Taiwan has become known as the 'Silicon Valley of the East' and, through the government's direct policy interventions, a number of Taiwanese IT-firms now rank among the world's largest manufacturers of IT-related products.

Owner of public enterprises

With a few exceptions, nation-states directly engage in *owning* and *managing* economic activities. In many ways, this state function contradicts conventional thought in economics because the ownership and management of business enterprises is seen as something best left to private capitalists. The state thus becomes the 'visible hand' that performs economic activities through direct ownership and management of firms and industries. It also challenges the notion that globalization is entirely driven by *private* corporations, as some of these state-owned enterprises are actively globalizing as well. Different countries, however, experience different levels of state involvement in such business enterprises that range from state-owned enterprises (SOEs) to government-linked corporations (GLCs).

SOEs are public enterprises that are directly owned and managed by the nation-state. CNOOC, in the example that opened this chapter, provides a good example. The Chinese state not only owns a substantial stake in CNOOC, but also directly appoints its top management. SOEs are commonly found in developing countries throughout the world, where institutions in favour of the market mechanism are not yet fully established and the state has immediate developmental goals that can be achieved through direct state ownership and management of business enterprises. According to the United Nations Conference on Trade and Development (UNCTAD), over 10 of the top 50 transnational corporations from developing economies in 2003 belonged to SOEs. These SOEs are found in such diverse developing economies as Brazil (e.g. Petroleo Brasileiro), China (e.g. China Ocean Shipping), Malaysia (e.g. Petronas) and Mexico (e.g. Cemex).

GLCs, on the other hand, refer to business enterprises in which the state has a direct or indirect stake and yet leaves the management to professional managers. State intervention in these GLCs is often less significant in comparison to SOEs. The state may own some equity in these enterprises although they are mostly operated as profit-driven businesses. GLCs are commonly found among the largest TNCs from many developing economies such as Singapore (e.g. Singtel), Taiwan (e.g. Taiwan Semiconductor), Hong Kong (e.g. China Resources Enterprises), and Malaysia (e.g. the MUI Group). Among advanced industrialized economies, some states continue to have substantial stakes in their national firms. For example, in France and Italy, the states have large ownership stakes in automobile manufacturers such as Renault and Fiat, respectively.

Provider of public goods and services

The nation-state is also a direct provider of a wide range of so-called *public goods* – for example, transport, education, health, infrastructure, and public housing – that are often seen as either too risky or unprofitable for individual private firms and/or fundamental to a nation's well-being. Throughout the world, all nation-states have at least to some extent taken up this role as a provider of public goods and services in lieu of the market and private firms, in many cases making the state the largest single employer in the country:

- *transport services*: the nation-state is involved in the provision of transportation at a variety of geographical scales ranging from rural to urban to inter-city to inter-national scales. For example, the German federal state continues to own and operate Deutsche Bahn, one of the most successful train systems in the world. Most developing and some developed countries own and manage their national airlines (e.g. Singapore Airlines). This state-owned approach to public transport contrasts with the UK experience where British Airways was privatized in 1987 and British Rail in 1993.
- *health and education services*: nation-states are usually heavily involved in the provision of health and education services for their citizens. For example, at the tertiary level, most universities throughout the world today continue to depend on state-level funding. In some countries, private sector health and education provision is available alongside public sector provision for those who can afford it. In the US, for example, the private college system funded by donations has thrived and produced some of the world's leading universities such as Harvard, Yale, and the University of Chicago. The state may also be important in providing research infrastructure and funding.
- *infrastructure services*: the nation-state is often tasked with providing basic facilities and amenities such as roads, highways, airports, ports, power grids, communications networks, and so on. In developing countries, the capacity of the nation-state to provide good public infrastructure is often critical to attracting inward investment. Some nation-states provide large public housing programmes in order to improve the living standards of their citizens and/or to maintain the cost competitiveness of its labour force. The latter experience has been argued to be particularly relevant for Hong Kong and Singapore, two city-states in Asia that link their highly successful public housing programmes to export-oriented industrialization policies.

To sum up, these multiple functions of the nation-state testify not only to its importance in economic management and governance, but also to the enormous diversity in state-economy relationships. As these political-economic relationships unfold in different historical and geographical contexts, the landscape of nation-states and national economies in today's global economy remains highly variegated, as we shall now move on to explain.

7.4 Types of states today

While the last section outlined the key economic functions of modern nation-states in general, it is important for us to bear in mind the enormous range of state formations in the global economy. In recognizing the institutional and geographical diversity of nation-states, we are challenging the 'end of the nation-state' thesis championed by ultra-globalists. This recognition is highly important precisely because of the uneven nature of the global economy (see Chapter 3). The different historical-geographical circumstances from which each nation-state emerged have produced a variety of *different* states rather than a homogeneous group of similar states. As alluded to in Figures 7.3 and 7.4, two of the world's most powerful states today, the US and China, emerged from rather different historical-geographical contexts. The US obtained its democratic statehood from Great Britain after Thomas Jefferson's Declaration of Independence was adopted by a meeting of statesmen held on 4 July 1776 in the Independence

Figure 7.3 Independence Hall, Philadelphia, the United States
Source: The authors.

Figure 7.4 The Forbidden City, Beijing, China
Source: The authors.

Hall in Philadelphia. The rise of the People's Republic of China (PRC) followed a bloody civil war and resulted in the inauguration of a drastically different communist political system at the Forbidden City in Beijing on 1 October 1949.

Different states therefore engage in a wide range of power relations with firms, markets, and international organizations. Whereas some states are no doubt in control of their domestic economies through strict economic regulation, other states have delegated such economic control functions to private firms, markets, and international institutions. In other words, the kind of state in question is crucial in determining its capacity to control the economy and to resist and shape globalization processes. In this section, we review briefly six groups of nation-states in relation to their different political-economic governance and power relations:

- *neoliberal states* in North America, Western Europe, and Australasia;
- *welfare states* in Nordic countries and some European countries;
- *developmental states* in Asia and South America;
- *transitional states* in post-socialist countries in Eastern Europe, the former Soviet Union, China, and Southeast Asia (e.g. Vietnam, Cambodia, and Laos);

Table 7.1 Types of states in the global economy

Type of states	Examples	Main characteristics	Political governance systems	Organization of economic institutions
Neoliberal states	North America (e.g. Canada and the US), Western Europe (e.g. the UK), and Australasia	Reliance on the market economy	Liberal democracy with multiple political parties	Strong role of capital markets and finance-driven investment regimes
Welfare states	Nordic countries (e.g. Sweden, Finland and Norway) and some European countries (e.g. Germany and France)	Coexistence of substantial provisions of state benefits and the market economy	Social democracy with multiple political parties	Bank-centric financial systems and strong interdependency between capital, labour and state
Developmental states	Japan and the newly industrialized economies (e.g. Brazil, Mexico, Singapore, Taiwan and South Korea)	Relative autonomy of the capitalist state from corporate interests and voters	Soft authoritarianism dominated by a single large political party	Direct involvement of state in economy through industrial policies and strategic investments
Transitional states	Post-socialist countries in Eastern Europe, former Soviet Union, China, and Southeast Asia (e.g. Vietnam, Cambodia, and Laos)	Former communist states that are moving towards capitalism and markets	Strong authoritarianism dominated by a single political party	Hybrid co-existence of state-owned enterprises and market economy
Weak and dependent states	African, Central/South American, and Middle East countries	Beholding of states to oligarchs, foreign corporate interests or international agencies	Unstable political systems and the frequent presence of dictatorship	Undeveloped or underdeveloped markets for investment and production
Failed states	Somalia, Congo, Afghanistan, Iraq, East Timor, and Bosnia	International agencies performing state roles on behalf of the host states	Disintegrating or embryonic form of political system	Lack of clearly defined market system

- *weak and dependent states* in African, Central/South America, and the Middle East;
- *failed states* such as Somalia, Congo, Afghanistan, Iraq, and Bosnia.

In Table 7.1, we summarize the main features of these six different types of states. Some of these states have experienced traumatic journeys to *nation*hood characterized by internal contests and civil wars (e.g. the former Yugoslavia and several African states). Still other states (e.g. Taiwan) face competing claims of *national sovereignty* from other nation-states (e.g. China). For simplicity, we therefore use the broad term 'states' (instead of 'nation-states') to describe these political entities. To distinguish these varieties of states, we use two broad criteria:

- *political governance systems*: these range from liberal democracy in neoliberal states and social democracy in welfare states to authoritarian political control in developmental states and outright dictatorship in some weak and failed states;
- *organization of economic institutions*: how firms, industries, state and non-state institutions relate to each other in these states.

The great diversity in state formations points to the highly *uneven* and *differentiated* relationships between these states and globalization tendencies. There is no single outcome of globalization in relation to these states. Instead, globalization processes occur in tandem with this diversity of states.

Neoliberal states

In *neoliberal states*, government institutions keep a distance from private firms and industries. The main role of the state in these economies is to establish market rules through legislation and to enforce these rules through their regulatory capacity. In the US, for example, competitive and antitrust or antimonopoly market rules have led to the development of a particular kind of investment regime that governs the market. In this investment regime, American firms tend to rely on capital markets for their investment needs and have their performance measured according to their ability to maximize shareholder value. Hence, short-term investors exert a great deal of influence on corporate decision-making. While we consider this aspect of corporate governance and national business systems in more detail in Chapter 11, it suffices here to note that this American-style investment regime has led to particular kinds of labour market practices to be described fully in Chapter 9 (e.g. labour flexibility and open market recruitment). In New Zealand, for example, labour market restructuring has gone in tandem with massive privatization programmes – all hallmarks of *neoliberalism* (Box 7.1). Overall, the economy in these neoliberal states is

Box 7.1 Neoliberalism: what's in a concept?

Neoliberalism is a set of free-market political-economic principles that have become the ideological 'common sense' of the twenty-first century in many parts of the world (Tickell and Peck, 2003). It initially emerged during the 1970s in the guise of *laissez-faire* economic policies designed to combat protectionism and excessive government intervention. By the 1980s, it had mutated into a series of state projects and restructuring programmes. In developed countries such as the US, the UK, and New Zealand, neoliberalism was the guiding ideology behind state-driven programmes of privatization and deregulation. Alternatively known as the 'Washington Consensus' – a term coined in 1990 to describe a series of policy prescriptions being applied in South America – neoliberalism quickly spread throughout the world. In developing countries, it became synonymous with the infamous 'structural adjustment programmes' championed by the International Monetary Fund and the World Bank (see Box 7.3).

More recently, during the 1990s and beyond, neoliberalism has grown into the most powerful economic policy 'fix' for all kinds of political-economic problems throughout the world. Although it exists in different versions and permutations, neoliberalism broadly encapsulates the following key policy propositions:

- trade liberalization;
- financial market liberalization;
- privatization of production;
- deregulation;
- foreign capital liberalization and the elimination of barriers to foreign investment;
- securing property rights;
- unified and competitive exchange rates;
- diminished public spending (fiscal disciplines): welfare cutbacks and financial austerity;
- restructuring of the state through privatization: changing public expenditure in health, schooling, and infrastructure;
- tax reform to broaden the tax base, cut marginal tax rates, and reduce progressive taxes;
- scaling back the 'social safety net': narrowly targeted and selective transfers for the needy;
- flexible labour markets.

underpinned by a well developed financial market and a flexible labour market. The role of the nation-state is to serve as the custodian of this system through complex sets of regulatory institutions and legislation.

Welfare states

In *welfare states*, labour unions and state institutions play a much more direct role in corporate governance and firm behaviour. Private firms do not necessarily have a free hand in labour management, let alone in controlling the national economies. Instead, there are stringent labour laws and other welfare provisions that shape the investment behaviour of private firms. The role of capital markets in driving the national economy is relatively less than in neoliberal states. The economy in both Germany and France, for example, is heavily funded by their national banks, many of which are state-controlled, and less so by the capital markets (e.g. national stock exchanges) than in neoliberal states. As a result, German and French banks have substantial ownership stakes and management input in large German and French firms. In this way, the welfare state is able to directly control the national economy through its well developed and regulated banking system. Through state taxation – another form of state control of economies – welfare states are also able to provide a significant range of national welfare services for their citizens ranging from unemployment benefits and medical treatments to retirement pensions and education.

Developmental states

In *developmental states*, the state is relatively autonomous from the influence of civil society, business and the populace. This autonomy – often achieved through authoritarian political control – is necessary in order for the state to pursue interventionist policies favouring economic development. Most developmental states thrived on newfound nation-building imperatives immediately after their independence from former colonial powers. To achieve their economic development objectives, these states exercise economic control through developing elite state-sponsored economic agencies and strategic industrial policies (e.g. Japan's former Ministry of International Trade and Industry and South Korea's Economic Planning Board). These agencies are heavily engaged in consultation and coordination with the private sector, and these consultations become an essential part of the process of policy formulation and implementation. Through these elite agencies, the state decides on the 'right' industries to nurture and the 'best' firms to promote, creating a range of 'national champion' firms, some of which are directly owned and managed by state institutions. The developmental state is also actively involved in regulating its domestic capital and labour markets in order to enhance the likely success of its strategic industrial policies. In capital markets, the ministry of finance in most developmental states takes

Box 7.2 Strategies of industrialization: import-substitution versus export-orientation

There are two main types of strategies for developing nations to industrialize over a period of time: (1) import-substitution; and (2) export-orientation. In reality, most developing nations have pursued a mixture of both industrialization strategies in different historical periods. Their policy choice depends very much on their internal political conditions and the prevailing climate in the global economy (e.g. geopolitics, global trade regimes, technological change, and cross-border investment in manufacturing activities). The import-substitution industrialization (ISI) strategy was commonly adopted by nation-states in their early post-colonial period when nationalistic sentiments were particularly strong. Such strategies have been pursued, for example, with varying degrees of success, in many economies across East Asia and South America. Under the ISI strategy, the focus of economic development was placed on the domestic national economy, as evidenced by the following:

- protectionist trade policies to reduce imports;
- an overvalued exchange rate and exchange controls;
- the nationalization of large export-oriented firms, particularly in the area of natural resources;
- price controls to check domestic inflation;
- a highly-regulated labour market;
- a fiscal policy featuring heavy dependence on trade taxes and high tax evasion;
- public expenditure focusing on subsidies to state-owned enterprises (SOEs) and domestically-oriented private firms;
- high concentration of state expenditure in urban areas.

On the other hand, most observers agree that export-oriented industrialization (EOI) strategies have worked very well in the newly industrialized economies (NIEs) of East Asia (e.g. Hong Kong, Taiwan, Singapore, and South Korea) and South America (e.g. Brazil and Mexico). Generally, an EOI programme includes the following:

- a strong trade policy focus on export of primary commodities and manufactured goods;
- devalued currencies to increase export competitiveness;
- explicit state incentives to promote large export-oriented firms;
- regulation of union activities to reduce labour militancy and to stabilize cheap labour supply;
- industrial policy featuring new industries with strong export propensities;
- public expenditure focusing on subsidies to state-owned enterprises (SOEs) and export-oriented private firms.

direct stakes in national banks and finances export-oriented industrialization programmes (Box 7.2) through export subsidies and generous grants to its national champions. In labour markets, developmental states are often actively involved in subordinating the interests and rights of their workers. For example, in Taiwan and Hong Kong, labour unrests and strikes are managed through tough laws that curtail labour union activity.

Transitional states

In general, *transitional states* are so called because they are moving away from their former existence as communist totalitarian states governed entirely by central economic planning to a hybrid mix of planned and market economies. They are usually neither entirely democratic nor dictatorial in their political system. In most of these transitional states, however, dominant political parties in the former regimes retain important political influence in the transition to market economies. National economies tend to be dominated by a mixture of state-owned enterprises (SOEs) and private firms (domestic and foreign). The precise ratio of this mix varies between different transitional states. What is clear is that the nation-state continues to exercise a great deal of control of domestic economies through owning stakes in SOEs and manipulating dense webs of state regulation covering private firms and industries. Transitional states are mostly found in Central and Eastern Europe where economic reform went hand-in-hand with political reform as democratization swept through the region in the late 1980s and early 1990s, leading to the break-up of the Soviet Union and the emergence of transitional states such as Hungary, Czech Republic and Poland. In Asia, the opening of China to the global economy since late 1978 marks another significant development in transitional states. Unlike the former Soviet Union, economic reform in China was not accompanied by significant political reform. As of 2005, the Chinese Communist Party remained the only legitimate political party in China, prompting a hybrid form of economic development characterized by strong communist control and expanding market freedom known as 'red capitalism' (Lin, 1997). In this system, the many state-owned enterprises are managed jointly by Party and firm managers.

Weak and dependent states

State domination of the domestic economy is much more pronounced in *weak and dependent states* throughout Africa, South America, the Caribbean and the Middle East. A *weak* state is one that has largely succumbed to the power of oligarchs and corporate interests – as distinct from the relative autonomy of the developmental state. A *dependent* state, meanwhile, would be one that is beholden to international forces including international financial institutions (e.g. the World Bank and the International Monetary Fund) or other powerful states (e.g. the US), to the extent that it has very little space for autonomous action. As most of these states rely heavily on natural resources for their economic development, state control of strategic resources becomes critical in preventing overexploitation and environmental degradation. Most of these dependent economies are highly vulnerable to cyclical fluctuations in global commodity prices. This dependency on primary commodity exports is apparent in most countries in Africa (e.g. Kenya and Ethiopia on coffee), South America (e.g. Venezuela and Ecuador on oil), the Caribbean (e.g. St Lucia on bananas) and the Middle East (e.g. the United Arab Emirates on oil).

In their efforts to industrialize, many weak and dependent states have engaged in import-substitution policies, often financed through lending from foreign banks. These debts, when combined with a lack of exports and therefore foreign reserves, and falling commodity prices, have left economies vulnerable to financial crises such as those that affected South America during the early-to-mid-1990s (e.g. Chile and Mexico) and 2002/2003 (e.g. Argentina). To help these South American economies out of their debt crises, international organizations such as the World Bank and the IMF were called in. The IMF's standard rescue package is commonly known as the Structural Adjustment Programme (Box 7.3). Under the strict conditions of these Programmes, states have lost influence over their domestic economies as they have become beholden to the policy advice of international organizations. From weak states, they have become dependent states insofar as they have succumbed to externally influenced programmes of financial management.

Failed states

A *failed state* is one where international agencies are actually performing state roles on behalf of the host government. Before they end up as failed states, economic institutions in these countries are usually subsumed to the state's political apparatus. As most of these states are subject to varying forms of dictatorship, the protection of property rights is weak and the existence of market rules is often ignored. This results in a form of economic organization that resembles central planning rather than a market economy. As the embryonic form of political system in these states disintegrates due to internal factors (e.g. civil wars or military coups) or external factors (e.g. international diplomatic or military pressure), there is pressure for international agencies such as the United Nations to step in and take over state functions described in the earlier section. Iraq and Afghanistan, in the post-US invasion periods, could be described in this way as failed states.

From this brief survey of how economic institutions are organized differently in different states (see Table 7.1), we can appreciate the highly differentiated power relations between state authority and market mechanisms. In neoliberal states, economic institutions are organized to reflect the centrality of *the market*,

Box 7.3 Structural adjustment programmes in South America, circa the late 1980s

Before the debt crisis hit South America in the 1980s, the role played by international financial institutions such as the International Monetary Fund (IMF) and the World Bank was mostly temporary and project-specific, designed to ease short-term cash-flow problems or fund specific infrastructure and public works schemes. With the onset of the debt crisis, however, many South American economies suffered from a sudden downturn in capital inflows, much greater debt servicing costs, massive devaluation of national currencies, and rapid inflation (known as hyper-inflation). Consequently, both the IMF and the World Bank began to focus more explicitly on the *structural reform* of South American economies. The conditions associated with these reforms were significantly more constraining than in their earlier involvement. In particular, the term *structural adjustment programme* (SAP) has been used to describe a series of conditions imposed by the IMF and, to a lesser extent, the World Bank on South American economies. Under SAPs, the international financial institutions would offer financial and development assistance to South American states such as Argentina, Bolivia, Brazil, Chile and Peru if they were willing to adopt and implement new market-oriented development strategies that eschewed the earlier ISI models of economic development (see Box 7.2). These new strategies included privatization, trade liberalization, the deregulation of prices and labour and financial markets, and civil service reform. As elaborated in Box 7.1, SAPs were part of a broader political-economic movement towards neoliberalism. While the implementation of such SAPs in part reflected the increased power and international institutions and the severity of the financial crisis, it also relied upon the support of domestic political coalitions. By the early 2000s, however, new populist leaders in Argentina, Bolivia, Chile and Peru were pursuing more nationalistic economic policies that sought to 'roll back' the SAPs implemented since the late 1980s, reasserting the importance of the nation-state in these processes.

resulting in high levels of flexibility and autonomy for private firms and enterprises. Notwithstanding these market-driven economies, there are still many other states in the global economy in which economic institutions are organized around *the state* in the form of either state-owned enterprises or government-sponsored business and industrial initiatives. To speak of the end of the nation-state in controlling domestic economic affairs is simply a gross generalization of a great geographical diversity of states and their approaches to national economic management. Our typology also shows very different institutional capacity in influencing and resisting forces unleashed through globalization processes. The explicit political ideologies adopted by different states result in different state-economy relationships, ranging from neoliberal free market systems to state-controlled planned economies.

7.5 Reconfiguring the State

Our recognition of the diverse functions and organization of nation-states in the preceding two sections does not imply that the state is a *static* feature in the global political economy, rather, we view the state as a dynamic set of institutions. In this section, we consider how the nation-state is being reconfigured in an era of globalization. First, we explain how the individual nation-state is experiencing changing structures of governance at both the international and sub-national scales – an economic-geographical process known as *rescaling*. Second, we note the rise of non-state forms of economic regulation through public–private partnership and private actors, a shift often described as being from govern*ment* to govern*ance*. In these subsections, we focus on the processes and dynamics of *institutional change* that enable nation-states to continue to exercise control over their national economies, albeit in different forms and in specific geographical contexts. Instead of delving in depth into the reconfiguration of a particular state, we offer a range of contrasting examples.

The rescaling of economic governance

As noted in Chapter 1, the global space-economy can be conceived in terms of different geographical scales that range from global, international, and macro-regional to national, regional, and local. Nation-states are actively reshaping their institutional structures of economic governance at both *international* and *sub-national* scales. In this *rescaling* process, the nation-state has to delegate some of its control of national economic affairs to authorities at higher geographical scales (international and macro-regional) or lower geographical scales (sub-national, regional, and municipal).

Historically, the nation-state executed its regulatory functions at the national scale. In other words, its policies and rules were applied throughout the nation, irrespective of local conditions. Globalization tendencies and changing international relations are increasingly challenging this nation-centric model of economic governance. In particular, the rise of international and macro-regional organizations has increased the importance of international cooperation and the coordination of economic policies among nation-states. This movement from national regimes of economic governance to authority at higher geographical scales can be conceived of as a process of *upscaling* (Swyngedouw, 2000). These international and macro-regional organizations exist primarily at two spatial scales:

1 *international organizations*: e.g. the World Trade Organization (WTO), the International Monetary Fund (IMF), the World Bank, the Organization for Economic Cooperation and Development (OECD), and the United Nations (UN);
2 *macro-regional groupings*: e.g. the European Union, Asia Pacific Economic Cooperation (APEC), North American Free Trade Agreement (NAFTA), the Association of South East Asian Nations (ASEAN), the New Partnership for Africa's Development (NEPAD), the South Asian Association of Regional Cooperation (SAARC), the Caribbean Community (CARICOM) and MERCOSUR (Mercado del Sur) in South America.

Among the *international organizations*, the United Nations is the oldest and broadest in terms of its mission and scope. However, it has only limited means and resources to influence economic governance at the global scale. In this sphere, we must consider the highly powerful IMF, the World Bank and the WTO and their influence upon the global financial system, development assistance and trade, respectively. The IMF was founded in 1944, and now has 184 member countries. It is the central institution of the international monetary system, and is charged with promoting stability and efficiency within the global system. The IMF is often called upon to resolve large financial deficits accrued by individual nation-states: as of August 2005, it had outstanding loan credits of US$71bn to 82 countries (http://www.imf.org, accessed 24 March 2006). However, as noted in Box 7.3, the IMF not only lends to these states to help them through cash-flow crises (e.g. a lack of sufficient foreign currency), but also seeks to impose conditions on how these economies should be restructured. Consequently, many IMF-assisted states are obliged to drastically reduce their state budgets and liberalize control of the domestic economy in line with the IMF's neoliberal policy prescriptions.

The World Bank was also founded in 1944, has 184 members, and provides development assistance – in terms of both knowledge and finance – to over 100 developing countries on an ongoing basis. This assistance covers a wide variety of activities including basic health and education provision, social development and poverty reduction, public service provision, environmental protection, private business development and macro-economic reforms. In 2005, it granted some US$22.3bn in development assistance, with Latin America and the Caribbean, and South Asia being the biggest recipients, with 24 and 22 per cent of total spending, respectively (http://www.worldbank.org, accessed 24 March 2006). As described in Section 7.4, some states have benefited from the World Bank's assistance and engaged in a virtuous circle of successful development pathways (e.g. post-war Germany, Japan, and the Asian NIEs). Other weaker and dependent states have benefited less from the World Bank's economic advices and development assistance (e.g. Chile and Peru). This uneven outcome of development interventions by the World Bank (and the IMF) has created discontent in certain developing economies, where some see the World Bank and the IMF as *causes* of the development gaps in today's global economy.

The WTO was created in 1995 as the successor to the General Agreement on Tariffs and Trade (GATT) established in 1947. As an organization it is responsible for the regulation of global trade, and operates a multilateral rules-based system derived by negotiation between its member states. The WTO has around 150 members, accounting for 97 per cent of global trade (http://www.wto.org, accessed 24 March 2006). In addition to administering trade agreements and

Box 7.4 The rise of macro-regional economic blocs

Regional economic blocs are a significant addition to the regulatory architecture of the global economic system. The initial stimulus for such formations comes from a desire to reduce barriers to trade and therefore enhance levels of *intra-regional* trade. Following Dicken (2003), we can identify four key types of bloc. As we move down the list, the level of economic and political integration increases:

- *the free-trade area*: trade restrictions between member states are removed, but states retain their individual trading arrangements with non-members e.g. the North American Free Trade Agreement (NAFTA) between Canada, the USA and Mexico since 1994;
- *the customs union*: members operate a free trade agreement between themselves and have a common trade policy for non-members e.g. the MERCOSUR customs union between Argentina, Brazil, Paraguay and Uruguay since 1991 (Venezuela has since joined);
- *the common market*: has the characteristics of a customs union, but in addition, allows the free movement of factors of production (e.g. capital and labour) between members e.g. Caribbean Community (CARICOM) since 1973;
- *the economic union*: harmonization and supranational control of economic policies, but only the European Union comes anywhere close to this form as evidenced by the adoption of the Euro as the sole currency of several member states in 2002.

Three further points should be made. First, numerically, the vast majority of agreements fall under the first two headings (i.e. free-trade areas and customs unions). Second, economic blocs may develop over time and 'move down' this list, as has been the case with what is now the European Union. Third, it is important to recognize that all these regional economic forms are initiated by, and derive legitimation from, their member states.

providing a negotiating forum, it handles trade disputes, monitors national trade policies, and provides technical assistance and training for developing countries. To its critics, the rise of the WTO and its enforcement of global trading rules have actually increased uneven development at the global scale. While developing countries find it hard to engage in protectionist trade policies because of WTO rules, some of the richest nation-states in the world continue to flout WTO rules by protecting their domestic producers in politically sensitive sectors such as steel, clothing, and agriculture (the US and European Union, respectively). Trade liberalization under the auspices of the WTO therefore does not necessarily reduce its uneven impacts, as different countries benefit differentially from free trade. The very slow progress of the Doha round of negotiations initiated in 2001, with agricultural tariffs proving to be a major sticking point, is reflective of these tensions.

Macro-regional groupings, on the other hand, have emerged primarily for member states to engage in economic integration and, to a limited extent, political integration. As shown in Box 7.4 and Table 7.2, these groupings come in many shapes and sizes. For example, the European Union is the oldest form of regional economic integration, first starting as the European Common Market in 1957, but now encompassing 25 European countries in an advanced form of regional integration (Figure 7.5), while NAFTA, inaugurated in 1994, brings together Canada, Mexico and the US into a free-trade zone. The basic rationale behind the formation of such groups is to reap the economic benefits of enhanced intra-regional trade and investment. However, as nation-states increasingly participate in such regional initiatives, some claim that they are ceding some of their power over their domestic economies. This claim is in part based on the experience of EU member states that have had to adjust their domestic budgetary expenditure in order to avoid deficits and thereby maintain the integrity of the single European currency system. Countries may also experience painful periods of adjustment as certain kinds of economic activity relocate to take account of the new regional context (e.g. manufacturing activity shifting from Western to Central and Eastern Europe, and from the US to Mexico).

Meanwhile, rescaling processes are also at work *within* nation-states, manifested in a tendency towards regional devolution. This process of *downscaling* of the nation-state is particularly evident in the heartlands of neoliberalism – the UK and the US (Jones et al., 2005). In the UK, the emergence of institutions such as Training and Enterprise Councils (from 1988 till 2001: now called Local Learning and Skills Councils) and Regional Development Agencies (since 1998) in the management of investment incentives, human resource development, and regeneration initiatives has reduced the role of the national state in domestic economic governance. This reconfiguration and rescaling of economic governance in the UK in favour of local states formed an integral part of an ongoing neoliberal project first initiated by Margaret Thatcher in the 1980s. Over time, local and regional authorities have been granted increased autonomy over

Table 7.2 Major regional economic blocs in the global economy

Regional group	Membership	Date	Type
EU (European Union)	Austria, Belgium, Cyprus, Czech Republic, Denmark, Estonia, France, Finland, Germany, Greece, Hungary, Ireland, Italy, Latvia, Lithuania, Luxembourg, Malta, Netherlands, Poland, Portugal, Slovakia, Slovenia, Spain, Sweden, UK	1957 (European Common Market) 1992 (European Union)	Economic Union
NAFTA (North American Free Trade Agreement)	Canada, Mexico, United States	1994	Free Trade Area
EFTA (European Free Trade Association)	Iceland, Norway, Liechtenstein, Switzerland	1960	Free Trade Area
MERCOSUR (Southern Cone Common Market)	Argentina, Brazil, Paraguay, Uruguay, Venezuela	1991	Customs Union
ANCOM (Andean Common Market)	Bolivia, Colombia, Ecuador, Peru, Venezuela	1969 (revived 1990)	Customs Union
CARICOM (Caribbean Community)	Antigua and Barbuda, The Bahamas, Barbados, Belize, Dominica, Grenada, Guyana, Haiti, Jamaica, Montserrat, St. Kitts and Nevis, Saint Lucia, St. Vincent and the Grenadines, Suriname, Trinidad and Tobago	1973	Common Market
AFTA (ASEAN Free Trade Agreement)	Brunei Darussalam, Cambodia, Indonesia, Laos, Malaysia, Myanmar, Philippines, Singapore, Thailand, Vietnam	1967 (ASEAN) 1992 (AFTA) 2005 (ASEAN +3, i.e. China, Japan and South Korea)	Free Trade Area

Source: Updated from Dicken (2003), Table 5.3. Reprinted by permission of Sage Publications.

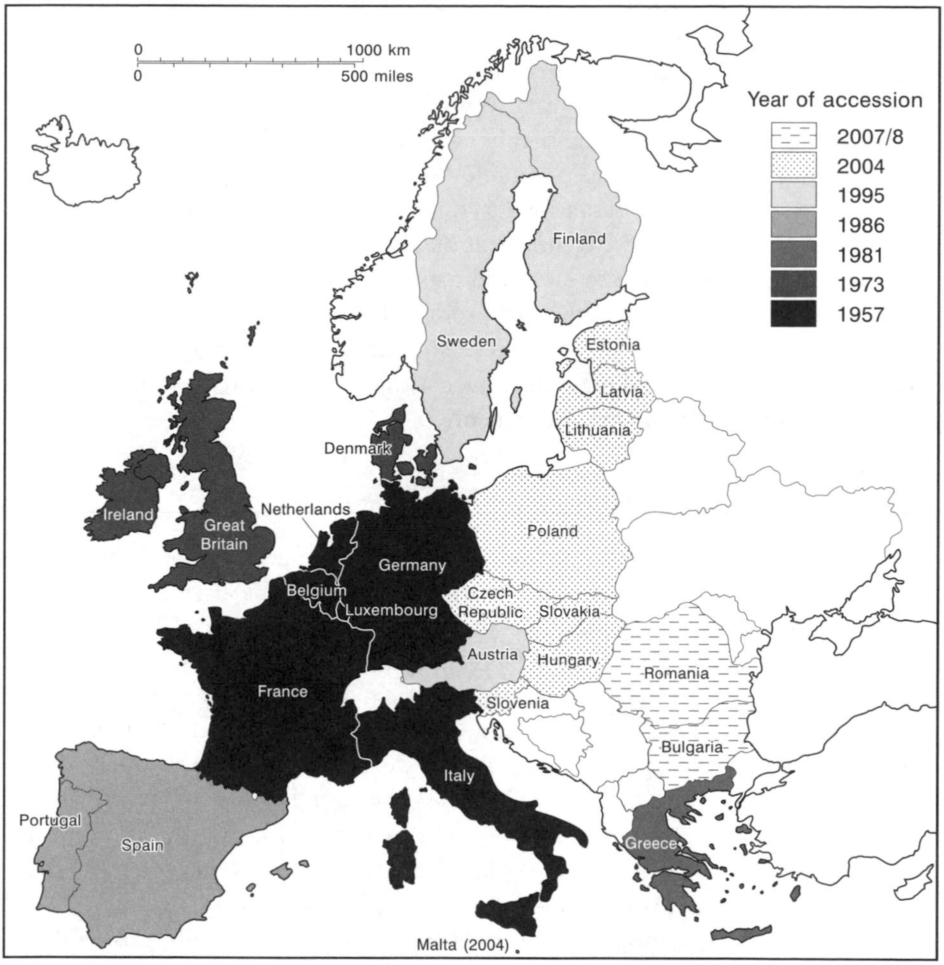

Figure 7.5 The expansion of the European Union since 1957

economic decision-making and investment initiatives. In some extreme examples (e.g. the Welsh Development Agency and OneNorthEast), RDAs have committed massive amounts of public funding to compete for specific foreign investors (South Korea's LG and Germany's Siemens).

In the US and Germany, the federal system of governance has historically granted individual state governments substantial autonomy and power in economic affairs within their own states (known as *Länder* in Germany). Since the 1990s, however, there has been a renewed interest in promoting major American cities as the 'growth poles' of the US economy. This phenomenon is broadly known as *urban entrepreneurialism*, reflecting the way in which metropolitan governments and city mayors are becoming more entrepreneurial in their economic policy initiatives, particularly those towards promoting inward investment. The process has led to stern inter-urban competition for trade and investment, a phenomenon widely observed throughout Western Europe as well.

While the US and the UK are well-known examples of devolution and downscaling, these processes have echoes in nation-states that are not necessarily neoliberal in their political-economic outlook. For example, the opening of China since late 1978 to international economic activities has been accompanied by a significant process of decentralization of economic decision-making from the central communist state to local and provincial state authorities. This decentralization process has subsequently produced a dramatic surge in economic activity, mostly in the coastal provinces. The local and regional governments in these regions (e.g. Guangdong and Lower Yangtze) are extremely aggressive in their efforts to attract foreign investment.

Hollowing out the state

Apart from relegating some of its authority over national economic governance to formal institutions at different geographical scales, the nation-state is also simultaneously undergoing another process, namely, that of *hollowing out*. In this process, state functions are taken over by public–private partnerships and private forms of regulation. This process represents the devolution of state power to non-elected QUANGOs (Quasi-Autonomous Non-Governmental Organizations) or even private entities, and clearly intersects with the rescaling of economic governance described above. As the state is increasingly privatizing its national economy, the management of this national economy becomes not just the responsibility of the nation-state, but also the private sector that effectively inherits this joint responsibility from the state.

To give an example, the *privatization* of former state-owned enterprises (SOEs) does not just represent a transfer of ownership from the public to the private sector. It also entails a transfer of economic management and governance rights from the public sector to the hands of new private shareholders of these former SOEs. In developing countries throughout the world (e.g. India, Brazil, and Thailand), the partial retreat of the nation-state from ownership of firms in various industries points to the rise of industry-specific private regulation. In the telecommunications industry, for example, the privatization of former state-owned telecommunications providers has led to new forms of service standardization and pricing strategies that are regulated by competitive market forces rather than state directives. The entry of foreign firms into these newly privatized industries has further strengthened the role of industry players in shaping the country-specific operation of these industries. In other words, the regulation and governance of telecommunications industries in many developing countries are as much shaped by the largest private telecommunications firms as by the ministries of communications in these national economies.

In the realm of *legal regimes*, we are also witnessing a certain degree of hollowing out of the state as national legal systems are increasingly intersecting with private mechanisms. This phenomenon is perhaps most obvious in the world of global finance. The regulation of the financial health of national firms used to be held in the hands of nation-states and their designated institutions – usually the central banks or ministries of finance. Today, however, the importance of private coordinating and evaluative agencies such as credit rating agencies (e.g. Moody's and Standard & Poors), institutional investors (e.g. Goldman Sachs and JP Morgan) and pension funds (e.g. the California Public Employees Retirement System and the Universities Superannuation Scheme), and accountancy firms (e.g. PricewaterhouseCoopers) has grown rapidly in an era of global finance. The role of the nation-state in ensuring order and transparency in their national financial markets is partially taken over by these private institutions. It is no exaggeration to say that many of the largest firms throughout developed and developing countries are more worried about their ratings with Standard & Poors and the buy/sell advice by Goldman Sachs than their standings with home country state authorities. This is because favourable credit ratings and 'buy' advice by Goldman Sachs can significantly enhance the share prices of these large firms that are constantly seeking cheaper forms of capital for their domestic and global business expansion.

In this regard, leading business media play a significant role in shaping how business practices are governed. Business magazines such as *Forbes*, *Fortune*, and *Business Week* and business newspapers such as *The Wall Street Journal* and *The Financial Times* often play a *de facto* monitoring role in the business world. Corporate scandals reported in these magazines and newspapers can have very damaging effects on a firm. In fact, the state often gets involved in investigating the wrongdoings of private firms *after* the media breaks the news (e.g. the role of the media in the 2001 collapse of Enron as the largest corporate bankruptcy in the US). *The Financial Times* and other media in the City of London were implicated, for example, in the pessimistic corporate discourses on the financial crisis that swept Asia in 1997 and 1998 (Clark and Wójcik, 2001). Negative assessments of the crisis published in *The Financial Times* may well have contributed to dwindling stock market performance in Asia as institutional investors rushed to sell their large holdings of Asian corporate shares and stocks.

Other than these *private agents*, there are many quasi-private institutions that operate at the *international* scale to regulate specific economic activities. These international institutions range from industry associations and 'watch dogs' to environmental agencies and private foundations. What these institutions have in common is that they are increasingly taking over some of the regulatory and coordination functions previously held by nation-states. While Chapter 4 offered some examples of these international institutions in regulating the production and consumption of commodities, we raise here the further example of global standards in accounting – the International Accounting Standards Committee (IASC), established in 1973, and the International Accounting Standards Board (IASB), established in 2001. As an independent private sector body, the IASC aims to create a set of global accounting standards that can be applied and implemented throughout the world. By the late 1990s, all stock exchanges in the world had adopted the reporting standards recommended by the IASC. This achievement should be viewed in a context where many state authorities (e.g. Japan and Germany) had previously found it very hard to implement a common standard among their domestic firms, let alone uniform international standards recommended by the IASC. In other words, one may argue that the IASC is much more effective in governing accounting practices throughout the global corporate world than individual nation-states.

7.6 Beyond the State?

Thus far we have considered the state as exercising influence and control in a uniform manner across its territory. In reality, however, state control is both socially and spatially uneven in its operation and effectiveness. In terms of social unevenness, certain portions of the population may be more or less excluded from state attempts to marshal the economy in the context of globalization. This may be voluntary on the part of some sections of the populace – for instance in the case of those deliberately working in the illegal or black economy beyond the reaches, or at the margins of state regulation – or involuntarily – as in the case of temporary migrant workers who may not be granted the same range of rights and opportunities as full citizens. With regards to spatial unevenness, we can think of certain spaces in which either a different form of state control is in existence – for instance in a tax-free export processing zone – or where the state simply cannot exert its controls – such as dangerous 'no-go' areas in certain cities. Rather than states exerting unproblematic control over their society and territory, most in reality operate a form of *graduated sovereignty* (Ong, 2000) in which different zones of economic control overlap and co-exist.

These ideas can be illustrated in the context of contemporary Malaysia, where it is possible to identify at least four zones of graduated sovereignty that are being created as the state has endeavoured to selectively integrate itself into the global economy (here we draw on Ong, 2000). First, there are the spaces of illegal migrants from Indonesia, the Philippines, Bangladesh and Myanmar, working as domestic servants, and on plantations and construction sites. Such workers have no rights to citizenship, and their short-term contracts are firmly enforced. Camps containing political refugees from across Southeast Asia are a related kind of 'non-citizen' space. Second, for Malaysia's diverse indigenous peoples – known in peninsular Malaysia as Orang Asli – differentiated government has meant displacement from traditional territories to increasingly environmentally and economically marginal land. Third, in the case of manufacturing workers, industrial sites and export processing zones (EPZs) are spaces exempted from a range of national labour and investment regulations, where labour unrest and potential strikes are quickly and firmly quashed, using force where necessary. Finally, the Multimedia Super Corridor (MSC) – a 750-sq km zone established in 1996 and stretching from Kuala Lumpur to the new international airport some 50 kilometres to the south – is a zone designed to meet the needs of the global knowledge economy, offering, among other benefits, advanced telecommunications, unrestricted employment of local and foreign knowledge workers, exemption from local corporate ownership requirements, a range of financial incentives and the promise of no Internet censorship.

Far from being a single territory governed in a uniform manner, contemporary Malaysia is in fact a fragmented space in which the experience of government depends on who and where you are. For example, a software worker enjoying the privileges of the MSC is seeing a very different side of Malaysia's efforts to integrate with the global economy than an electronics assembly worker in Penang. While the lines between these zones are particularly stark in the context of developmental states such as those in Southeast Asia, these arguments have a much wider resonance. In all kinds of states it is possible to think about how different portions of the population are governed differentially and the graduated zones or spaces that are created as a result. In some cases – as in the Malaysia example – these zones are deliberately created as part of state strategies, while in others – such as contemporary Russia, where crime, nepotism and bribery are widespread within the economy – they may reflect the state's inability to exert regulatory control over large portions of society.

7.7 Summary

In this chapter we have shown that it is far too simplistic to claim the irreversible demise of the nation-state in managing national economies. As we have seen, nation-states continue to profoundly shape the economic activity within, and across, their borders in a wide range of ways: from the assurance of basic laws and property rights, through the provision of basic infrastructure and education, to direct ownership of companies and a range of financial and tax incentives. At the same time, nation-states are not all the same: there is a tremendous range of different state forms in today's world, from neoliberal variants through to marginal states that constantly teeter on the brink of disarray and lawlessness. These different states in turn, have widely differing abilities both to control their economies, and to exert influence on international institutions. The US and Japan, for example, together contribute over 40 per cent of the operating budgets of the United Nations, the World Bank, and the IMF. They can thus wield substantial power in these international organizations through their voting rights, financial contributions, and the appointment of key personnel.

Overall, however, the nation-state continues to an extremely important actor in the global economy. Ongoing processes of rescaling and hollowing-out, while extremely significant, have not reduced its significance as perhaps the *primary* shaper of political-economic activity worldwide. Importantly, the state itself is always implicated or involved in directing these processes through its conscious decisions whether to engage with international organizations or to devolve power and authority to local states. In short, nation-states remain critical institutions through which international, regional, and local economic issues are evaluated and acted upon. Together with the transnational corporations that we will describe in the next chapter, they are key drivers of economic globalization.

Further reading

- Coe and Yeung (2001) is a useful supplementary reading for understanding different geographical perspectives on globalization and the nation-state. Dicken et al. (1997), Yeung (1998), Kelly (1999), Amin (2002), and Dicken (2004) offer some of the best geographical analyses of globalization.
- O'Neill (1997) gives a good overview of the nation-state in economic geography. Dicken (1994) analyzes specifically the interaction between nation-states and transnational corporations. Glassman (1999) shows how the nation-state is rapidly internationalizing to project its power.
- Brenner and Theodore (2003) bring together an excellent collection of important geographical work on neoliberalism and its global geographies.
- Brenner et al. (2003) offer a state-of-the-art review on the rescaling of governance. MacLeod and Goodwin (1999) offer a critical summary of the geographical debates on this topic.
- Mansfield (2005) demonstrates the importance of understanding the relationships between the nation-state and all other geographical scales of governance.

Sample essay questions

- Why is the end of the nation-state thesis flawed?
- How are nation-states different in their approaches to economic governance?
- How does neoliberalism influence state behaviour?
- What does the rescaling of governance mean in geographical terms?
- To what extent is the nation-state still capable of managing its national economy?

Resources for further learning

- http://globalpolicy.igc.org/nations: the website of the Global Policy Forum provides some useful insights into the history and formation of modern nation-states.
- http://www.englandsrdas.com: the English Regional Development Agencies website has very useful information on the functions and activities of the nine RDAs in England.
- http://www.imf.org and http://www.worldbank.org: the websites of two of the world's most powerful financial and economic institutions offer a wide range of information on how these international organizations can shape a whole variety of political-economic processes in specific nation-states.
- http://europa.eu: the European Union portal contains a whole host of information on the genesis and working of the world's largest macro-regional organization.

References

Amin, A. (2002) Spatialities of globalisation, *Environment and Planning A*, 34: 385–99.

Brenner, N., Jessop, B., Jones, M. and MacLeod, G. (eds) (2003) *State/Space*, Oxford: Blackwell.

Brenner, N. and Theodore, N. (eds) (2003) *Spaces of Neoliberalism*, Oxford: Blackwell.

Clark, G.L. and Wójcik, D. (2001) The City of London in the Asian crisis, *Journal of Economic Geography*, 1: 107–30.

Coe, N.M. and Yeung, H.W-C. (2001) Geographical perspectives on mapping globalisation, *Journal of Economic Geography*, 1: 367–80.

Dicken, P. (1994) Global-local tensions: firms and states in the global space-economy, *Economic Geography*, 70: 101–28.

Dicken, P. (2003) *Global Shift: Reshaping the Global Economic Map in the 21st Century*, 4th edn, London: Sage.

Dicken, P. (2004) Geographers and 'globalization': (yet) another missed boat?, *Transactions of the Institute of British Geographers*, 29: 5–26.

Dicken, P., Peck, J. and Tickell, A. (1997) Unpacking the global, in R. Lee and J. Wills (eds) *Geographies of Economies*, London: Arnold, pp. 158–66.

Friedman, T.L. (1999) *The Lexus and the Olive Tree*, New York: HarperCollins.

Glassman, J. (1999) State power beyond the 'territorial trap': the internationalization of the state, *Political Geography*, 18: 669–96.

Jones, M.R., Goodwin, M. and Jones, R. (eds) (2005) Special issue on devolution and economic governance, *Regional Studies*, 39: 397–553.

Kelly, P.F. (1999) The geographies and politics of globalization, *Progress in Human Geography*, 23: 379–400.

Lin, G.C.S. (1997) *Red Capitalism in South China: Growth and Development of the Pearl River Delta*, Vancouver: University of British Columbia Press.

MacLeod, G. and Goodwin, M. (1999) Space, scale and state strategy: rethinking urban and regional governance, *Progress in Human Geography*, 23: 503–27.

Mansfield, B. (2005) Beyond rescaling: reintegrating the 'national' as a dimension of scalar relations, *Progress in Human Geography*, 29: 458–73.

Ohmae, K. (1995) *The End of the Nation State: The Rise of Regional Economies*, London: HarperCollins.

O'Neill, P. (1997) Bringing the qualitative state into economic geography, in R. Lee and J. Wills (eds) *Geographies of Economies*, London: Arnold, pp. 290–301.

Ong, A. (2000) Graduated sovereignty in South-East Asia, *Theory, Culture & Society*, 17: 55–75.

Swyngedouw, E. (2000) Elite power, global forces, and the political economy of 'glocal' development, in G. Clark, M. Feldman and M. Gertler (eds) *The Oxford Handbook of Economic Geography*, Oxford: Oxford University Press, pp. 541–58.

Tickell, A. and Peck, J. (2003) Making global rules: globalization or neoliberalization?, in J. Peck and H.W-C. Yeung (eds) *Remaking the Global Economy: Economic-geographical Perspectives*, London: Sage, pp. 163–81.

Yeung, H.W-C. (1998) Capital, state and space: contesting the borderless world, *Transactions of the Institute of British Geographers*, 23: 291–309.

CHAPTER 8

THE TRANSNATIONAL CORPORATION

How does the global firm keep it all together?

Aims

- To question the claim that transnational corporations are really 'global'
- To understand how firms organize complex global activities
- To explore the variety of organizational forms used by transnational corporations
- To appreciate the inherent limits to the global reach of firms.

8.1 Introduction

In 1984, the now-defunct Barings Bank decided to buy a small brokerage firm from Henderson Crosthwaite and create Baring Far East Securities in order to start trading in the burgeoning stock markets in Asia. This was certainly not Barings' first overseas venture. In fact, as the world's oldest merchant bank – founded in the City of London in 1762 – Barings' international expansion started early. It played a crucial role in the sale of the French territory of Louisiana to the United States and the funding of the British government loans to finance the Napoleonic wars. The real significance of the 1984 acquisition rested with its radical departure from Barings' historical conservatism in merchant banking. For the first time in its corporate history, Barings engaged in the securities business – a high risk and high return business in an era of global finance. At its peak in 1991, Baring Securities had 19 offices worldwide and 1,100 staff, contributed over half of the Group's profits, and dwarfed the size of its parent bank. The global potential of the Barings Group had never looked better and brighter.

Just four years later in February 1995, the entire bank collapsed with trading losses amounting to £860 million (US$1.4 billion). What went wrong? One can easily point the finger at Nick Leeson, the 27-year-old rogue trader operating out of Baring Securities in Singapore. Once dubbed 'the Michael Jordan of trading', it was revealed that Leeson had consistently misled the bank by covering huge trading losses in an unauthorized secret 'error' account and gambling with the bank's money on Japan's Nikkei Index in the hope of recovering his massive losses. At the peak, on 23 February 1995, he lost Barings £143 million (US$233 million), the largest losing bet in the history of the Japanese stock market, and he disappeared from Singapore. In the same year, ING Group, a Dutch insurance and banking giant, acquired Barings for a token £1 and assumed its debts of over US$1 billion.

The Barings example vividly illustrates the enormous *organizational complexity* and *operating risks* faced by today's transnational corporations (TNCs) when they expand internationally. The lack of adequate management supervision and control systems, the failure of top management in London to understand Leeson's trading activities in new financial instruments such as futures and options, and the 'irrational exuberance' of the culture of global financial markets were all partly responsible for the collapse of the world's oldest transnational merchant bank. The physical distance between London and Singapore itself played a role in reducing the likelihood of Leeson being socialized into the conservative banking culture of Barings' home office in London (see also Chapters 11 and 12 for elaboration on the role of culture and gender relations in global finance). The influence of this home office culture clearly diminished with the number of time zones traversed. What we have here then is a case of a global firm failing to adequately control and coordinate its various parts with disastrous consequences.

On the whole, though, Barings is an exception. Many firms successfully coordinate immensely complex global operations, some with hundreds of thousands of employees and turnovers the size of the economies of small countries (e.g. Wal-Mart in Chapter 4 and Dell in Box 5.3). How do they do it? How does the global firm keep it all together? In Chapter 7, we have already seen the continued relevance of nation-states in today's global economy. This chapter aims to further strengthen this argument for understanding the dynamic interplay between nation-states and TNCs by unpacking the nature and organization of the latter's global activities. A TNC can be defined as 'a firm that has the power to coordinate and control operations in more than one country, even if it does not own them' (Dicken, 2003: 17). In particular, we adopt an *economic-geographical perspective* that explicitly takes into account the role of *place*, *space*, and *scale* in shaping the strategic organization of these TNC activities. By questioning the false claim that global corporations are truly 'global' in their reach, we argue that TNCs are not 'placeless' economic institutions of capitalism. Rather, they are deeply spatial entities in the sense that they are constantly looking for new ways of organizing space in order to stay ahead of their competitors. In doing so, TNCs organize their economic activities across space in a variety of different ways, with important implications for the places that 'plug in' to these corporate networks.

In illustrating these conceptual arguments, this chapter is organized into five main sections. Section 8.2 introduces a commonly held view of the TNC as an all-powerful entity that effortlessly straddles the globe in an ongoing search for profits. We believe that such a view clearly both exaggerates and simplifies the corporate power of TNCs. In Section 8.3, we briefly revisit the ideas of *production chains* and *business networks* for understanding TNC activities. The section serves to reinforce our understanding of the structure and organization of commodity chains introduced in Chapter 4. The next two sections (8.4 and 8.5) seek to explain the *internal* and *external divisions of labour* in which TNCs are involved. In Section 8.4, we introduce the idea that the diverse activities coordinated and controlled by TNCs can be understood as *intra-firm networks*: managing different TNC functions across borders forms the core issue in this section. In Section 8.5 we examine the different modes of *inter-firm relationships* that are increasingly important in sustaining the competitive advantage of TNCs. Here, we argue that space plays an integral role in these inter-firm networks because managing across different territories can often involve conflicting objectives and cultures. In section 8.6, we consider the different obstacles faced by TNCs in globalizing their economic activities, arguing that TNCs are not an invincible force operating beyond, and effectively immune to, competitive pressures and social and political regulation.

8.2 The Myth of Being Everywhere, Effortlessly

On its global website, HSBC claims categorically that it is the 'world's local bank' (http://www.hsbc.com, accessed 15 February 2006). Headquartered in London, HSBC (Hongkong and Shanghai Banking Corporation) is one of the world's top three banking and financial services groups, operating a global business network of over 9,700 offices in 77 countries. Its shares are listed and traded on stock exchanges in London, New York, Hong Kong, Paris, and Bermuda. Some 200,000 individuals or institutions from over 100 countries are its shareholders. Despite its humble origin as a British bank founded in Hong Kong and Shanghai in 1865 to finance the growing trade between China and Europe, the HSBC has seemingly developed into a global corporation that operates in almost every corner of today's global economy.

Just like the Barings Bank story described in the previous section, HSBC's global reach is not just a matter of its geographical locations. More importantly, it gives substance to the so-called 'myth' of the global corporation. In this myth, TNCs are conceived as having the managerial and organizational power to easily control and command their worldwide operations from one central corporate headquarters. This interpretation of TNCs is not inconsequential as it leads to all kinds of misleading conceptions about what TNCs do and how they organize their business operations on a global basis. The myth of the global corporation is underpinned by its apparently enormous financial capital, massive productive capacity, huge levels of employment, and privileged access to technology and knowledge competencies.

In reality, most TNCs are indeed significant corporations that have the necessary capital to fund their international expansion. Their huge financial muscle often allows them to bargain hard with host country governments. Their ability to achieve both economies of scale and scope means that TNCs are able to outcompete most, if not all, local competitors in host markets. In developing countries that are desperately looking for ways to industrialize their economies, the dependence on TNCs for job creation means that TNCs have substantial power to negotiate and bargain with these developing countries. As TNCs gain experience through globalization, they develop relatively more sophisticated marketing and distribution know-how and technological activities in different host countries.

Upon closer inspection, however, the notion of the effortless global reach of the TNC does not withstand scrutiny. Global presence does *not* necessarily mean that organizing on a global basis is easy. On the contrary, as TNCs become more globalized, they face much greater difficulty in holding everything together within a single corporate entity – as the Barings case aptly shows. TNCs not only need to carefully configure their functions and roles in order to adapt to different local economies and places – an idea known as 'placing firms' (Dicken, 2000); their entry into diverse host markets also necessitates different organizational forms and processes explained later in this chapter. To make sense of the diversity of TNCs and their operations, we must understand the following three highly important and yet interrelated issues:

- How do TNCs organize their own diverse activities in different parts of the world?
- How do TNCs engage and interact with other firms when they operate internationally?
- How do TNCs keep everything together in the face of intense market competition, civil society opposition, and political barriers?

8.3 Revisiting Chains and Networks: the Basic Building Blocks of TNCs

In this section we will build on the ideas of commodity/production chains and networks introduced in Chapter 4 as a necessary first step towards an understanding of contemporary TNCs. How, for example, might we make sense of Nokia and the geography of its mobile phone business? Arguably the best-known Finnish TNC today, Nokia first started as a humble forest industry enterprise in southwest Finland in 1865, only seriously moving into mobile communications at the beginning of the 1980s. By 2005, it was undoubtedly the market leader in mobile communications and had a worldwide market share of almost 32 per cent in the mobile phone business. It has spent hundreds of millions of dollars on research and development (R&D) for mobile communications; it has invested considerable money and effort on marketing its products; and it no longer manufactures in its home country but has instead established worldwide production facilities. For every Nokia mobile phone that reaches the final consumer, numerous *actors* are involved: the *designer* responsible for the phone's 'look' and 'feel', the *engineer* for the technological specifications, the *financier* for funding the project, the *worker* for assembling it, the *driver* for handling logistics, the *sales person* for facilitating the act of buying, the *government official* for ensuring safety and product standards, and so on. Following this line of thinking, we can describe the mobile phone commodity chain and unpack the entire production chain starting with Nokia, the lead firm that coordinates the phone's global production chain.

Organizing the production chain of a commodity such as a mobile phone, is, therefore, far from a simple process. Two key issues are at stake here, namely the *configuration* and *coordination* of the production chain. *Configuring* the production chain is a matter of deciding where each value-added activity along the production chain should be located. This configuration is important in the formation of the various spatial structures of commodity chains described in Chapter 4. Nokia, for example, has to decide if it wants all handset manufacturing to take place in one country from which it can ship products to the rest of the world. This would be a highly geographically *concentrated* production chain configuration. On the other hand, it may decide to locate production in its major markets in different regions, thereby following a more *dispersed* production chain configuration. Geographical outcomes can therefore vary.

Coordinating the production chain, however, is mostly an *organizational* matter that shapes the chain's governance (see also Chapter 4). Depending on the nature of the commodity chain (producer- or buyer-driven), a TNC has to tackle the issue of how to coordinate a wide range of activities spanning different countries and locations. Nokia, for example, may choose to centralize most of its corporate decisions at the headquarters located in Helsinki, with its regional headquarters located in various major regions implementing these decisions within the respective markets. Equally, hypothetically at least, it might decide to operate as a much looser confederation of autonomous national subsidiaries, which meet the needs of the various markets in which they are located with only a light guiding hand and financial oversight being provided by central headquarters. Different TNCs may thus coordinate their production chains via different mechanisms that range from being highly *centralized* to highly *decentralized*. Taken together, each TNC will develop its own particular way of configuring and coordinating its production chain(s).

The above discussion has assumed that all value-added activities are conducted by the same TNC and its subsidiaries. What if the TNC is only engaged in *some* parts of the production chain and develops contractual relationships with other firms as part of a wider system? A distinction should therefore be made between two types of transactions along the production chain:

1 *internalized transactions*: these are activities that occur within the legal and organizational boundaries of a particular TNC. We use the term 'hierarchy' to describe these intra-TNC transactions as they are subject to the internal governance system of the TNC. For example, when Motorola's mobile phone division 'buys' microprocessor chips from the semiconductor division, this is an internalized transaction *within* Motorola and it is not necessarily subject to the open market mechanism of price competition;
2 *externalized transactions*: these are relationships that exist between independent firms, some of which may be TNCs. We describe these inter-firm relationships as 'markets' because they are subject to pressures of price competition and other factors beyond the governance system of a particular firm. For example, when Nokia buys lithium batteries for its mobile handsets from its Japanese supplier Sanyo, this transaction is market-based and both Nokia and Sanyo are considered to have engaged in an *inter-firm* transaction.

In short, there are many ways of organizing the production chain of a particular product or service. A large TNC such as Hewlett Packard (HP) has to coordinate the production chains of hundreds of products on a global basis. These chains often overlap in terms of corporate functions (e.g. common technologies between personal computers and electronic medical devices) and geographical locations (e.g. China) so that we can use the term *networks* to better describe these complex sets of internal (intra-firm) and external (inter-firm) relationships. In the next two sections we look at these different kinds of relationships – graphically illustrated in Figure 8.1 – in turn.

8.4 Organizing Transnational Economic Activities 1: Intra-firm Relationships

Building on the notion of the production chain as a basic building block of the global economy, we will now explore how TNCs organize their economic functions and divisions of labour *internally* (drawing extensively on the ideas in Dicken, 2003). We argue that the internal organizational structure of TNCs varies according to different corporate strategies being pursued. To support our argument, we draw on examples from both manufacturing and service sectors, and pay special attention to the role of *space* and *place* in shaping diverse

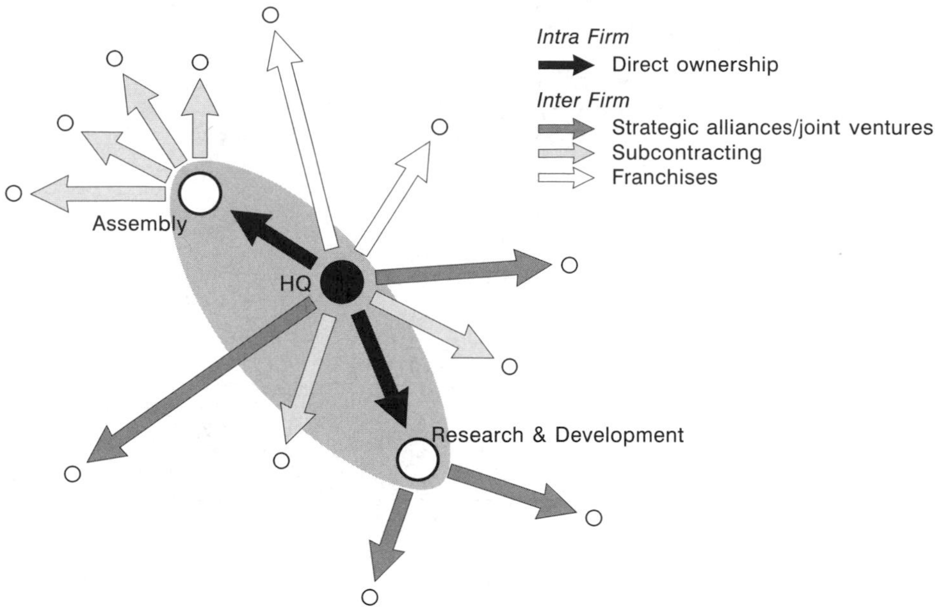

Figure 8.1 Different forms of organizing transnational operations

organizational formations. We leave the role of culture, gender relations and ethnicity in shaping TNC activities to Part IV. In Chapters 11–13, we will analyse how TNCs are not internally coherent and homogeneous precisely because of the diversity of cultures, gender relations and ethnic tensions that constitute their worldwide operations. In this chapter, however, our primary intention is to illustrate the *spatiality* of the organization of TNC activities.

The exact internal divisions of labour within TNC networks are often the combined results of organizational/technological forces, and location-specific factors. Different parts of the TNC may have different location needs, and their geographical outcomes have to be considered on a case-by-case basis. Generally, three key kinds of TNC organizational units can be discerned (see Figure 8.1):

- *corporate and regional headquarters*: these are the 'nerve centres' of TNCs where important strategies are formulated and decisions are made. These decisions apply to a wide range of corporate functions such as corporate finance and investment decisions, market research and development, product choice and specialization, human resource development, and so on.
- *research and development (R&D) facilities*: these activities are found in both manufacturing and service TNCs. R&D facilities encompass activities such as product development, new process technologies, operational research, and so on. They provide important knowledge and expertise to keep the TNC competitive in the global marketplace.

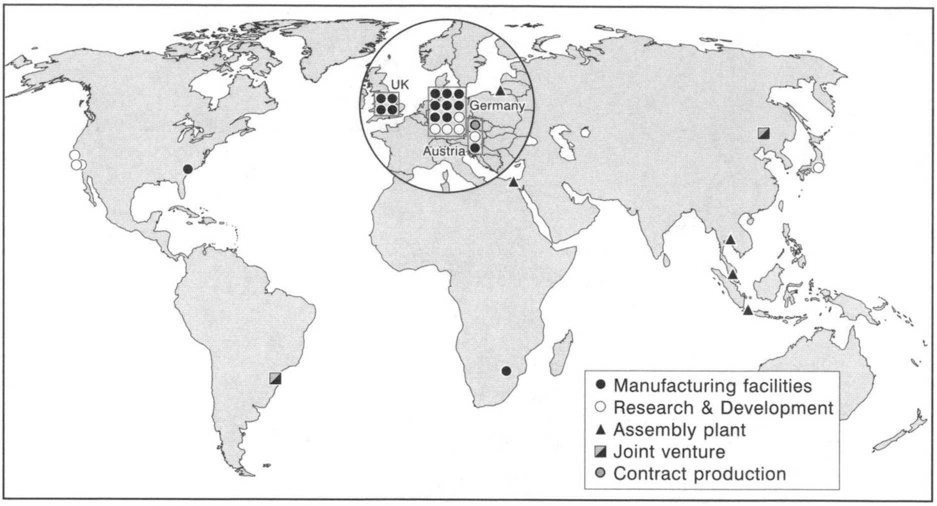

Figure 8.2 BMW's global production networks
Source: Adapted from http://www.bmwgroup.com/e/nav, accessed 22 March 2006.

- *transnational operating units*: these units cover a wide range of activities from manufacturing plants and facilities to sales and marketing offices, fulfilment centres and after-sale service centres.

We can use the example of BMW to illustrate the complex *spatial organization* of intra-TNC networks. First, *headquarters* plays a highly important role as the apex of management, the financial controller, and the processor and transmitter of market and production information within the TNC. Such corporate functions thus require a strategic location in major cities which provide access to high-quality external services (e.g. management consulting or advertising), and the presence of skilled labour and excellent infrastructural and communications support. As shown in Figure 8.2, BMW is headquartered in Munich, the third largest city and a major business and financial centre in Germany. Leading global cities such as this act as important control and command centres in the global economy by virtue of their hosting of the headquarters of major TNCs. Most of the *Fortune Global 500* TNCs are headquartered in a small group of global cities including Chicago, Frankfurt, London, New York, Paris, Shanghai, and Tokyo (see Box 5.5).

Second, the location of *R&D facilities* can be critical to the success of TNCs because of the importance of continuously developing new products and/or services. And yet despite decades of partial internationalization of R&D activity, the most significant R&D activities remain located in TNC home countries. Different types of R&D activities have different locational requirements, however. In general, there are three main types of R&D activities (after Dicken, 2003):

- *internationally integrated R&D lab*: this centre normally supplies the core technologies and knowledge for the entire TNC's operations worldwide. It is thus the 'brain' behind a TNC's innovative products and/or services that contributes to its competitive advantage.
- *locally integrated R&D lab*: these laboratories apply the fundamental technologies and knowledge developed in the internationally integrated R&D lab in order to develop new innovative new products for the host market in which the lab is located. Such labs are thus oriented towards local market and regulatory requirements that are not necessarily found in other markets.
- *support lab*: this is the lowest level of R&D facility and is primarily concerned with adapting parent company technologies for the local market and providing technical back-up.

In the case of BMW (see Figure 8.2), the most important R&D centres for the Group's innovation, technology, and car IT research continue to be located in, or near to, its Munich headquarters. It has also established other R&D labs in California, to engage in design work and emission tests, for example. A Tokyo lab supports the Group's R&D efforts by tapping into the technological and innovative environment of Japanese carmakers. A newly opened design centre in Singapore reflects the Group's anticipation of Asia's emerging influence on design. What is clear in the case of BMW is that there are different R&D patterns for different market orientations. The firm's Munich-based R&D activities cater to the global market whereas its California, Singapore, and Tokyo centres have more specific local market and regulatory orientations.

Third, the most spatially mobile part of a TNC's global production chain is often its *production operations* (with respect to either tangible products or services). Drawing upon Dicken (2003), we highlight in Figure 8.3 four main ways of organizing transnational production units among TNCs. Each cell in the figure represents a geographical unit, e.g. a country:

- *globally concentrated production*: this mode of transnational production applies to some resource and manufacturing industries. It does not work well for service industries, as services are much harder to export. In this mode, a TNC can exercise tight control over its subsidiaries and tends to follow a spatial pattern of geographical concentration. Most TNCs start with this strategy in their early years when they are likely to locate all production in their home countries and export to the rest of the world.
- *host-market production structure*: this mode of organization is preferred where there are considerable barriers to trade in host countries meaning that exports may not be the most efficient channel to reach host markets. In service industries, the demand for a local presence also explains why service TNCs, ranging from professional services (e.g. law and accounting) to consumer services (e.g. retailing and hotels), often set up operations in each host

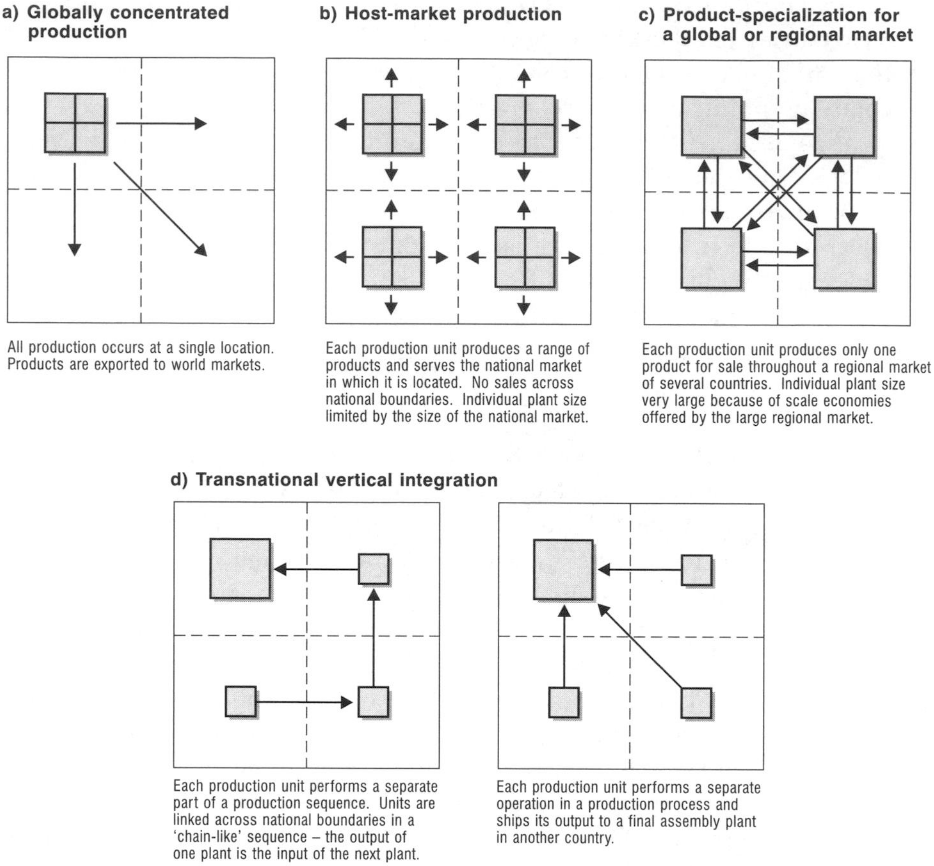

a) Globally concentrated production
All production occurs at a single location. Products are exported to world markets.
b) Host-market production
Each production unit produces a range of products and serves the national market in which it is located. No sales across national boundaries. Individual plant size limited by the size of the national market.
c) Product-specialization for a global or regional market
Each production unit produces only one product for sale throughout a regional market of several countries. Individual plant size very large because of scale economies offered by the large regional market.
d) Transnational vertical integration
Each production unit performs a separate part of a production sequence. Units are linked across national boundaries in a 'chain-like' sequence – the output of one plant is the input of the next plant.
Each production unit performs a separate operation in a production process and ships its output to a final assembly plant in another country.

Figure 8.3 Geographies of transnational production units
Source: Dicken (2003), Figure 8.4. Reprinted by permission of Sage Publications.

market. In this case, centralized corporate control of local subsidiaries and operations is likely to be rather difficult as considerable autonomy needs to be granted to each national unit (see also the case of Barings Bank above). This is because each operation in the host country is predominantly 'local' in its business orientation and very sensitive to local demand. No sales occur across national boundaries as products and services are highly customized to the tastes and preferences of local markets.

- *product-specialization for a global or regional market*: this mode of transnational production tends to be applicable mainly to manufacturing industries, for example, the electronics, automobile, and petrochemical industries. As product mandates are given to macro-regional production units, a great deal of autonomy rests with these operations supplying global or regional markets. Hewlett Packard's regional headquarters in Singapore, for example, coordinates the global mandate for a HP printer. Its high-end computer servers, however, remain coordinated by its global headquarters in Palo Alto, California.

- *transnational vertical integration*: this mode is the most developed and coordinated organizational structure, although it is very demanding in terms of management expertise and control in order to ensure that production responsibilities are well specified and coordinated across the TNC's global operations.

One of the main advantages of this transnational production mode is to exploit spatial variation in production costs. A TNC can establish offshore production for some of the more labour-intensive manufacturing operations. For example, many American TNCs in the electrical, electronics, textiles, and garment industries have set up assembly operations in the *maquiladoras* in Mexico (see Box 8.1) to take advantage of the latter's lower production costs and tax-exemptions under the North American Free Trade Agreement (NAFTA). As assembly work represents the most labour-intensive and the least technologically sophisticated part of the production chain, it can be easily transferred to low-cost locations. However, assembly operations in these overseas subsidiaries often have little autonomy to develop new products or market strategies. Instead, they tend to have to strictly follow the production plan set by the corporate headquarters.

In reality, we should note that these four organizational structures are ideal-typical modes of transnational production. A TNC may use one or more ways to configure its worldwide operations. BMW, for example, has moved a long way from vertically integrated production based in Munich during its early days. Figure 8.4 illustrates the complexity and large number of components of its modern 7 Series sedan that necessitates the globalization of its production networks. BMW has therefore developed sophisticated production networks spanning different regions in the European Union and Southeast Asia. While its plants in Bavaria, Germany, continue to manufacture the full range of BMW cars for the entire European Union market (Figure 8.3c), BMW has a full production plant in the UK to serve the right-hand-drive British market (Figure 8.3b). Its parts and components manufacturing takes place in its supplier and innovation park in Wackersdorf in Eastern Bavaria. The establishment of an international logistics centre in Wackersdorf is critical because it distributes BMW's international parts and components to its foreign plants in the US, South Africa, Russia and East and Southeast Asia (see Figure 8.2). Managed by a third party logistics provider, the Wackersdorf centre sends out daily shipments of 2.5 million parts.

More specifically, BMW's assembly operations in East and Southeast Asia resemble the transnational vertical integration mode in Figure 8.3d. In this mode of transnational production, its core parts and components are produced in

Box 8.1 Transnational production in the maquiladoras *of Northern Mexico*

The term *maquiladora* refers to an assembly plant in northern Mexico that manufactures export goods for the US market. It was instituted by the Mexican government in 1965 when it started the Border Industrialization Program to try and tackle the severe economic and social problems of the towns along the Mexico-US border. Under the auspices of the North American Free Trade Agreement (NAFTA) described in Box 7.4, which came into effect in 1994, US firms manufacturing in the *maquiladoras* across the Mexican border can benefit from much cheaper labour costs in Mexico, virtually nonexistent taxes and custom fees, and less stringent environmental regulations. The *maquiladoras* have thus become offshore factory spaces for American firms and TNCs from other countries (e.g. Japan and Western Europe) selling into the US market. *Maquiladoras* are prevalent in Mexican cities such as Tijuana, Ciudad Juarez, and Matamoros that lie directly across the border from the US cities of San Diego (California), El Paso (Texas), and Brownsville (Texas). Factories in the *maquiladoras* primarily produce electronic equipment, clothing, plastics, furniture, appliances, and automobile parts. By 2005, at least one million Mexicans worked in over 3,000 *maquiladoras*. Dicken (2003) estimates that some 60 percent of the workforce in the *maquiladoras* is employed in electrical assembly and a further 30 percent in textiles and clothing. Some 90 percent of the goods produced in the *maquiladoras* are shipped to the US.

Many *maquiladoras* might be considered 'sweatshops', employing young women for as little as 50 cents an hour, for up to ten hours a day, six days a week. The cost of living in border towns is often 30 percent higher than in southern Mexico and many of the *maquiladora* women are forced to live in shantytowns, that lack electricity and water, surrounding the factory cities (see also Box 12.2 on 'devaluing the third world woman'). While some of the companies that own the *maquiladoras* have sought to increase their workers' living and working conditions, much of the workforce is non-unionized. Some *maquiladoras* are responsible for significant industrial pollution and environmental damage to the northern Mexico region. Competition from China has weakened the attraction of *maquiladoras* for inward investors in recent years and reports suggest that more than 500 plants have closed since the beginning of 2000, causing a loss of several hundred thousand jobs. The future of the *maquiladoras* remains unclear, as they are heavily dependent on the US market and NAFTA.

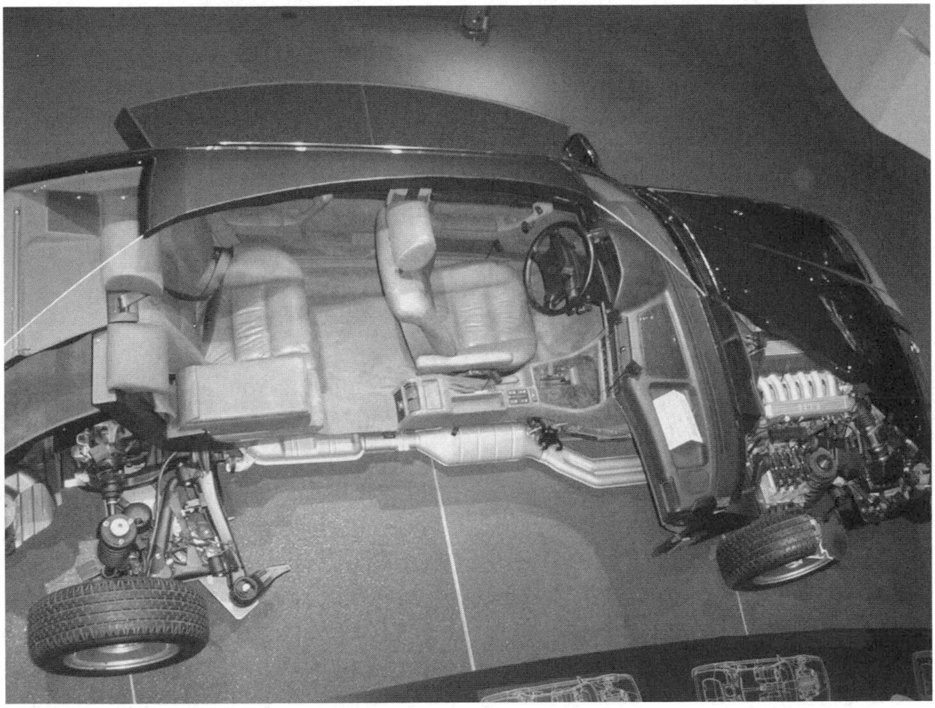

Figure 8.4 Making a BMW: Munich and beyond
Source: The authors.

Germany and shipped to its plants in Asia for assembling with some local parts and components provided by either local firms, or foreign suppliers that have 'followed' BMW's move to these overseas assembly locations. The local content of a BMW car assembled in Thailand, for example, was about 40 per cent in value-added terms in 2003. However, this included labour costs and sourcing within Southeast Asia, not just Thailand. To sum up, the example of BMW represents well the global production strategies, locational impacts, and organizational characteristics of all major automobile manufacturers.

Even though the automobile industry depends on a large number of independent parts and component suppliers that may be TNCs in their own right, the major car makers continue to organize their transnational production networks with a great deal of *vertical integration* – a pattern of industrial organization in which the parent TNC directly owns and controls most of its operations along the production chain. Compared to other modern industries such as electronics and consumer products (e.g. garments and shoes), tightly coordinated and controlled intra-firm networks are extensive within the automobile industry. This occurs despite the extensive use of independent suppliers by major carmakers through inter-firm relationships, the theme of the next section.

8.5 Organizing Transnational Economic Activities 2: Inter-firm Relationships

As products and services become increasingly sophisticated, few TNCs today can produce everything by themselves. The BMW sedan in Figure 8.4 is made up of at least 20,000 to 30,000 parts, an Apple personal computer can be broken down to a few thousand parts, a Nokia mobile phone has hundreds of components, and even an ordinary Nike sports shoe has at least 10–20 parts. In the service sector, an advertising project requires creative talents, financial consultants, technologists, public relations experts, and clerical workers among many others. Running a hotel business may bring together property owners, international hotel management companies (e.g. Marriott), independent food and beverage operators, service suppliers, and so on.

In short, most TNCs are simultaneously controlling their own *intra-firm networks* of foreign subsidiaries and affiliates at the same time as managing a dense web of *inter-firm networks* of independent contractors, suppliers, business partners and strategic allies. When these intra-firm networks overlap with inter-firm networks in different *geographical contexts*, we can imagine each TNC as being part of a complex set of organizational patterns and processes (see Figure 8.1). In this section, we focus on three main modes of inter-firm relationships that lie somewhere between market (buying goods and services from independent firms) and hierarchy (intra-firm relations): (1) subcontracting and outsourcing; (2) strategic alliances and joint ventures; and (3) franchising and cooperative agreements. Our decision to focus on these three forms reflects both their significance in organizing TNC activities and in contributing to international trade and investment patterns. First, the rapid rise of subcontracting and outsourcing during the 1990s and the 2000s has helped many TNCs to ease competitive pressures and to focus on their core businesses. Second, the role of strategic alliances and joint ventures is critical in industries where investment costs are high and business horizons are uncertain. Third, franchising and cooperative agreements (e.g. licensing) have emerged to be very successful modes of organizing international expansion in the service industries e.g. retailing, hotels, and fast-food chains.

In particular, we are interested in the various strategies and governance mechanisms through which TNCs engage in these inter-firm relationships. Viewed from an economic-geographical perspective, we pay special attention to the role of place and space in shaping the formation and dynamics of inter-firm relationships. In the context of TNCs, the discussion focuses primarily on the *international dimensions* of these relationships. As in the preceding section, we draw upon various empirical examples to illustrate our discussion.

International subcontracting

Subcontracting (also known as outsourcing) involves engaging independent firms to produce goods specifically for the principal firm. There are two types of international subcontracting: commercial and industrial. *Commercial subcontracting* occurs when the principal firm (the buyer) subcontracts most, if not the entire, production chain to another firm (the supplier) in another country (see Figure 8.1). In the beginning, the buyer may supply product specifications in what is called original equipment manufacturing (OEM) so that the finished products are exactly the same as if the buyer had produced them. Over time, the supplier may learn and develop new technology and expertise so that it can move up the production chain by engaging in some original design manufacturing (ODM). In this latter mode of subcontracting, the supplier will discuss product specifications with the buyer and design and make the product in question to meet those technological and marketing requirements. For buyers that subcontract their entire production, they will specialize in brand management and the marketing of products bearing its brand name.

This mode of international subcontracting is most common in today's electronics industry, particularly the manufacturing of personal computers, consumer electronics, and household appliances. Most standard personal computers, televisions, refrigerators, digital devices and the like are no longer manufactured by the TNCs whose brand names appear on the product. Instead, a large group of independent contract manufacturers or electronics manufacturing service (EMS) providers are designing, producing, and, sometimes, distributing *for* these brand name TNCs. These independent subcontractors are mostly based in newly industrialized economies and emerging developing countries, whereas the brand name TNCs are still mostly from advanced industrialized economies.

Take the example of notebook or 'laptop' personal computers. Table 8.1 reveals that in 2003, the top ten global brand name TNCs were subcontracting the production of large proportions of their notebook products through original design manufacturing (ODM) agreements with leading Taiwanese contract manufacturers such as Quanta, Compal, and Wistron. In 2004, these relatively unknown Taiwanese contract manufacturers accounted for some 68 per cent of the world's total notebook shipments of 49 million units. This figure included notebooks manufactured in China by Taiwanese-owned production facilities. As Taiwan is rapidly emerging as the world's leading centre of electronics manufacturing, the same phenomenon of subcontracting to Taiwan is occurring with electronics products such as computer monitors, flat panel and plasma televisions, mobile telephone handsets and personal digital assistants (PDAs).

To illustrate how this international subcontracting works, we can look at the example of a customized Apple computer notebook. A customer in the US may order an Apple MacBook or MacBook Pro via Apple's website or through retail channels (e.g. local distributors). This order will be received by Apple Computer

Table 8.1 Subcontracting of the world's top ten notebook brand-name companies to Taiwan, 2003

Company	Proportion of outsourcing (%)	Taiwanese OEM/ODM manufacturers
APPLE	100	Quanta, Elitegroup
DELL	90	Quanta, Compal, Wistron
HP	90	Inventec, Arima, Quanta
IBM	90	Wistron, Quanta
NEC	80	Arima, FIC, Wistron, Mitac
SHARP	50	Quanta, Mitac, Twinhead
SONY	20	Quanta, ASUS
Fujitsu-Siemens	15	Quanta, Compal, Wistron
Toshiba	15	Compal, Inventec

Source: Yang and Tsia (2007), Table 1.

Inc., the brand name buyer, which in turn will subcontract the notebook production to a system manufacturer or EMS provider in Taiwan (e.g. Quanta). Upon receiving its daily order from Apple and other brand name customers, Taiwan's Quanta will regularly place its order of parts and components with its key suppliers, e.g. IBM or Intel (chipsets), Seagate (hard disks), Samsung (memory chips), Delta (power supplies), Chunghwa Picture Tubes (computer screens). As Quanta has already moved its entire notebook manufacturing facility to Shanghai, China, it will ask its key suppliers to ship parts and components directly to its warehouse and assembly plants in Shanghai. Once the notebook is made in China, it will be sent via express service to the customer in the US. Typically, this process – from the initial order by the consumer to receiving the finished product – takes no more than seven days within the Asia-Pacific region and 14 days worldwide. This manufacturing system is a form of the integrated production shown in Figure 8.3d: in short, the Apple MacBook thus embodies a complex geography of international production and trade flows spanning many different locations in the global economy.

Unlike commercial subcontracting, *industrial subcontracting* takes place when the suppliers carry out only OEM production on behalf of its key customers, i.e. manufacturing only. Industrial subcontractors do not engage in ODM and final distribution on behalf of their customers. In the sports shoes industry, Nike's international subcontracting practices provide an excellent example (Donaghu and Barff, 1990). Headquartered in Beaverton, Oregon, Nike specializes in the design, R&D, and marketing of its sports products. Its production in Asia is handled entirely by 'developed' partners in Hong Kong, Taiwan, and South Korea. These partners have established production facilities throughout Asia, e.g. Bangladesh, China, Indonesia, Malaysia, Sri Lanka, Thailand and Vietnam. A Japanese trading company or *sogo shosha* (see Box 4.3) handles the financial and logistical aspects of Nike's production in Asia. Nike's subcontractors in South America (e.g. Brazil, El Salvador, Ecuador, and Mexico) and Eastern Europe (e.g. Bulgaria, Turkey, Tunisia) are managed respectively by its international buying offices in the US and Europe. As of April 2005, Nike had contracts with over 700 such factories throughout the world. Its 124 active contract factories in China alone employed over 200,000 workers and its 34 active contract factories in Vietnam employed another 84,000 workers.

The *geographical implications* of international subcontracting are highly variegated. First, subcontracting may lead to the development of *highly localized clusters* in developed countries (see Chapter 5). In the automobile industry, for example, tightly-knit inter-firm relationships between Japanese automobile manufacturers and their key first-tier suppliers led to a wave of Japanese suppliers following their key customers to open plants in North America in the mid-1980s and Europe in the early 1990s. These transplants often clustered in specific locations to form flexible production clusters. In the US, Honda, Nissan, and Toyota had developed closely integrated supplier networks at their production bases in Ohio, Tennessee, and Kentucky by the early 1990s (Figure 8.5).

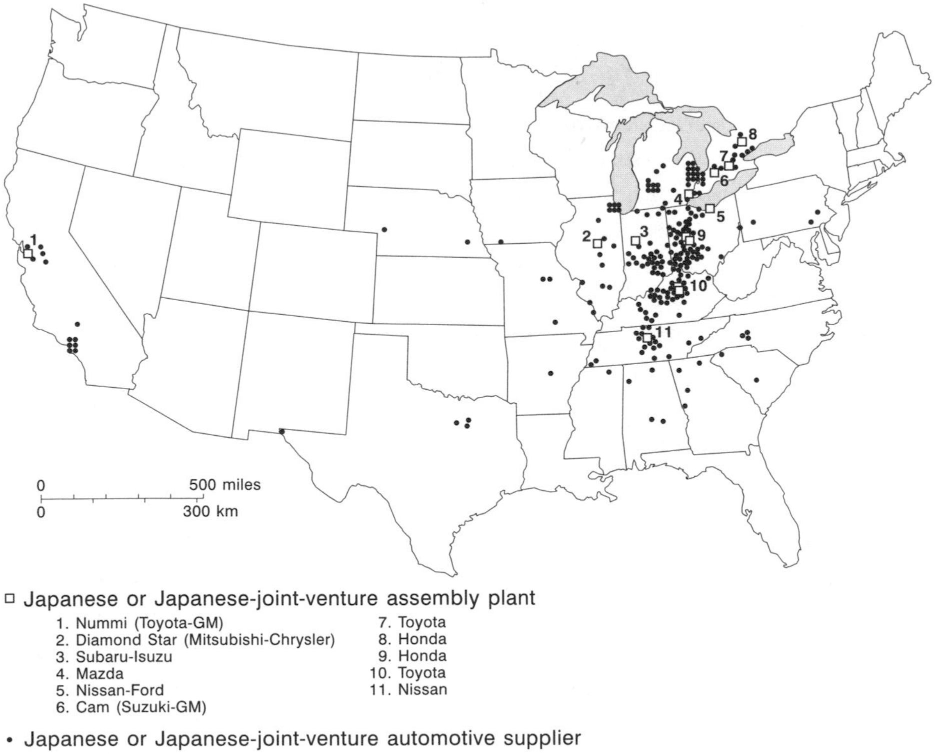

Figure 8.5 Japanese transplant networks in the US in the early 1990s
Source: Redrawn from Mair (1994), Figure 5.3. With permission of Palgrave Macmillan Publications.

Second, international subcontracting has led to the development of enclaves of export-oriented production space in developing countries – a spatial phenomenon described in Box 8.2 as the new international division of labour. These *export enclaves* or satellite production clusters (see Chapter 5) host labour-intensive manufacturing activity that forms an integral part of the global supply chain of most TNCs. Collectively, these enclaves constitute the manufacturing base of the global economy. Prime examples are the *maquiladoras* in Mexico (electrical assembly and clothing), Dongguan in China (one of the world's largest shoe production centres), Penang in Malaysia (electronics), Rayong in Thailand (automobiles) (Figure 8.6), and Slovakia's Presov (clothing). In the service industries, extensive outsourcing in the software industry has led to the rise of Bangalore in India and Dublin, Ireland, as centres of software production and data processing. The outsourcing of back-office operations has also contributed to the global coordination of call centre activities (see also Chapter 5.3).

Box 8.2 Transnational corporations and the new international division of labour

During the nineteenth century, the world economy was dominated by the industrialized economies of Western Europe and the US who specialized in the manufacturing of goods on the basis of raw materials extracted from developing countries. Supported by imperialism and colonial relations, this 'traditional' form of international division of labour prevailed until at least the 1960s. From the 1960s onwards, a 'new' international division of labour (NIDL) supposedly started to emerge in which certain developing countries took on new roles within the global economy through the investment strategies of TNCs. The NIDL described the establishment by European, North American and Japanese TNCs of a global manufacturing system based on establishing labour-intensive export platforms in so-called 'newly industrializing economies' (NIEs). The NIDL theory, proposed most explicitly by Fröbel *et al.* (1980) in their study of garment manufacturing, argued that falling rates of profitability in manufacturing production in core countries were combining with market saturation and under-consumption as market growth lagged behind productivity increases. In response, TNCs used their global reach to relocate production from the industrial core to low-cost production sites in the NIEs; the manufactured goods were then in turn exported back to core markets from the offshore branch plants. Crucially, the system depended on new technologies that allowed production fragmentation, thereby creating tasks that could use young and female, semi-skilled or unskilled workers in the periphery. While the more capital-intensive part of the production processes continued to be located in advanced industrialized economies, NIEs across South America, Eastern Europe, Africa, and Asia became bases for labour-intensive phases of production. The appearance of designated export processing zones (EPZs), often offering a range of tax incentives, from the 1960s onwards were the most obvious physical manifestation of the emerging NIDL.

The NIDL thesis, however, has been extensively critiqued. In particular, it overlooks the following important concurrent trends in the global economy from the 1960s onwards:

- the continued importance of the 'traditional' form of international division of labour through significantly more investment in raw material extraction and agribusiness;
- foreign direct investment flows remain dominated by investments between developed countries;
- a rapid growth of outward investment and TNCs from the newly industrialized economies has occurred, complicating the simple NIDL model;
- much of the world's foreign direct investment now occurs in the service sector, primarily aiming to serve host markets rather than being for re-export;
- the significance of cheap labour in attracting manufacturing functions is clearly overstated. Many other institutional factors such as conducive government policies and appropriate business climates are equally, if not more important.

The reality of the global economy today thus reflects overlapping divisions of labour, some of which are newer than others. The geographies of these divisions of labour are necessarily more complicated and messy than the simply core–periphery model underlying the NIDL thesis.

Third, extensive *macro-regional integration* can take place if the production networks of TNCs span different countries within the same region and yet strong inter-firm relationships serve as the 'thread' to stitch together different places within the region. Box 8.3 explains how this geographical process of macro-regional economic integration works in relation to the presence of TNC production networks in Southeast Asia.

Figure 8.6 The automotive cluster, Rayong Province, Southern Thailand
Source: Martin Hess, with permission.

Box 8.3 TNC production networks and macro-regional integration in North America and Southeast Asia

Most commentators on the global economy agree that the globalization of TNC activities is producing a *disintegrated* global production system and an *integrated* global trading system. However, on a closer look, it becomes apparent that many TNCs are increasingly organizing their production activities within specific macro-regions of the global economy. This growing *macro-regionalization* of the global economy can be explained by three interrelated factors. First, the scope and complexity of many TNCs' operations mean that they seek to coordinate and integrate their activities at a macro-regional, rather than truly global, scale. Second, economic resources such as labour, capital, and technology are increasingly mobile within specific regional contexts, for example, the European Union. Third, macro-regional free trade arrangements have been growing in number and significance at the same time as globalization processes have been accelerating (see also Box 7.4).

We can illustrate these dynamics by looking at the example of macro-regional integration in Southeast Asia, an area covered by the Association of Southeast Asian Nations (ASEAN) grouping. Here, integration processes are primarily being driven by TNCs from Japan and, more recently, South Korea, and have been given further impetus by the implementation of the ASEAN Free Trade Agreement (AFTA) since the mid-1990s. Evidence of macro-regionalization can be seen in two sectors in particular (UNCTAD, 2004):

- *the automobile and automotive component industries*: Japanese and other automakers are consolidating their production in Southeast Asia and are adopting macro-regional production network strategies and plant specialization to serve the AFTA market. Toyota and Nissan, for example, have networks of operations linking different functions (e.g. regional HQ, finance, training, parts suppliers and assembly) in different ASEAN countries. Ford also adopts a regional strategy of specialization to serve the ASEAN market: its plant in Thailand makes pick-up trucks, and its plant in the Philippines produces cars.
- *the electronics industry*: similar dynamics can be observed in this sector. Samsung (South Korea), for example, has developed extensive macro-regional production networks in ASEAN. Samsung Corning (Malaysia) provides tube glass as a major input to Samsung Display's factory for colour picture tubes in Malaysia. It also sells intermediate products and components to Samsung Electronics (Thailand) and affiliates in Indonesia (colour TVs), Malaysia (computer monitors), and Vietnam (colour TVs). Sony from Japan has also developed similar regional production networks for producing electronics goods.

To sum up, these production networks serve to strengthen macro-regional integration through production and supply linkages and intra-firm sourcing of key parts and components from affiliates located in the same macro-region.

Strategic alliances and joint ventures

Apart from subcontracting, inter-firm networks can also be formed when TNCs engage in strategic alliances and joint ventures with other firms (see Figure 8.1). These two organizational forms are different from mergers and acquisitions (M&As). Firms participating in strategic alliances or joint ventures do not experience *ownership change* and thus remain independent of the alliance or joint venture partner. In contrast, firms in M&As necessarily undergo ownership change. When Daimler-Benz AG from Germany merged with Chrysler Corporation from the US to form DaimlerChrysler in 1998, both original TNCs ceased to exist. Instead, a new transnational company known as DaimlerChrysler emerged. Similarly 'merged' TNCs are Royal Dutch Shell (oil), Fuji-Xerox (electronics), GlaxoSmithKline (pharmaceuticals) and AOL-Time Warner (media and entertainment).

In some fiercely competitive industries, firms tend to engage in a very specific type of inter-firm collaboration – *strategic alliances*. Only some business activities are involved in these function-specific strategic alliances and no new equity capital will be involved. This cooperative approach to competition can be thought of as 'coopetition', i.e. a mix of cooperation and competition. In the semiconductor and pharmaceutical industries, for example, strategic alliances are common because of intense competition, the high costs of R&D and new product development, and the rapid rate of technological change. These pressures raise the investment stakes in these two industries beyond the financial means of any individual TNC. Cooperation through strategic alliances becomes the most effective means of competing on a global basis. Many of the TNCs in these strategic alliances are mutual competitors in some product segments and allies in other.

In service industries, function-specific strategic alliances are also common. Most major airlines, for example, tend to participate in one of the following strategic alliances: Star Alliance, One World Alliance, and SkyTeam Airline Alliance. Through these alliances, participating airlines can, for example, enjoy code-sharing in their computer reservation systems that allows for cross-loading of passengers and thus reduces the excess capacity of any individual airline.

When two or more firms decide to establish a separate corporate entity for a specific purpose, *alliance joint ventures* are often the organizational outcome. Here, partners need to invest new equity capital in the joint venture. Sometimes, a partner may use other assets (e.g. land or goodwill) to substitute for capital investment and this is known as a *cooperative joint venture*. As joint ventures are formed for partners to share financial risks, to benefit from inter-firm synergy, and to develop new products or markets, they are a popular form of inter-firm relationship that can be found in most industries. In many developing countries, government regulations requiring a minimum shareholding to be held by host country citizens or firms further promote the use of equity or cooperative joint ventures as the preferred entry mode for foreign TNCs. Clearly many automobile TNCs engage in joint ventures with each other. In South America, Africa, Eastern Europe and Asia, equity joint ventures are very common among foreign automobile manufacturers seeking to develop new markets.

Just as in international subcontracting, the *geographical implications* of international strategic alliances and joint ventures are manifold. We are witnessing the rise of project-based teamworking in many industries in which strategic alliances occur (see Box 5.6). As face-to-face interaction within these project teams tends to take place between cities and science hubs located in different countries, certain kinds of places become more strongly interconnected to the exclusion of others. For example, in the semiconductor industry there are intense transnational flows and interactions between Taiwan's Hsinchu and Silicon Valley (see Section 9.4). As most of the activities in these alliances and ventures are developmental and high value-added, places hosting these activities become more prosperous and competitive (e.g. R&D clusters in Europe and North America). Spatial uneven development is thus exacerbated.

Franchising and cooperative agreements

Franchising refers to an organizational form in which the TNC owner of a registered trademark or intellectual property rights agrees to let a *franchisee* (often outside the home country) use that trademark or rights provided that the franchisee follows the guidelines and requirements laid down by the TNC. There is thus an inter-firm relationship established between the franchisor (a TNC) and the franchisee (a local firm). Familiar examples are McDonald's, KFC and Burger King fast food restaurants (Figure 8.7), Starbucks coffee outlets (see Box 4.1), and 7-Eleven convenience stores. We are all aware of the global presence of these registered trademarks because of the extent to which the TNCs have used franchising as a preferred method of internationalization. Their outlets in many countries are *not* necessarily owned by these TNCs, but rather are often operated by local franchisees.

This form of inter-firm network is particularly popular in the service sector for two key reasons. First, most service TNCs may not have sufficient capital or may not want to incur the substantial costs of expanding into many markets at the same time. This often occurs in the retail and fast-food industries. Second, some service TNCs may not want to be exposed to risks arising from unfamiliarity with local cultures, social relations and practices of local customers. Franchising provides a convenient and low-cost alternative for service TNCs to have a local presence through their franchisees that are likely to be local entrepreneurs familiar with the host country markets.

Figure 8.7 Fast food chains in the Caribbean
Source: The authors.

On the other hand, *cooperative agreements* encompass a wide range of inter-firm relationships that range from licensing agreements to non-equity forms of cooperation. These agreements can be found in both manufacturing and service industries. In the manufacturing sector, a TNC may decide to license out its patented technology in return for royalty payments. For example, through owning the DVD format patents, Philips (the Netherlands), Sony (Japan), Pioneer

(Japan), and LG (South Korea) will receive royalties for every DVD player manufactured by their licensees (http://www.licensing.philips.com, accessed 15 February 2006). In service industries, for example the hotel sector, two firms may come together to form a cooperative agreement in training, combining their training teams to offer human resource development programmes to their combined employees. Both hotels can thus reap the benefits of combining human resource practices.

Through both franchising and cooperative agreements, TNCs can rapidly internationalize their market presence and promote the consumption of their products and/or services in these host markets (see Chapter 10 for more on consumption). In the retail and restaurant business, the global presence of certain restaurant outlets (e.g. McDonald's and KFC) and consumer products (e.g. Coca-Cola) has often been taken as direct evidence for the globalization of economic activity, alternatively known as 'McDonaldization' or 'Coca-Colonization'. The pervasive presence of iconic franchised activities throughout the world leads us to an obvious question: is there any barrier to the global reach of TNCs? This brings us to the penultimate section on the limits to global reach.

8.6 The Limits to Global Reach?

Our discussion of the production networks has focused on the corporate logics of TNCs and the different organizational structures and strategies that they may pursue as they expand internationally. Table 8.2 summarizes the costs and benefits associated with the different organizational forms we have discussed. We want to argue, however, that contrary to the common perception of TNCs as all-powerful economic institutions, there are significant limits to their global reach in today's highly competitive world economy. These limits go well beyond the internal organizational limits illustrated in the case of Barings Bank. In particular, we discuss three such limits to the global dynamics of TNC activity: the ongoing time–space compression of competition, the rise of anti-globalization movements, and social and cultural barriers to expansion.

First, increased levels of global competition have had the effect of 'shrinking' both time and space: the pace of new product and service development has increased dramatically and 'distant' competitors are no longer that far away (Harvey, 1989). This relative *time–space compression* has impacted on the spatial extent of TNC activities. As we saw in Chapter 5, this may necessitate spatial clustering, thus placing limits on the potential spread of TNC production activities. In the petrochemical industry (Figure 8.8), for example, clustering of upstream refineries and downstream specialty chemical facilities serves to reduce possibility of supply disruption and enhances production efficiency – all competitive strategies to decrease the time-to-market of final products. Moreover,

Table 8.2 Different forms of organizing transnational operations – costs and benefits

Organizational form	Costs to the TNC	Benefits to the TNC	Geographical impacts
1. Fully owned subsidiaries	• Heavy capital investment • Potential conflicts with host governments and local firms • More cross-border managerial and organizational issues	• Full managerial control • Protection of trade secrets and proprietary technology • Consistency in production and market services	• Potential benefits to local economies through spin-offs and technology transfer • Spatial transfer of organizational cultures
2. International subcontracting	• Disruption in supply chains • Leakage of proprietary technology and intellectual property rights (IPRs) • Lack of direct managerial control of production	• Cost competitiveness • Flexibility in inventory management • Reduction in investment risks	• Development of export-processing zones in developing countries • The rise of the new international division of labour • Potential upgrading in some host regions
3. Strategic alliances and joint ventures	• Lack of control of technology and IPRs • Problems in managing partners	• Sharing of risks • Access to technology and knowledge of partners	• Deepening of linkages in local economies • Development of extra-local network ties
4. Franchising and cooperative agreements	• Potential infringement of IPRs • Costs of monitoring franchisees and managing ties	• Low or no capital investment required • Rapid expansion of market penetration • Financial returns to trademarks, brands and IPRs	• Rapid spatial development of franchising outlets • Diffusion of cultural norms and practices to franchisees • Homogenization of consumption

Figure 8.8 A petrochemical cluster in Jurong Island, Singapore
Source: The authors.

many product-based TNCs have adopted just-in-time (JIT) production methods in order to enhance efficiency, market-responsiveness and innovative capacity (see also Table 5.2). In the automobile industry, JIT production necessitates proximity to key suppliers for major automobile manufacturers. For example, Denso, a major Japanese automobile supplier, has to duplicate many of its manufacturing functions in individual East and Southeast Asian countries. Its subsidiaries in Australia, Indonesia, Malaysia, the Philippines, Taiwan, and Thailand are all involved in manufacturing air-conditioning systems for both local and foreign automobile manufacturers located in the host economies. This local production of a key component is critical to its main customers. Curiously, then, while time–space compression may help TNCs achieve global reach in sales of their products and services, at the same time it can create an imperative for selectively organizing activities at the national and macro-regional scales in addition to global structures. Relatedly, as we shall see in the next chapter, the inherent complexity of contemporary TNC production systems and the exposure of the 'weakest link' through JIT practices make them increasingly vulnerable to worker/union actions.

Second, the global reach of TNCs has increasingly faced serious and organized resistance in recent years. Since the Seattle ministerial meeting of the World Trade Organization (WTO) in 1999 the *anti-globalization movement* has gathered momentum with subsequent organized demonstrations against globalization, global capitalism and global corporations in Prague (2000), Quebec City (2001) and Genova (2001) among many others. The rise of the World Social Forum since 2001 as a critical response to the World Economic Forum, a pro-TNC organization of business and political leaders, has further politicized the global reach of TNCs. These organized movements have exerted counter-pressures on the unlimited expansion of TNC activity to an extent never seen before. TNC behaviour has thus become much more cautious and careful in order not to be targeted by these social movements. One of the best examples is Nike. As a consequence of intense criticisms from social and labour movements on the working conditions in its subcontractors' factories, Nike has attempted to develop extensive subcontractor monitoring and assessment systems. Through its partnership in the Global Alliance for Workers and Communities, it aims to give greater voice to its contract factory workers through regular field studies and focused interviews. As we saw in Chapter 4, severe doubts remain, however, about the ability of such schemes to improve dramatically the conditions of factory workers.

Third, there remain significant *social and cultural barriers* to the globalization of TNC activity. Clearly, the notion of the 'global' economy masks huge variation in cultural norms and patterns of social behaviour. The continued vitality of localized cultures, consumption practices and different forms of kinship/social relationships, for example, can pose serious problems for the globalizing efforts of TNCs. A vivid example was Nike's failure to appreciate local culture in its December 2004 worldwide advertising campaign. The commercial showed the Cleveland Cavaliers' basketball player LeBron James defeating a Chinese kung fu master, a pair of Chinese dragons and two women in traditional Chinese attire. What was presumably meant to be a sleek and dynamic advertisement showing the meeting of a modern basketball player bedecked in Nike sportwear with ancient elements of Chinese culture was, unsurprisingly, deemed to be insulting and insensitive in China. This lack of cultural sensitivity proved costly in this instance for Nike, as the advertisement generated negative publicity and was banned in China, one of its major markets for sports products. This incident is symptomatic of many other corporate failures in managing joint venture partners, suppliers, marketing, and so on that can arise out of insufficient appreciation of social and cultural differences (see more analysis in Chapter 11). Conversely, TNCs that appreciate and can effectively adapt to these local differences tend to reap the benefits. The success of the UK retailer Tesco in South Korea, for example, can in large part be put down to its effective localization to meet the specific needs of the Korean consumer (see Coe and Lee, 2006).

8.7 Summary

This chapter has systematically unpacked the transnational corporation, traditionally something of a 'black box' in neoclassical economics. We can usefully understand the TNC as a system of both internal (intra-firm) and external (inter-firm) *production networks*. Interpreting TNCs in this way has many advantages. First, it allows us to appreciate the great variety and diversity in the global organization of production systems orchestrated by leading TNCs. Our analysis has demonstrated various ways in which these intra- and inter-firm networks can be organized. Second, it shows how different localities and places are articulated into globally extensive production networks that operate across different national and macro-regional formations. This approach thus allows us to be sensitive to geographical variations in the global economy that are both a cause and an outcome of TNC globalization strategies. It also reveals how value is produced, captured, and retained in, or extracted from, different places within specific global production networks. Third, conceptually, the network approach brings together a great range of actors and stakeholders beyond conventional economic focus on simply firms and workers. We will examine these actors systematically in the next few chapters.

With this great diversity in organizational forms, geographies of transnational production, and active stakeholders, can the TNC really keep everything together? This chapter has argued that effective globalization is not easy for the TNC. To set up extensive global operations requires the TNC to take into account *geographical differences* that exist in physical, political, economic, and social-cultural realms. As TNCs globalize into different regions of the global economy, these geographical differences are further accentuated to produce greater organizational challenges and problems of maintaining legitimacy. This in turn increases the vulnerability of TNCs to a wide range of potential risks such as industrial action, sabotage, consumer boycotts, punitive regulation, financial exposure, logistical bottlenecks, and so on. Many of these risks are increasing in tandem with global reach, and are also beyond the control of any individual TNC.

Moreover, while many TNCs emphasize the importance of globalization, still others appreciate the strategic imperative of macro-regionalization and localization. Organizing production chains at the macro-regional and local scales may just be as important a strategic consideration to many TNCs. This increasing attention to the specificity in different regions and localities is not merely about expanding the market for TNC products and reducing the operating risks mentioned above. Equally importantly, it also relates to *the people* who constitute the very human fabric of any TNC. It is to the topic of labour and its power that the next chapter turns.

Further reading

- Dicken (2003) offers the most authoritative account of TNCs as movers and shapers in the global economy from an economic-geographical perspective.
- For a full account of the global production networks perspective, see Henderson et al. (2002) and Coe et al. (2004).
- Yeung (2000) and Grabher (2006) provide extensive reviews of the geographical literature on networks and organizations.
- For some recent geographical studies of the automobile industry, see Dicken (2003: Chapter 11), Park (2003) and Depner and Bathelt (2005). For a service sector example, see Wrigley et al. (2005) and Coe and Lee (2006) on retail TNCs.
- See Tickell (1996) for a detailed account of the collapse of Barings Bank.

Sample essay questions

- Why do TNCs organize transnational activities differently across different industries and host regions?
- What is the distinctive role of geography in the global configuration of TNC production chains?
- What factors may lie behind a TNC's decision to engage in inter-firm relationships?
- What are the constraints that may limit the global reach of TNCs?

Resources for further learning

- http://www.unctad.org/wir: the World Investment Report website contains the most accurate data and comprehensive analysis on the global reach of transnational corporations.
- http://www.hsbc.com: the HSBC global corporate website gives a good indication of what it means to be 'the world's local bank'.
- http://www.bmwgroup.com: the BMW corporate website contains valuable information about its global production networks.
- http://globalization.about.com: this website provides very useful information on and links to anti-globalization organizations, institutions, and individuals.
- http://www.sed.manchester.ac.uk/geography/research/gpn: the Global Production Networks website at the University of Manchester contains relevant research papers for this chapter's main discussion.

References

Coe, N.M., Hess, M., Yeung, H.W-C., Dicken, P. and Henderson, J. (2004) 'Globalizing' regional development: a global production networks perspective, *Transactions of the Institute of British Geographers*, NS, 29: 468–84.

Coe, N.M. and Lee, Y.S. (2006) The strategic localization of transnational retailers: the case of Samsung-Tesco in South Korea, *Economic Geography*, 82: 61–88.

Depner, H. and Bathelt, H. (2005) Exporting the German model: the establishment of a new automobile industry cluster in Shanghai, *Economic Geography*, 81: 53–81.

Dicken, P. (2000) Places and flows: situating international investment, in G.L. Clark, M.A. Feldman and M.S. Gertler (eds) *The Oxford Handbook of Economic Geography*, Oxford: Oxford University Press, pp. 275–91.

Dicken, P. (2003) *Global Shift*, 4th edn, London: Sage.

Donaghu, M.T. and Barff, R. (1990) Nike just did it: international subcontracting and flexibility in athletic footwear production, *Regional Studies*, 24: 537–52.

Fröbel, F., Heinrichs, J. and Kreye, O. (1980) *The New International Division of Labour*, Cambridge: Cambridge University Press.

Grabher, G. (2006) Trading routes, bypasses, and risky intersections: mapping the travels of 'networks' between economic sociology and economic geography, *Progress in Human Geography*, 30: 163–89.

Harvey, D. (1989) *The Condition of Postmodernity: An Enquiry into the Origins of Cultural Change*, Oxford: Basil Blackwell.

Henderson, J., Dicken, P., Hess, M., Coe, N.M. and Yeung, H.W-C. (2002) Global production networks and the analysis of economic development, *Review of International Political Economy*, 9: 436–64.

Mair, A. (1994) *Honda's Global Local Corporation*, New York: St. Martin's Press.

Park, B-G. (2003) Politics of scale and the globalization of the South Korean automobile industry, *Economic Geography*, 79: 173–94.

Tickell, A.T. (1996) Making a melodrama out of a crisis: reinterpreting the collapse of Barings Bank, *Environment and Planning D: Society and Space*, 14: 5–33.

UNCTAD (2004) *World Investment Report 2003*, Geneva: United Nations.

Wrigley, N., Coe, N.M. and Currah, A. (2005) Globalizing retail: conceptualizing the distribution-based transnational corporation (TNC), *Progress in Human Geography*, 29: 437–57.

Yang, Y-R. and Hsia, C-J. (2007) Spatial clustering and organizational dynamics of trans-border production networks: a case study of Taiwanese IT companies in the Greater Suzhou Area, China, *Environment and Planning A*, 39, in press.

Yeung, H.W-C. (2000) Organising 'the firm' in industrial geography I: networks, institutions and regional development, *Progress in Human Geography*, 24: 301–15.

CHAPTER 9

LABOUR POWER

Can workers shape economic geographies?

Aims

- To recognize the ways in which capital's mobility gives it bargaining power over labour
- To appreciate the range of mechanisms used by states and firms to control labour
- To understand the different geographical strategies that workers may use to improve their position
- To reflect on the possibilities for alternative or non-capitalist labour geographies.

9.1 Introduction

In the mid-1990s, the Argentinian economy was being widely lauded as a success story. In a series of economic reforms initiated in 1989, the economy had been opened up, state industries privatized, and the currency was fixed to the US dollar. The result was a huge influx of foreign investment and growth rates of around 8 per cent in the early 1990s. By late 2001, however, the situation was very different. The fixed exchange rate had become a hindrance, making Argentinian goods expensive. Poverty levels were at an all-time high, and the unemployment rate was estimated to be topping 30 per cent. Argentina was struggling to repay its private sector debts, and, on 14 November 2002, failed to make a debt repayment to the World Bank. While it repaid the interest – about US$80 million – it could not repay the entire debt, which stood at around US$800 million. The country was in a state of political and economic turmoil.

In the midst of this confusion, however, a new social movement became evident. Across the country, workers started to take control of abandoned or struggling factories, and put them back into operation. In October 2001, the Zanón Ceramics factory was occupied by its laid-off workers. In December 2001, in perhaps the best-known case, the Brukman garment factory in Buenos Aires – the capital of Argentina – was abandoned by its owners and taken over by its workers. In some cases, the occupation was peaceful, while others resulted in direct confrontation with the authorities. In the Brukman case, the workers were forcibly evicted by police officers in April 2003, and did not return to the factory until later that year. By mid-2004, it was estimated that there were some 200 worker-run factories in Argentina across a wide range of sectors from bread-making to metal-smelting. In the most successful cases, sales and profits have increased, and worker salaries have seen substantial increases. The future of the movement is not entirely certain, however: while some have achieved official status as worker cooperatives, others remain in a state of legal limbo, facing challenges from former owners and local government. Some cooperatives have started collaborating together, and they have become part of a wider movement across Latin America – in mid-2004, there were an estimated 100 such factories in Brazil and 20 in Uruguay. In Brazil, since 1994, the factories have been formally organized through the National Association of Worker-Managed Enterprises (ANTEAG). The Argentinian movement proved the inspiration for a critically acclaimed 2004 documentary by Naomi Klein (the well-known anti-globalization author and journalist) and Avi Lewis entitled, *The Take: Occupy. Resist. Produce* (Figure 9.1).

This example illustrates the key arguments that run through the chapter. It shows how workers in particular places are increasingly exposed to the forces of global competition, and policy choices made by governments and supranational organizations. The actions of the factory workers, however, reveal that even within the confines of a global capitalist economy, workers have the ability to intervene and alter the course of events. The emergence of coordinated networks of factories both in Argentina, but also more formally in Brazil, illustrates the potential for worker groups to organize their activities at various geographical scales other than the workplace and the locality. Finally, the advent of worker-owned and managed factories hints at alternative ways of organizing labour other than the standard employer–employee relationship inherent to capitalism.

In developing these arguments, the chapter is structured into four main sections. Next, we explore the general account of how capital's spatial mobility in the contemporary era gives it a strong bargaining position with respect to place-bound workers (Section 9.2). While there is a large element of truth in this story, it is actually an over-simplification of real processes. Second, we review the range of mechanisms and policies that firms and states can use to extract concessions from workers (Section 9.3). These labour controls come together in different places in very different ways, and with different outcomes. Third, and moving on from viewing labour in this rather passive way, we then turn to look at the active role of workers within the economy (Section 9.4). We will look

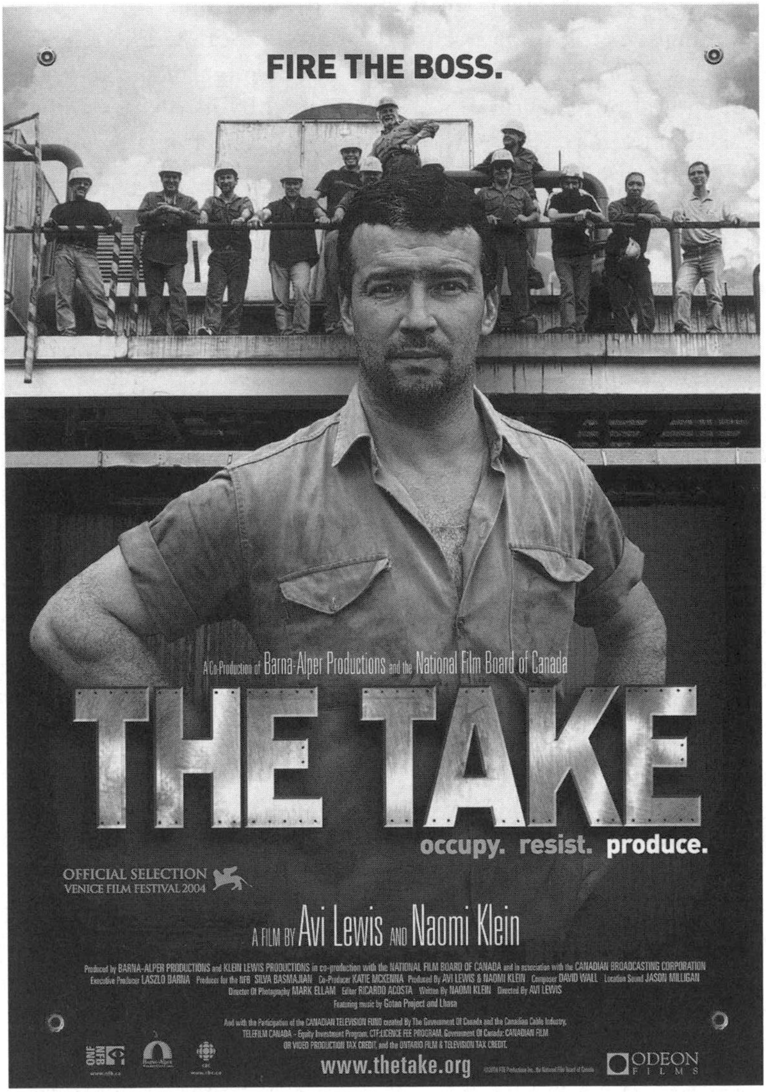

Figure 9.1 *The Take*
Source: *Occupy. Resist. Produce.* A film by Avi Lewis & Naomi Klein (© 2004).

at examples of how labour can assert itself through a number of inherently geographical strategies: by moving between places, by acting in place, and by connecting across places. Fourth, we finish by considering the potential of alternative ways of working that are not based around free market exchange and waged labour (Section 9.5). The central argument that drives the analysis is that in working collectively, labour can become an important and active agent helping to shape contemporary economic geographies.

9.2 Global Capital, Local Labour?

We start by reviewing the popular account of the position of labour in global capitalism. The basic argument is that a defining characteristic of contemporary capitalism is the *relative* geographic mobility of capital compared to that of labour, which in turn gives the former tremendous bargaining power when negotiating with the latter. In particular, the persistent threat that corporations may relocate production becomes a powerful tool for employers when determining wages and benefits, contracts, investment strategies and the like with employees. In this way, the *potential* reorganization of the activities of transnational corporations on a global scale (see Chapter 8) becomes a critical tool for firms negotiating with workers and local interest groups. Workers, in turn, become increasingly pressured into having to defend their interests in particular places, and in the process may enter into direct wage competition with workers in other places in a bid to secure jobs. This inter-place competition fosters a progressive ratcheting-down of labour terms and conditions as mobile capital seeks out the best rates of return, a process sometimes known as the 'race to the bottom'.

This competition can occur at the inter-national and/or sub-national scales. The UK and Spain may compete, for example, to host a new Japanese car manufacturing plant to serve the European market. Equally, locations within those countries may also be drawn into competition with one another in order to attract much-needed investment. National, regional and local governments may become embroiled in this competitive dynamic, offering up large sums of public money in the form of incentives in a bid to secure investments. To give one example, in 1993, the US state of Alabama offered what was then Mercedes-Benz some US$250m in incentives to secure a car plant in competition with other Southern states; this amounted to US$167,000 per job created. Overall, this competitive bidding for employment is put forward by commentators on both the left and right of the political spectrum. From the perspective of neoclassical economics, this competition-driven system is a route to efficiency gains through globalization that will generate benefits for all. From a more critical, Marxist perspective, it constitutes a neoliberal economy in which labour is subordinated to employers in a process that undermines labour standards and salaries (see Box 9.1 for a range of viewpoints on labour and labour markets).

The general thrust of this account is persuasive. The geographic mobility of capital certainly *has* increased over the last few decades, due to a number of overlapping factors. *Deregulation*, as seen in the lowering of barriers to international trade and foreign direct investment, increasingly enables the transfer of both finished goods and the factors of production (i.e. materials, technologies, people, and capital) (see Chapter 7). Dramatic improvements in productive,

Box 9.1 Different perspectives on labour and labour markets

Three broad approaches to understanding labour and labour markets can usefully be identified here. It should be noted that each approach is, in reality, a cluster of related approaches rather than the homogeneous entities we depict here. Contemporary economic geography tends to be critical of the neo-classical worldview and instead combine facets of the Marxist and institutional perspectives (see Martin, 2000, for further details):

1 The *neo-classical approach* emphasizes the free market as the key mechanism for matching the supply of, and demand for, labour as with any other 'commodity'. In this idealized world, workers simply 'sell' their services to employers. Employers benefit from the workers' capacity to labour for a finite period, while workers get wages that allow them to purchase the commodities they need to live and prosper. Equilibrium is reached and the labour markets 'clear' (i.e. everyone finds a job) through price adjustments and the basic laws of supply and demand. Not all workers are equal, however. Different skill levels, for example, are reflected in different wage levels as determined by the open market. When markets do not clear, market imperfections (such as unemployment and job vacancies) are attributed to various barriers to the efficient working of the labour market (e.g. government intervention). Individual workers, meanwhile, are seen as rational individuals seeking to get the best deal from their employers.

2 The *Marxist approach* is highly critical of the neo-classical perspective. A Marxist viewpoint sees the emphasis upon unregulated and apolitical labour markets as highly unrealistic. In capitalism, the worker–employer relationship is, in principle at least, a contested affair. The labour market is a site of struggle, not a forum of free exchange. Building upon this, Marxists view capitalism as an inherently unequal system in which profits are derived through workers being paid less than the value of the goods they make. This means that the neo-classical idea of the market as a neutral means of exchange hides the exploitation that goes on daily in the workplace, and says little about social power and resistance within society. For Marxists then, exploitation pits workers against capitalists in both theory and practice. Contrary to the neo-classical vision of sovereign individuals buying and selling labour power, Marxists see a more collective politics of labour where class interests shape individual interest and actions.

3 The *institutional approach* focuses on the role of institutions in economic life. An advocate would argue that, in both neo-classical economics and to a lesser extent in the Marxian approach, economic life tends to be abstracted from its socio-political and cultural context. Institutions are best thought of as generally accepted rules and regulations that shape human behaviour, and which are either policed by individuals themselves, or by some kind of external authority. It is helpful to distinguish between the institutional *environment* (rules, customs, routines, etc.) and institutional *arrangements* (organizational forms such as markets, firms, unions etc.). Labour markets are thus seen as a set of social institutions which – importantly for economic geographers – will vary in nature from place to place due to differences in employment practices, workplace norms, work cultures and labour traditions. Work in this area is concerned with revealing how both local and non-local institutions intersect to shape local employment, wage and welfare conditions.

transport and communications technologies have facilitated increasingly complex geographic arrangements of production and the rapid global circulation of financial capital (see Chapters 5 and 8). Together, these deregulatory and technological dynamics have increased the intensity of inter-place competition. Technologies, for example, have contributed to productivity growth that has often outstripped the expansion of markets, thereby enhancing cost-competition. Deregulation, on the other hand, may both increase competition in a firm's home market by facilitating the entry of foreign firms, and increase opportunities abroad for domestic enterprises.

At the same time, we also can think of a number of ways in which working and labouring are attached to particular places:

- *the daily commute*: the majority of workers do not have the time or financial resources to travel far to work. While in advanced capitalist economies the average commute can be as high as 100km or more per day, the vast majority of the world's labour lives in much closer proximity to the workplace. Put simply, 'labour-power has to go home every night' (Harvey, 1989: 19).
- *labour reproduction*: the reproduction of labour is also necessarily local. The various institutions of everyday life and communities such as the family, church, school, clubs and so on, all develop over time in particular places.
- *place attachments*: as labour is unlike other commodities, it has the propensity to develop place-attachments or place-identities. The notion of 'home-place' points to this intensely local dimension to many people's sense of themselves.

Places may become sites of familiarity, routine, affection and friendships. These emotional ties to place can be hard to break.

- *the grounding of production*: all production activities are necessarily local. For the vast majority of paid workers, production occurs at fixed sites. Even the largest TNCs, while *relatively* more mobile than most small businesses, for example, must still combine labour, materials and technology in particular work*places*.
- *regulation*: the regulation of workers is always expressed and experienced locally. Even where the regulatory institutions are national (e.g. a national trade union) or international (e.g. the European Union or the North American Free Trade Agreement) their regulatory mechanisms must ultimately be articulated at the local scale.

In sum, there are clearly a number of strong place attachments for workers and their families to be contrasted with the enhanced mobility of capital.

However, the argument we will develop in the remainder of this chapter is that this general thesis is somewhat overstated. Accepting it as a blanket characterization of the position of all workers in contemporary capitalism has two problems. First, it obscures the social and spatial variations in the incidence of these conditions. Socially, we can think of inter-sectoral differences. While harsh inter-place wage-based competition may characterize the garment industry, it clearly bears little resemblance to the dynamics of the legal profession. Even within individual firms or commodity chains, it is usually only certain kinds of workers (e.g. manufacturing or routine administration staff) who are subject to this kind of competition. Spatially, we can think of different scales of variation. At the scale of the *macro-region*, regional blocs differ in the extent to which they institute protections for labour rights and protections. For example, the European Union (EU) is generally seen to be more progressive than the North American Free Trade Agreement (NAFTA) in this regard. *National* labour systems with different institutional and cultural histories will vary greatly in the extent to which they safeguard workers. Table 9.1 contrasts the deregulated, neoliberal American model with Japanese and European systems that offer more support to workers and worker groups. Taking the example of employment systems, we can see how these range from short-term employment relationships forged outside of the firm in a competitive market context (the American way) to long-term employment relationships dominated by slow progression within individual firms (the Japanese way). Within national territories, there is considerable *place variation* in the character of labour-capital relations. Places represent unique coalescences of firms, workers and institutions: no two local labour markets are exactly alike, and the precise problems and possibilities facing workers vary accordingly. Second, it serves to obscure the potential for worker intervention in the global economy, an issue we shall return to shortly.

Table 9.1 Different national labour conditions: three ideal types

	The American Way	The Japanese Way	The European Way
Economic structures	Low-skilled, price competitive approach Downward pressure on wages from cost sensitivity to international competition	Flexible system, combining scope and scale economies High, but firm-specific skills	High-skill, high-wage economy, based on non-price competitive goods and services Extensive human capital investment
Employment practices	Short-term employment relationships Market-led employment adjustment through external labour market Individualised and competitive employment relations	Long-term employment relationships Confinement of core workers to internal labour markets Slow, internalised labour market adjustment	Medium-term employment relationships High standards of employment protection Macroeconomic stabilisation of employment levels
Economic and social norms	'Market' work ethic Increasing working time combined with low productivity growth, and falling investments in human capital High levels of labour market inequality	'Corporate' work ethic Acceptance of long working hours Submission to corporate authority Implicit socially encoded rules of managerial dominance	'Social' work ethic Declining working time Low inequality Universalised basic income levels
Industrial relations	Market based Individualistic ethos De-unionized, human resource management (HRM) approach	Ethos of culturally and institutionally embedded trust, cooperation and compliance Enterprise unionism	State-articulated regulation of economic and political relations Institutionalized rights of employment Unions conferred with public status

Source: Abridged version of Peck (1996: 246–7).

9.3 Geographies of Labour: Working under Pressure

Notwithstanding the above discussion, it is undoubtedly true that workers in contemporary capitalism are faced with a broad range of labour control mechanisms that either threaten to erode the terms and conditions under which they work, or worse still, remove their opportunity to work at all. In this section we review the range of different mechanisms that can be used by both firms and states in this regard, before moving on to look at how they work out on the ground in various localities in Southeast Asia.

Firm and state strategies for labour control

Many of the most detrimental dynamics for workers come about as a result of firm restructuring and reorganization. Firms seek to manage strategically their workforce in pursuit of profits. Indeed, as we saw in Chapter 3 through the notion of spatial divisions of labour, firms often initially establish operations in particular places to access certain types of labour, with labour costs, levels of skills and productivity, and the relative degree of militancy and collective representation, being important variables in such decisions. In the call centre industry, for example, employers may seek out pools of cheap and weakly organized female workers in relatively peripheral regions (part of a broader pattern that we will examine more closely in Chapter 12). In some instances, groups of migrant workers may be similarly exploited due to their need for work and relative lack of local social contacts.

Some of the restructuring strategies pursued by firms relate to firms altering the *scale* of operations in particular places. Most obviously perhaps, firms may simply decide to relocate production to another locality. Increasingly, as part of broader globalization processes, production is being moved offshore. Attempts by displaced workers to find jobs on similar pay and conditions locally may be complicated by sectoral shifts – such as that from manufacturing to service employment – which may result in mismatches between workforce skills and the demands of the local economy. In reality, however, the decision to 'up and leave' a place completely is one very rarely taken by firms. Instead, firms are more likely to change the scale of operations in a particular place, or, in the case of multi-plant firms, alter the balance of activity *between* various locations. Again, such decisions are increasingly taken at the global scale.

Another method which firms may use to downsize operations in a locality is to *externalize* certain activities that are subsequently secured from suppliers or through subcontractors. In such situations, workers may find themselves working for worse pay and conditions due to increased competition between suppliers. In some instances, such as in garment manufacturing in both developed and developing world cities, retrograde forms of weakly regulated labour such as home-working and informal work can emerge.

Other firm strategies, however, involve changing the *nature* of operations in particular places. As suggested in Chapter 5, technological change is crucial here. New production technologies may lead to the replacement of workers and result in job losses (e.g. automation of an assembly task) or they may alter the labour process for the workers involved (e.g. moving from paper to computer-based systems). While some view this process positively, suggesting that repetitive tasks are replaced and skilled positions are created, critics suggest that some new technologies have simply increased the potential for worker surveillance and monitoring in a more mundane way. Worker performance in call centres, for example, is often monitored by recording phone conversations with clients. Closely related to the implementation of new technologies, firms may seek to intensify the labour process of their employees in other ways. Examples include moving to new and extended shift systems, the increased use of overtime, the lowering or ending of barriers between different jobs (known as multi-skilling), and the use of a range of more flexible employment contracts such as part-time, short-term and temporary working.

These technology and work intensification strategies may also be accompanied by attempts to constrain the activities of unions within the workplace. Employers may simply ban unions completely in the workplace, they may establish one-union rather than multi-union workplaces, they may insist on workplace bargaining and dispute resolution as opposed to collective bargaining at larger spatial scales, and they may establish worker-management councils within the workplace to try and negate the need for union representation. Moreover, in their negotiations with workers and worker groups, they may pursue concessionary bargaining strategies in which they seek to freeze, and in some cases reduce, wages. Important discursive strategies come into play here as firms can emphasize the perils of global cost competition and the threat of relocation during negotiations. The term *lean production* has been used to describe the nature of the production system in contexts where most or all of these various mechanisms coincide (Box 9.2).

These various processes of corporate restructuring may be supported – or not, as the case may be – by state and quasi-state bodies at a range of scales. In particular, as we demonstrated in Chapter 7, the national state remains of critical importance in any analysis of global capitalism. One group of strategies relate to how the economy is regulated, with a dominant trend in many contexts being towards deregulation. As we mentioned earlier, by removing or lowering regulatory barriers to trade, foreign direct investment and the movement of people, technologies and capital, states (and quasi-state bodies such as the WTO and IMF) are creating conditions that allow capital to be switched between places at the international scale. At the same time, the re-scaling of economic development and place marketing functions to the subnational or local level

Box 9.2 Lean production

In reality, many different corporate practices are used together to control the workforce. This coalescence of strategies has resulted in a new form of production system often termed *lean production*. The essence of lean production – which originated in Japan in the post-war period and has subsequently spread widely – is that it combines economies of scale with economies of scope derived from various practices of *flexibility*. Three types of flexibility are encapsulated in lean production. First, *functional* flexibility concerns increased levels of job rotation and multi-skilling. Second, *numerical* flexibility reflects a growing use of contracting-out and part-time and casual labour. Third, *temporal* flexibility describes the move to new forms of shift patterns (such as 24-hour working) and overtime. These changing working patterns are often combined with the very latest productive technologies. While proponents choose to emphasize that these processes enhance product and process innovation, critics argue that flexibility is most often implemented to facilitate cost reductions. Despite seemingly positive descriptors such as 'quality management' and 'team-working', these changes are not simply about worker empowerment, but also about the micro-management of workers' time and activities, a system that might be described as 'management-by-stress' (Moody, 1997: 87). New forms of groupings designed to increase worker input to the production process such as teams and job circles can actually serve to bypass unions, or avoid them altogether. In some situations, workers and their representatives are forced to develop proposals to reduce costs and improve quality in competition with other plants. The usual outcome of lean production is a combination of job losses, and increased work and time pressure on the primary workforce. For more, see Moody (1997).

(explained in Chapter 7) means that places within the same country are often competing for inward investment.

More specifically, states may choose to alter the nature of their regulatory intervention in labour markets. For example, they may seek to enhance the flexibility of labour markets through removing constraints on labour mobility between jobs or by using toughened legislation to constrain the activities of unions, for example. Another component of a move to labour market flexibility may be to change the nature of support available to those without work. Some industrialized countries such as the USA and the UK are seeing welfare systems replaced by so-called *workfare* systems in which the unemployed are enrolled into a variety of temporary training schemes and low-end jobs in order to 'earn' their benefit payments and hopefully secure long-term employment.

Finally, we must not forget the fact that the state remains a massive employer in most countries, with three important implications. First, any moves to shrink the public sector may result in redundancies (e.g. rationalization of the civil service). Second, the contracting-out of certain activities (e.g. catering, security or IT services) to external providers brings into play the same range of dynamics considered above in the context of firms. Third, the wholesale privatization of previously government-owned industries (e.g. utilities such as water, gas and electricity) can similarly expose more workers to the forces of inter-firm competition.

Labour control in action: new industrial spaces in Southeast Asia

Workers, then, face a wide range of corporate and state strategies in their bid to make a living in the global economy. Figure 9.2 summarizes these various overlapping dynamics and their impacts on workers and labour markets. But

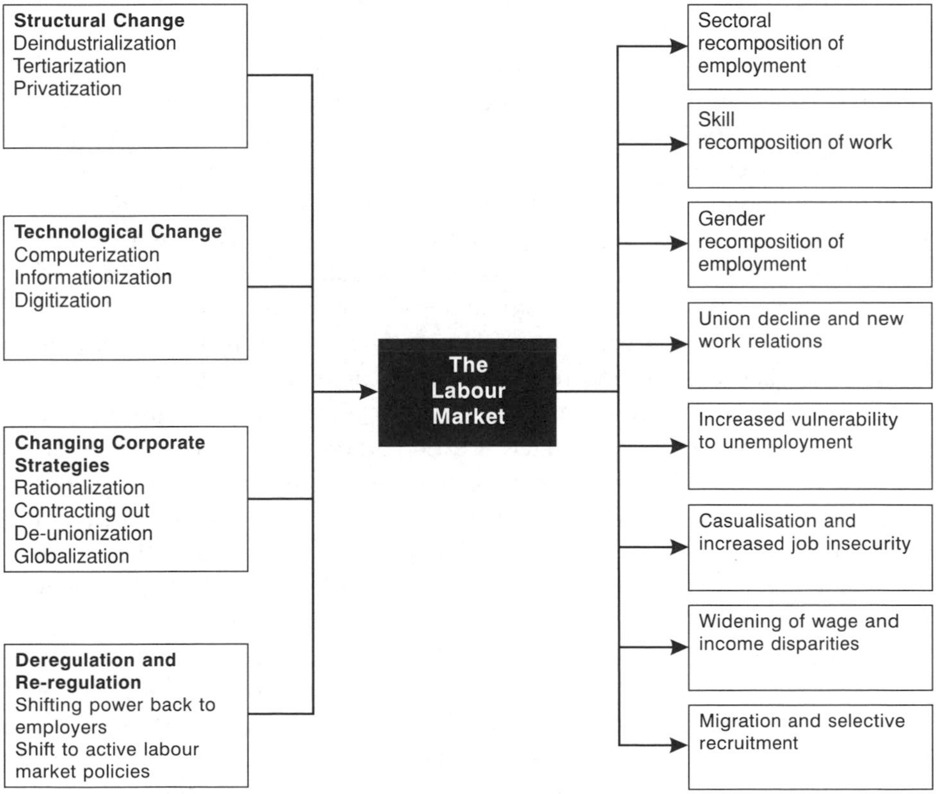

Figure 9.2 Forces of labour market restructuring
Source: Adapted from Martin and Morrison (2003), Figure 1.1.

how do these forces interact and play out on the ground? This question can only really be effectively answered from an economic-geographical perspective. Geography is important here in two ways. First, as we shall see shortly, strategies of labour control are innately spatial and operate at multiple spatial scales. Second, different strategies come together in distinct combinations in particular places to create unique local labour control regimes (Box 9.3). These two dimensions can be illustrated by considering labour control in the industrial estates and export processing zones of three localities in Southeast Asia (Kelly, 2002).

Box 9.3 Local labour control regimes

The *local labour control regime* (LLCR) is a useful concept for understanding the inherently place-based nature of economic activity. The development of LLCRs is driven by one of the basic contradictions of capitalism, namely that which exists between the *potential* spatial mobility of many firms, and their associated bargaining power, and the need for firms to extract profits from concrete investments in particular localities, which requires a certain measure of stability. An LLCR is the place and time-specific set of mechanisms – social relations, norms, rules and habits – that coordinate the links between production, work, consumption and reproduction in particular localities (Jonas, 1996). Importantly, this formulation extends beyond the workplace to incorporate the domains of consumption (housing, recreation, household consumption etc.) and reproduction (education, training, health care, welfare etc.). It naturally follows that the full range of worker, household, firm, civil society and state institutions are – or at least may be – involved in shaping the LLCR. In part, a LLCR constitutes the unique, place-specific relations between firms, workers, unions and regulatory institutions that enable workers to be integrated into a production system. At the same time, it is also shaped by multi-scalar processes, and there are a broad range of extra-local worker, employer and regulatory connections that will influence its character. Put another way, every local regime is 'nested' within labour control regimes operating at larger scales that will influence, *but not determine*, the nature of employer–worker relations in that particular locality. Of particular importance is the way in which the regime is integrated into production systems organized at the national and international scales that actively seek to take advantage of the differences between various regimes (e.g. in terms of labour costs, skills, unionization rates). The global capitalist economy is characterized by a bewildering variety of different LLCRs. Even within the same country and the same industry, the differences may be profound. In the US computer industry, the highly flexible innovative labour regime of Silicon Valley, California, is reputedly very different from the regime dominated by large firms found in the Route 128 area near

Boston, Massachusetts (see Chapter 11). A corollary of this great variability is that the problems and possibilities facing workers will differ dramatically between different places. After all, it is only certain kinds of businesses that benefit from inter-locality flows of investment and aggressive spatial restructuring. Others serve to benefit far more from becoming progressively involved in the LLCR on a long-term basis.

Cavite and Laguna (The Philippines), Penang (Malaysia) and Batam (Indonesia) share several important features: high levels of female employment in light manufacturing and assembly work in the electronics and garments industries; substantial inward foreign direct investment from North America, Western Europe and leading East and Southeast Asian economies; strong state provision of locational and export incentives; and very low levels of unionization. Five distinct spatial strategies of labour control are apparent in these industrializing localities. First, labour control is evident at the level of *the individual*. This can be seen in the way that employers try to establish individual relations with workers rather than allowing collective representation – through unions, for example – and also in the use of dormitories and hostels to provide a tightly controlled and regulated non-work environment in which workers may be offered counselling and a range of religious, educational and sporting activities. Second, *the workplace* is a site of labour control. Where collective organization is allowed, labour relations are usually confined to the factory level through the use of in-house unions and labour-management councils. This is to prevent independent trade unions from operating across different workplaces and firms, which may facilitate the comparison of wages and working conditions.

Third, *the industrial estate* is another scale of labour containment. Estate management companies provide services to tenants that go way beyond infrastructural provision, including tight security, establishing and maintaining worker databases identifying non-union employees, undertaking employer–worker mediation, and fostering links and contacts to local politicians. In this way, estate management companies actively seek to maintain their jurisdictions as union-free spaces. Fourth, as described earlier, *national regulatory policies* can also serve to constrain worker organization, through trade union legislation. This may not be through legislation that bans unionization outright, but more likely through highly bureaucratic and sector-specific registration processes that present an effective barrier to worker organization. Fifth, different *spaces of migration* can be identified which allow employers flexibility and promote labour cooperation. Employing workers either from other regions of the country or from abroad can provide employers with a flexible workforce that is often disinclined to change jobs (or may be prevented from changing jobs through the conditions of their work visa) and is insulated from the distractions of family or social life. Such workers may also be highly motivated by the desire to send

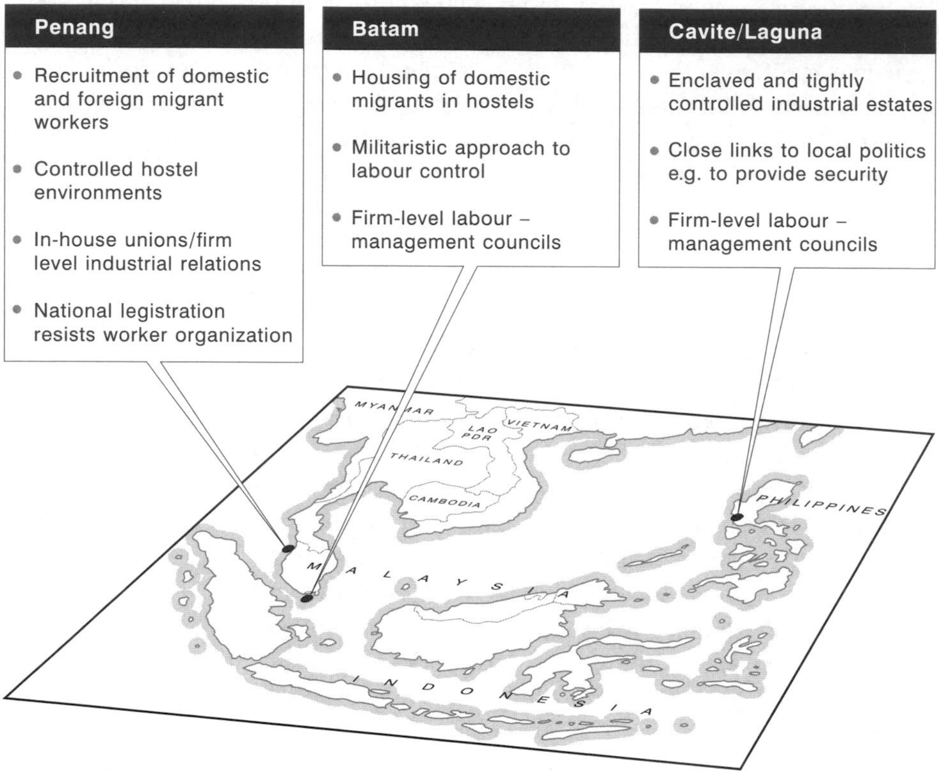

Penang
• Recruitment of domestic and foreign migrant workers
• Controlled hostel environments
• In-house unions/firm level industrial relations
• National legistration resists worker organization
Batam
• Housing of domestic migrants in hostels
• Militaristic approach to labour control
• Firm-level labour – management councils
Cavite/Laguna
• Enclaved and tightly controlled industrial estates
• Close links to local politics e.g. to provide security
• Firm-level labour – management councils
MYANMAR
LAO PDR
VIETNAM
THAILAND
CAMBODIA
PHILIPPINES
MALAYSIA
INDONESIA

Figure 9.3 Labour control regimes in Southeast Asia

money home to their families. In sum, in these localities a compliant workforce is delivered through the combination of different corporate and state spatial strategies of labour control.

The precise configuration of these different strategies, however, is quite distinct in each of the three localities, thereby illustrating our second geographical dimension to labour control (Figure 9.3). In Penang (Malaysia), the use of domestic and foreign migrant workers is common, as is their accommodation in carefully controlled hostel environments. In-house unions and firm-level industrial relations combine to keep disputes at the factory level, while national legislation provides a strong barrier to worker organization at larger scales. In Batam (Indonesia), the housing of domestic migrants in hostels, supported by a militaristic approach to preventing organization, has provided an effective system of control (Figure 9.4). In Cavite and Laguna (Philippines), it is the enclaved nature of industrial estates as carefully controlled spaces combined with strategic links into local politics that facilitate labour control, rather than migrant labour, hostel spaces or national legislation. There is clearly geographic variability in the relative importance of different strategies of labour control in each of the

Figure 9.4 Worker dormitories in Batam, Indonesia
Source: The authors.

localities. By inference, this variability can be extended to export processing zones in Mexico, China, India and beyond. Recognizing and understanding these differences is the first step towards a more progressive understanding of the role of labour in the global economy. It is to this topic that we now turn.

9.4 Labour Geographies: Workers as an Agent of Change

Arguably what is needed is a shift from thinking about *geographies of labour* – i.e. how workers are distributed through space and how they are affected by the dynamics of the global economy to a notion of *labour geographies* that enables us to conceptualize workers not merely as a factor of production similar to capital, equipment or raw materials, but as thinking beings who actively shape economic geographies through their interventions (Herod, 1997). This is not to argue that labour is the only force shaping contemporary economic geographies: such a stance would clearly be nonsensical in the light of the various powers of labour control available to both firms and governments illustrated above. Rather, it is to recognize that workers do have the ability (or *agency*) to work to improve their relative position and at the same time re-shape the economic geographies around us.

Indeed, two inherent characteristics of contemporary capitalism offer potential for such interventions. First, however mobile capital may be in theory, all economic activity is ultimately grounded in particular workplaces, and there are a number of prohibitive costs that firms will incur when leaving a locality (e.g. redundancy and pension payments to workers, losses on immovable plant and equipment). Second, the world economy is characterized by a tremendous variety of inter-place connections and interdependencies, many of which are highly time- and space-sensitive due to the application of new information and communication technologies. This creates sometimes unexpected vulnerabilities on the part of employers, and the possibility for geographical action by workers. In the remainder of this section, we illustrate four different spatial strategies: proactive migration, resistance within the workplace, acting in a locality both alone and in coalitions, and mobilizations of workers across different places.

On the move: proactive migration strategies

One strategy through which some workers can assert their agency to try and improve their working and living conditions is to migrate. Migration comes in many varied forms: temporary or permanent, internal to a country or international, legal or undocumented, skilled or unskilled, and proactive or reactive. Here we choose to focus on a form of migration that it is a salient characteristic of the contemporary global economy, namely the voluntary or proactive international movement of high-skilled workers over a variety of timeframes (we will look at some less progressive forms of international migration in Chapter 13). Professional labour migration of this kind has expanded dramatically over the past 20 years as part of broader globalization dynamics (Figure 9.5). In some cases, this migration is the result of organized overseas secondment and assignment schemes within transnational corporations, as are common, for example, in the financial services sector. In other cases, migration is the result of individual and family livelihood strategies, and may be linked to the pursuit of overseas education opportunities.

We can illustrate different forms of professional migration through looking at the information technology (IT) industries of the Pacific Rim. Skilled immigrants play an increasingly significant role in the economic development of Silicon Valley, California: currently over one-third of scientists and engineers in Silicon Valley's high-technology workforce are foreign-born, with the majority being of Asian descent (Saxenian, 1999). By 1998, immigrant-run technology companies (started since 1980) accounted for more than US$16.8bn in sales and 58,000 jobs. There were 2,001 firms led by an ethnic Chinese manager and 774 by an ethnic Indian manager, together accounting for 24 per cent of the total number of firms. The continued success and dynamism of Silicon Valley is thus increasingly based on trans-Pacific flows of highly skilled migrants – particularly from

Figure 9.5 Skilled workers on the move: 'foreign talent' in Singapore's financial district
Source: The authors.

Taiwan, India and China, but also in much lesser numbers from Vietnam, the Philippines, Japan and Korea – who contribute both directly to the economy, as engineers and entrepreneurs, and indirectly, as traders and middlemen linking California to the technologically advanced areas of Asia. While the success of Asian migrants in Silicon Valley in parts rests upon their ability to integrate themselves into local technology and business networks, there has also been an ongoing 'globalization' of Silicon Valley's ethnic networks. Many of the new firms set up by immigrant entrepreneurs are 'global' actors from the outset, with the ability to use ethnic network connections to access Asian sources of capital, development skills and markets.

Such links are especially apparent between the technology communities of Silicon Valley and Taiwan's predominant technology region, Hsinchu (Hsu and Saxenian, 2000). The key actors are a cohort of US-educated Taiwanese engineers who have the experience and language skills to operate effectively in both regions. This community is constituted by three international migration flows. First, there is the continued flow of Taiwanese migrants to the US, and Silicon Valley in particular, seeking educational and economic opportunities. Second, there are now increasing numbers of returnees to Taiwan, who have become an integral part of the technology industries of Hsinchu. Many immigrants have returned to Taiwan to start businesses there while keeping close links with former colleagues and classmates in Silicon Valley. Return migration grew appreciably in the 1980s (200 engineers and scientists per year on average) and accelerated in the 1990s (1,000 per year in mid-decade). Third, there is a growing population of circulatory migrants or 'astronauts' that work in *both* regions. With families based on either side of the Pacific – most commonly in California due to lifestyle factors – these engineers travel between the two areas as often as once or twice a month. They include Taiwanese 'angel' investors and venture capitalists as well as managers and engineers from firms with operations in both regions. The 'bridge' between Silicon Valley and Hsinchu is thus constituted by a combination of permanent, temporary and transient highly skilled international migrants.

'Office politics': action within the workplace

Some forms of labour agency manifest themselves at the scale of the individual (and/or groups of workers within a workplace) and are shaped by the particular family and household configurations of which those workers are a part. All employer–worker relations inherently embody tensions concerning a wide range of issues including worker autonomy, training and skills development, wages, working hours, promotion opportunities, job security, job status, non-wage benefits and workplace facilities. We have already seen the various strategies that employers may use to extract concessions from workers *in situ*. Yet, workers also have their own resistance strategies which can be used to either support or contest employer strategies. These strategies can vary along three key dimensions. First, they may either *proactively* initiate debate with employers or *reactively* respond to employer actions. Second, they may be either *official* or *unofficial*, depending on whether they operate within approved industrial relations frameworks and laws or not. Third, they may be either *physical* (e.g. striking or working to rule) or *non-physical* (e.g. using persuasion and bargaining). In many cases, and mirroring the interconnecting labour control strategies implemented by states and firms, workers will use a combination of resistance strategies at the same time.

An excellent example of such activities is provided by the coping strategies employed by women working in the Jamaican information-processing industry (Mullings, 1999). The industry exports a range of services associated with the collection, transmission, storage and processing of information, using information and communications technologies, to mainly US clients. Often working out of the Export Processing Zones of Kingston and Montego Bay, the activities undertaken by these Jamaican subcontractors range from simple data entry work for magazine subscriptions to software localization. Growth in the sector – as in other Caribbean states – has been driven by the wage savings of up to 90 per cent that can be secured by US firms using data entry clerks, secretaries and telephone operators located in Jamaica. Some 90 per cent of the workforce is female – with the highest concentrations in the data entry rather than management strata – and the average age is only 25. In short, the Jamaican data processing industry mobilizes a workforce of non-unionized, low-wage female workers. As in the Southeast Asian examples recounted earlier, labour control is enacted through a range of practices including premises designed to obscure workers from the public gaze and prevent interaction, strict control on toilet and drinks breaks, and close electronic monitoring of worker productivity and accuracy.

The worker–manager relation in this context is clearly one of constant tension. While few workers have overtly resisted work conditions, many have engaged in acts of everyday resistance that may hamper the ability of firms to meet contractual deadlines for their clients. Examples include absenteeism, working to rule, rejecting overtime when offered, and 'finger dragging' (that is, deliberately slowing the work-rate when basic targets have been met). These are activities that can actually reduce the wages that workers receive, and may indeed put their employment in the industry in danger. The viability of these strategies – and the psychological boost that workers may receive from them – can only be understood in terms of the role of the extended family in the Jamaican context. Again, geography clearly matters – seen here in place-specific forms of household organization – in shaping worker actions. The basic wages earned in the information processing industry are not enough to meet the income needs of a family. Hence, many families survive by pooling a variety of resources: money received from overseas family members (known as *remittances*), formal wages, and by exchanging unpaid services within and between family groups. It is these collective household economies (see also Chapter 13) that cushion the effects of the low wages and harsh discipline of the information-processing industry, and make viable a series of workplace resistance and coping strategies that at first glance seem illogical.

Defending place: worker actions in situ

Despite the apparent global mobility of capital, workers still have considerable potential to protect jobs and livelihoods in particular localities. Indeed, the geographical mobility and spread of large employers can make them especially vulnerable to certain kinds of worker action. An excellent example of such *in situ* action occurred at the General Motors (GM) plants in Flint, Michigan, in 1998 (Herod, 2000). Flint, like many other places in the so-called North American rustbelt, lost several thousand manufacturing jobs during the 1980s and 1990s, placing great strain on local communities. The large employers that remained sought to implement more flexible systems of lean production (Box 9.2) in their factories, often with negative impacts on working conditions, overtime pay and shift patterns. In particular, they sought to impose low inventory just-in-time production systems (see Chapter 5) to cut costs and improve efficiency. Workers at two GM plants in Flint were unhappy both with these changes and with the low levels of investment made by GM in the factories, and they decided to take strike action. The *local* nature of strike action was critical for the United Auto Workers (UAW) union to which the workers belonged: a national strike would have been illegal under US labour law, and as such the UAW may have become liable for any costs incurred by GM. As a result, the UAW was clear from the outset that it was local workplace terms and conditions that were at stake.

The first strike began on 5 June 1998, when 3,400 UAW members stopped working at a metal stamping plant in Flint. Six days later, on 11 June, 5,800 UAW members at Flint Delphi Automotive Systems followed suit. This plant was the only source for certain components in North America. Within just two weeks, over 190,000 workers at 27 General Motors plants across North America had been sent home for the simple reason that there was no work for them to do. In addition, 117 component and supplier plants owned by subsidiaries of GM had either closed or reduced production. The impacts were also felt *globally*, with production at GM's plants in Mexico and Singapore being affected. Within a matter of days, the strike action had crippled General Motor's highly complex Just-In-Time production system across North America (Figure 9.6). The impacts on GM were significant – it lost production of some 500,000 vehicles and posted a loss of US$2.3bn for the second and third financial quarters of 1998. By the end of the strike action, local workers had extracted a number of concessions from their employer related to preserving existing forms of work organization, investment in the metal stamping plant in return for productivity increases, the removal of the threat of legal action against the UAW, and assurances about the short- to medium-term future of both Flint plants.

This example illustrates how workers acting in a particular place may be able to exert their agency to good effect. By choosing strategically important nodes in GM's global production network, and stressing that local issues were behind the strike, the UAW was able to gain the upper hand over GM and extract real concessions during subsequent negotiations. Workers and worker groups taking action on their own is not the only *in situ* option available, however. Workers may also choose to collaborate with other local interest groups through either *cross-class alliances* or *community unionism*. In the former case, workers join forces with local employers and government to defend existing investments in a particular place, or to try and attract new ones. In the cities of the industrialized world, these kinds of alliances have come to be known as *growth machines*. In the case of community unionism, workers collaborate with a variety of local non-work-based groups to pursue particular issues (see Box 9.4 for more on community unionism).

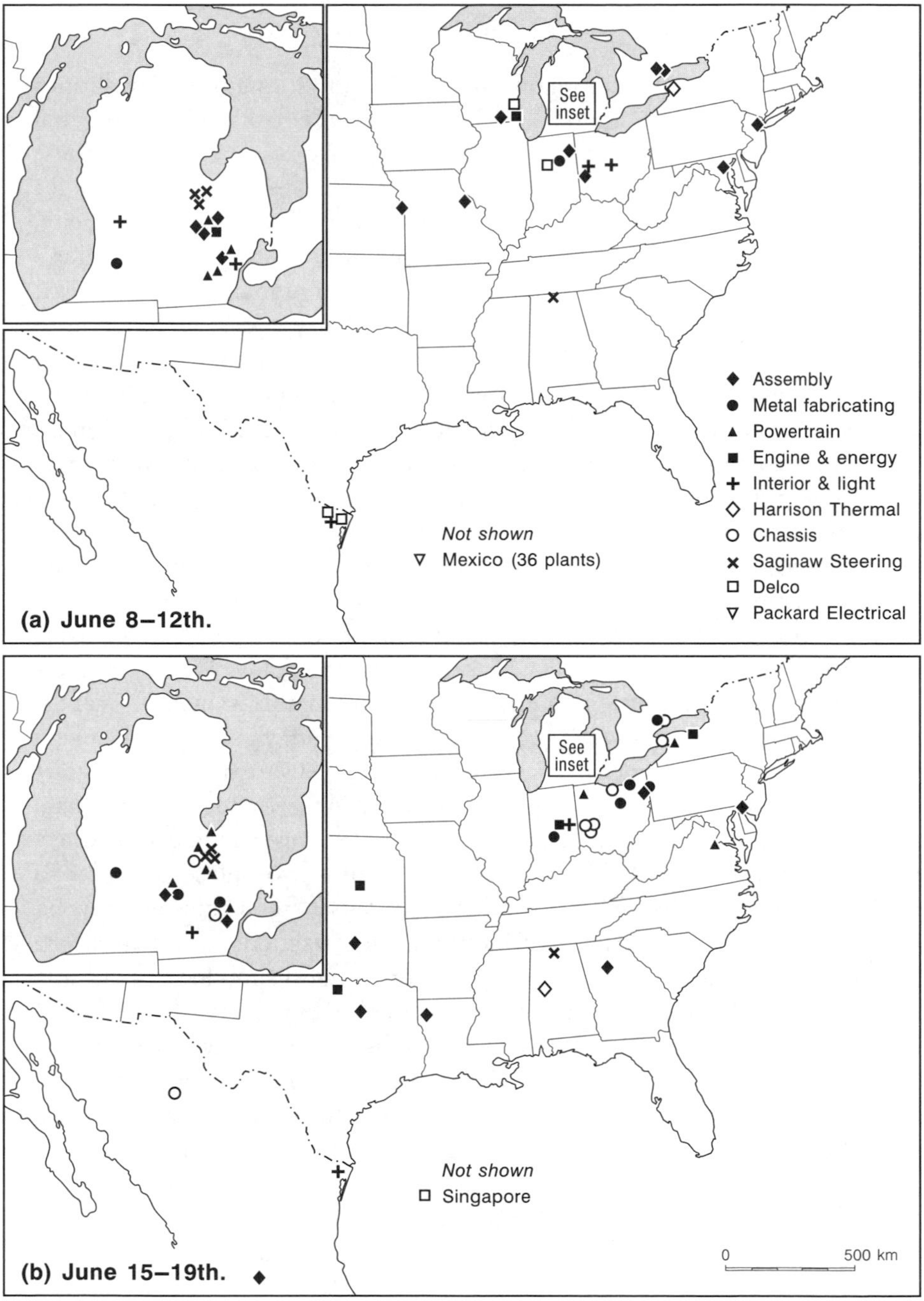

Figure 9.6 The spread of the GM UAW dispute
Source: Redrawn from Herod (2000), Figures 1 and 2. With permission of Blackwell Publishing.

Box 9.4 Community unionism

Community or reciprocal unionism refers to the formation of coalitions between unions and non-labour groups in order to achieve common goals, such as a minimum Living Wage for public sector service workers. In other words, it is about unions forging alliances with other groups whose primary identity might be based on affiliations of religion, ethnicity, gender, disability, environmentalism, neighbourhood residence or sexuality. Community union initiatives aim to achieve a number of things. First, by increasing the scale of organizing activity, they can deal with questions of economic justice beyond particular workplaces. Second, by working with community groups, unions are able to reach workers in traditionally non-union environments. Third, they can enable unions to help organize fragmented workforces split across large numbers of small workplaces (e.g. security and cleaning staff). Fourth, extensive links into the local community may help unions defend terms and conditions within their traditional workplaces. One pioneering example of community unionism is the Battersea and Wandsworth Trades Union Council (BWTUC) in Southwest London. BWTUC is a multi-union body made up by local representatives of unions affiliated to the national Trades Union Congress (TUC). In 1999, a centre was established with the aim of making Wandsworth a 'Respect at Work Zone'. The centre's workers are tasked to help organize non-union employees in Wandsworth to ensure five key principles are upheld for workers: (1) dignity – a secure job with adequate training; (2) decency – a say in the future of their job; (3) fairness – a decent wage; (4) equality; and (5) justice – a safe and healthy workplace. In order to do this, BWTUC has forged a number of links both within and beyond the trade union movement, working with organizations as varied as the West London Business Link (a support organization for local small and medium-sized businesses), the Citizen's Advice Bureau (CAB), the local Further Education College, after-school clubs and local environmentalists. These various links have enabled BWTUC to reach a large number of people in the locality, and it is estimated that the centre brought at least 3,500 people into trade unions during the first two years it was in operation. Workers in poorly unionized sectors such as retailing, fast food and private manufacturing firms have been particularly targeted. For more, see Wills (2001) and http://www.respectatwork.org.uk/.

'Upscaling' worker action: organizing across localities

The final kind of geographical strategy we will consider is organizing across localities, or *upscaling* worker activity. This upscaling can take place at different spatial scales: it might be between workers in neighbouring towns; it might be between workers in distant parts of the same region or country; or, it might take the form of transnational connections between workers in different countries. Equally, it might be organized formally through local, national or international trade union structures, or it might be relatively organic and grassroots in nature, connecting across localities outside of formal worker organizations. Translocal activities of these kinds have a number of advantages: first, inter-place cooperation can prevent different groups of workers being played off against each other; second, a simple strength in numbers can be achieved by bringing more workers into a particular action; and third, a greater level of resources can be tapped by involving national and international organizations. In many countries, the formation of national trade unions (for example, on occupational or sectoral grounds) has traditionally been the favoured means of coordinating worker representation and action. As globalization dynamics have gathered pace in recent decades, however, and national labour agreements and unionization rates have come under pressure from both governments and corporations, increased emphasis has been placed on worker internationalism, that is, worker activism coordinated at the international scale to try and tackle the global reach of firms and capital.

A compelling example of a globally organized worker campaign is provided by the 20-month struggle between 1,700 members of the USWA (United Steelworkers of America) Local 5668 and the multibillion dollar metals trading empire of the international financier Marc Rich (Herod, 1995). The dispute started in 1989 when an aluminium plant near Ravenswood, West Virginia, was purchased by investors and became the Ravenswood Aluminium Corporation (RAC). Almost immediately, the RAC sacked 100 workers, increased the pace of work in the factory, and disbanded a joint worker-manager safety committee. Over the course of the next 18 months, five workers were killed and several injured at the Ravenswood plant. Employer–worker relations deteriorated still further when, in September 1990, the Ravenswood branch of the USWA (Local 5668) came to renegotiate its members' contracts with the RAC. In late October, RAC suspended the negotiations and hired 1,700 non-union replacement workers, arguing that Local 5668 was delaying the resolution of talks and was therefore technically on strike. The union workers were effectively locked out of their workplace. They decided to take up the fight through the USWA and the general US trade union, the American Federation of Labor-Congress of Industrial Organizations (AFL-CIO).

The campaign developed momentum when Local 5668 discovered who the owners and shareholders of RAC were. Most importantly, the investors were found to be led by Marc Rich, a notorious financier living in Switzerland and wanted in the US on tax fraud charges. Armed with this information, and in addition to pursuing labour law resolutions that would help their position, the USWA and the AFL-CIO organized an openly transnational campaign to get the Ravenswood workers reinstated. First, the two unions lobbied over 300 end-users of RAC metals worldwide to stop buying Ravenswood products. Second, the unions strove to reveal the links between Rich and the events at Ravenswood. In a series of actions involving unions in Western Europe, attempts were made to disrupt economic activities associated with Rich. For example, organized by the International Union of Food and Allied Workers, some 20,000 trade unionists protested against Rich's attempted purchase of a luxury hotel in Bucharest. By the spring of 1992, anti-Rich activities had been organized in some 28 countries on five continents, with the result that in June of that year, RAC was forced to concede defeat and to reinstate the workers of Local 5668. Coordinated in this instance by two large US unions, and by focusing on the various transnational operations of the Rich business empire, thousands of disparate workers worldwide were able to secure a favourable outcome for their colleagues at Ravenswood. What this reveals is that, in some instances at least, the power of corporations can be challenged by up-scaling worker action and connecting across different places.

9.5 Beyond Capital Versus Labour: Towards Alternative Ways of Working?

So far, this chapter has examined the role of workers *within* the capitalist system. However, it is possible to think of a wide range of non-capitalist or *alternative* economies within which the nature of work might be different to the dominant capitalist system. As we saw earlier in the case of the Jamaican information-processing sector, economic geographers need to recognize the diverse range of ways of working that *together* sustain livelihoods in communities across the globe. In the context of workers, it is possible to identify two alternative kinds of employment spaces that are being carved out within contemporary (capitalist) economies (Leyshon et al., 2003):

- alternative *formal employment spaces*: the emergence of 'third sector' community-owned and run enterprises – for example health, youth or arts centres – intended to fill the gaps in welfare provision between the state and private sectors (Box 9.5 profiles a well-known example of a workers' cooperative, the Mondragon Cooperative Corporation in Spain).
- alternative *informal employment spaces*: encompassing a variety of informal non-market work activities, which may be paid work that is hidden from the state (e.g. a builder paid cash-in-hand to evade taxes), unpaid

Box 9.5 The Mondragon Cooperative Corporation (MCC)

The Mondragon Cooperative Corporation (MCC) was formed in the Basque area of Spain in 1955. Unhappy with the close relations between large businesses and the Spanish government in the aftermath of Spanish Civil War, Father José María Arizmendiarrieta set up the first cooperative in order to establish a degree of economic and political independence for the area. Since its creation, the cooperative has greatly expanded, and now encompasses over 200 separate companies and bodies, divided into three industrial groups and a series of research and training centres (including a university). By 2003, the MCC employed just over 68,000 staff – 49 per cent in the Basque region, 39 per cent in the rest of Spain, and 12 per cent abroad – and had assets of €16.3bn and total sales of €9.6bn. Underpinning the whole endeavour are ten guiding principles:

1. open admission
2. democratic organization
3. the sovereignty of labour
4. the instrumental and subordinate nature of capital
5. participatory management
6. payment solidarity, i.e. just and equitable salaries
7. inter-cooperative cooperation
8. social transformation, i.e. reinvestment in community
9. universality, i.e. involvement in international organizations
10. importance of education and lifelong learning.

From its formation until the early 1990s, the MCC was able to use these principles to inform a model of growth that was widely lauded. Since then, however, it has gone through a period of expansion and restructuring driven by the logics of global competition that has challenged its founding principles, thereby illustrating the difficulty of instituting truly 'non-capitalist' forms of working. Traditional labour-intensive production methods have been steadily replaced by the greater use of technology. In addition, the MCC widened its growth strategy to include joint ventures – often with non-cooperatives – and offshore production and markets. By 2003, the MCC had 38 overseas plants and exports of €2.5bn. Overall, just over half of employees are now cooperative members. While the figure is 80 per cent in the Basque region, it is far lower in the rest of Spain, and negligible overseas. Another indicator of change is that the MCC has had to relax its rule that the ratio of the highest to the lowest wage should be at most 3:1. In sum, growing involvement in global markets has 'diluted' the coverage of MCC ideals. Managers now find themselves wrestling with depressingly familiar dilemmas, such as whether to keep production in the Basque region, where pay rates are higher, and risk job losses and firm closures, or establish joint ventures in regions where wages are low (Clamp, 2000). For more, see: http://www.mondragon.mcc.es/ing/index.asp.

work undertaken for one's own household (e.g. childcare and cleaning), or unpaid work for another household (e.g. work exchanged within family networks).

The importance of these alternative ways of working will clearly vary both socially and spatially. While in many developed societies they exist in addition to, and towards the margins of, mainstream capitalist activity, in many developing world cities they may constitute the chief forms of daily economic exchange for large sections of the population. For example, estimates suggest that between 40 and 60 per cent of total employment within the cities of Latin America, Africa and Asia is informal in nature.

We should not assume that these other forms of economy are unproblematic alternatives that might replace the capitalist system simply if more people engaged in them. Rather, the challenge for economic geographers is twofold. First, it is necessary to explore geographical variations in the prevalence and significance of these activities, and the ways in which these ways of working connect to, and interact with, the dominant capitalist system. Second, we need to consider how such activities might offer a potentially important starting place for thinking about how capitalist labour relations might be resisted, challenged and even changed – even if change is gradual and incremental. In reality, however, these initiatives quite often reveal as much about the difficulties of disconnecting from capitalist dynamics as they do about the potential for change. The Mondragon case (Box 9.5), for example, clearly reveals the problems inherent in combining a successful cooperative venture with the imperatives of the global marketplace. Equally, some third sector schemes can be read as state-driven initiatives designed to conceal gaps in formal welfare provision. Alternative economic geographies are seemingly always open to incorporation by capitalist activity. Equally, they may simply supplement the mainstream while remaining only notionally separate from it, or may support it without really providing a challenge to its dominance. While alternative economies need to be read critically then, this in no way detracts from the fact that there *are* different ways of living and working. If economic geographers believe in progressive change to the capitalist system, they will surely have to foreground and engage with these activities in order to complement understandings of labour geographies based on formal wage labour.

9.6 Summary

This chapter has moved from discussing *passive* geographies of labour to exploring the *active* geographies of labour. Under the former heading, we have reviewed a range of mechanisms and controls used by both corporations and states to place workers under pressure in the context of a global economy where capital is *relatively* mobile with respect to labour. This does not mean that workers are powerless, however. Considering the agency of labour, we have explored a range of spatial strategies that workers may use to try and resist, and indeed change, the dynamics of global capitalism. Workers may act alone, or in coalition with other groups in society, and they may do so in place, by moving between places, and by connecting across places. Various examples have been given to demonstrate these different strategies in action. The chapter concluded by considering the potential for alternative non-capitalist ways of working, albeit within the context of a global capitalist system.

Returning to the question which frames this chapter – can workers shape economic geographies? – our answer is a qualified 'yes'. As noted in Chapter 3, capitalism is an economic system that depends on the activities of waged workers at the same time as it tries to extract concessions from them. Furthermore, it is a system that depends on complex connections both within, and between, different places. As we have seen, these two attributes offer considerable potential for labour to exert its collective agency in different ways. Importantly, however, the ability to act varies both socially and spatially. What is possible in some workplaces, sectors, and localities may not be viable in others. It is the task of the economic geographer to reveal this on-the-ground complexity and the implications that stem from it. What is clear though, is that workers, in addition to states (see Chapter 7) and firms (see Chapter 8), are extremely important agents for change in the development of the global space economy.

Further reading

- Castree et al. (2004) provide a student-friendly and explicitly geographical approach to the study of labour.
- Herod (2001) charts the shift from geographies of labour to labour geographies and offers a range of compelling case studies of the latter.
- Jonas (1996) offers a challenging but rewarding conceptualization of local labour control regimes and the ways in which capital is necessarily attached to place.
- Kelly (2002), Mullings (1999), Saxenian (1999), Herod (1995, 2000) and Wills (2001) offer further details on the various case studies covered in this chapter.
- Examples of different 'ways of working' are described and evaluated in Leyshon et al. (2003).

Sample essay questions

- In what ways are business- and state-imposed mechanisms of labour control interrelated?
- How and why do the mechanisms of labour control employed vary between different places?
- How can labour work with others in the local community to assert its agency?
- What are the main geographical strategies through which workers can assert their agency?
- What are the possibilities and limits of workers organizing themselves transnationally?

Resources for further learning

- http://www.ilo.org: the website of the International Labour Organization, a United Nations agency, has a wealth of information of labour and working conditions worldwide.
- http://www.icftu.org/default.asp?Language=EN: the International Confederation of Free Trade Unions (ICFTU) is one of the most important international union federations (see also the World Federation of Trade Unions: http://www.wftucentral.org/).
- http://labornotes.org/: the home of Labour Notes, a regular report on labour activism worldwide.
- http://www.unglobalcompact.org/: the website of the Global Compact, a UN initiative to bring together companies, labour and civil society groups to promote universal principles in the areas of labour, human rights, the environment and corruption.
- http://www.etui-rehs.org/: the research arm of the European Trade Union Confederation (ETUC) which offers a range of reports and information on labour issues in Europe.
- http://www.labourstart.org/: this is a trade union website that contains up-to-date stories about worker struggles worldwide.

References

Castree, N., Coe, N.M., Ward, K. and Samers, M. (2004) *Spaces of Work: Global Capitalism and the Geographies of Labour*, London: Sage.

Clamp, C. (2000) The internationalization of Mondragon, *Annals of Public and Cooperative Economics*, 71: 557–77.

Harvey, D. (1989) *The Urban Experience*, Oxford: Blackwell.

Herod, A. (1995) The practice of international labor solidarity and the geography of the world economy, *Economic Geography*, 71: 341–63.

Herod, A. (1997) From a geography of labor to a labor geography: labor's spatial fix and the geography of capitalism, *Antipode*, 29(1): 1–31.

Herod, A. (2000) Implications of just-in-time production for union strategy: lessons from the 1998 General Motors-United Auto Workers dispute, *Annals of the American Geographers*, 90: 521–47.

Herod, A. (2001) *Labor Geographies*, New York: Guilford Press.

Hsu, J-Y. and Saxenian, A. (2000) The limits of guanxi capitalism: transnational collaboration between Taiwan and the USA, *Environment and Planning A*, 32: 1991–2005.

Jonas, A.E.G. (1996) Local labour control regimes: uneven development and the social regulation of production, *Regional Studies*, 30: 323–38.

Kelly, P.F. (2002) Spaces of labour control: comparative perspectives from Southeast Asia, *Transactions of the Institute of British Geographers*, 27: 395–411.

Leyshon, A., Lee, R. and Williams, C.C. (eds) (2003) *Alternative Economic Spaces*, London: Sage.

Martin, R.L. (2000) Local labour markets, in G. Clark, M. Feldman and M. Gertler (eds) *The Oxford Handbook of Economic Geography*, Oxford: Oxford University Press, pp. 455–76.

Martin, R.L. and Morrison, P.S. (eds) (2003) *Geographies of Labour Market Inequality*, London: Routledge.

Moody, K. (1997) *Workers in a Lean World*, London: Verso.

Mullings, B. (1999) Sides of the same coin? Coping and resistance among Jamaican data-entry operators, *Annals of the American Geographers*, 89: 290–311.

Peck, J. (1996) *Work-place*, New York: Guilford Press.

Saxenian, A. (1999) *Silicon Valley's New Immigrant Entrepreneurs*, San Francisco: Public Policy Institute of California.

Wills, J. (2001) Community unionism and trade union renewal in the UK: moving beyond the fragments at last? *Transactions of the Institute of British Geographers*, 26: 465–83.

CHAPTER 10

CONSUMPTION

Is the customer always right?

Aims

- To recognize the position and importance of the consumption process within the capitalist system
- To appreciate the changing geographies of consumption and in particular, retailing
- To understand how consumption spaces are actively designed and used
- To reflect on the ways in which consumption, place and identity are interrelated.

10.1 Introduction

Not many people in the UK have heard of 'dunnhumby'. Initially established in 1989 by Clive Humby and Edwina Dunn, by 2005, the firm employed 400 staff across offices in the UK, Ireland and USA, and had a turnover of £130 million. Dunnhumby has become a market leader in collecting and analysing information about consumers and their shopping patterns, in turn, using this information to advise clients on new ways of interacting with customers and enhancing 'brand value'. By contrast, almost everyone living in the UK will have heard of – and indeed visited the stores of – dunnhumby's majority shareholder and main client, the food retailer Tesco. Since 1995, the companies have worked together to introduce and develop Tesco's *Clubcard*, the card-based loyalty scheme that monitors members' shopping patterns in return for a one per cent rebate on money spent in Tesco stores. The scheme, used regularly by some 12 million shoppers, is widely credited with fuelling the retailer's explosive growth in the past decade – quite an accolade when one considers Tesco's current size and status. One in every eight pounds spent in a British retailer goes to Tesco, and almost one in three in the grocery market. The company made £2 billion profits on turnover of £37 billion in 2004, employing 250,000 people across 1,800 stores in the UK, and another 100,000 across 600 stores in 12 overseas markets.

Dunnhumby's latest *Crucible* database is designed to profile every customer in the UK, whether they shop at Tesco or not. Pulling together a variety of data sources, it classifies individual consumers according to variables such as their wealth, travel patterns, credit levels, amount of free time, lifestyle, environmental ethic, charity contributions and susceptibility to promotions and new products. Promotional material for the company describes a 'Mrs Pumpkin', for example, who is careful about money when shopping, mostly uses cash, buys a steady range of goods but experiments with new products, shops at various times, buys some eco-friendly goods, is involved in charitable giving, travels rarely, and likes buying promotional products (*The Guardian*, 2005). Combining this detailed customer information with Clubcard data enables Tesco to develop individually targeted promotional and incentive schemes. The information can also be analysed geographically, of course. First, it allows the retailers to gauge the most appropriate locations for new stores by comparing local social and economic conditions with the profile of their wider customer base. Second, it enables them to adjust their product mix both spatially and temporally. For example, in Brixton, London, an area settled by immigrants from the Caribbean, Tesco sells plantains, a type of savoury banana. Tesco stores in central London do not, but instead are based on selling sandwiches to office workers in the middle of the day, and pre-prepared 'ready' meals to them in the evening. In short, technologies such as the Clubcard and Crucible help retailers 'to combine the local knowledge of the village shop with a multinational's economies of scale in buying and logistics' (*The Economist*, 2005: 23).

The example of dunnhumby and Tesco serves to introduce us to several important aspects of contemporary consumption dynamics. The sheer number and coverage of Tesco stores alert us to the fact that consumption is a necessarily geographically more extensive activity than production. It illustrates the growing influence of large transnational retailers on the range of products we buy, and where we are able to buy them. It reveals how this power is in part based upon access to highly detailed information about how and where we shop. It speaks to processes of retailer globalization and the development of similar retail experiences in many countries around the world. And yet, as the comparison of Tesco in Brixton and central London shows, the consumption experience remains highly geographically variable, even between the stores of the same retail corporation in the same city. Consumption spaces such as supermarkets are shaped by the economic, social and cultural spaces of which they are part. This chapter seeks not only to explore these geographies of consumption, but also to consider where the power to shape these spaces lies – are our consumption spaces and decisions shaped for us, or 'is the customer always right'?

The argument proceeds in four stages. First, we consider different interpretations of who controls consumption dynamics, arguing for a perspective that takes the roles of both corporations and consumers seriously (Section 10.2). Second, we look at the changing geographies of consumption in the contemporary era, with a particular focus on shifting patterns of retail activity (Section 10.3). Third, we zoom in to look at how particular consumption spaces – the store, the mall, the theme park – are designed and used (Section 10.4). Fourth, we switch focus to consider how consumers use the commodities they buy as part of *place-specific* processes of identity construction (Section 10.5). In short, this chapter aims to reveal how consumption is as geographically varied and complex as processes of production.

10.2 The Consumption Process

What exactly do we mean when we talk of 'consumption'? As we saw in Chapter 4, the commodity chain concept allows us to see how consumption is the final step in a series of material transformations and value-adding activities. Importantly, consumption refers to the sale, purchase *and use* of commodities. Hence it is not only about the interface between those offering products/services for sale, and those making decisions about which products/services to buy, but is also about what people do with commodities after buying them. Consumption is thus not just about simple economic transactions at the point of sale. Rather, it is a *process* that includes socio-cultural aspects of commodities and their use, encompassing a range of activities: purchasing, shopping, using, discarding, recycling, reusing, wearing, washing, eating, leisure, tourism and home provisioning and renovation among others. Consumption is therefore underpinned by a wide range of activities such as shops, restaurants and hotels, and repair, servicing, cleaning and recycling operations, which are significant economic sectors in their own right.

The nature of the consumption process means it possesses an inherently different economic geography to production operations. Put simply, while not every large city or region will have a car or plastics factory, all will have supermarkets, cinemas, laundrettes, restaurants and the like. Due to the frequency and nature of the way in which we use such services, consumption activities necessarily have a more extensive and disparate geography than the other parts of the production chain. One key task, then, is to understand how this economic geography is structured by different kinds of capitalist firm as part of their profit-making strategies. Equally, however, we need to comprehend the ways in which the *nature* of consumption varies geographically, due in large part to its socio-cultural dimensions. In this regard, sustained research into consumption has been a key characteristic of the 'new' economic geography that has emerged since the mid-1990s (see Box 1.2).

Before going any further, however, we need to consider the wider significance of consumption within the production chain. What role do final consumers play in shaping the chain and the nature of its operation? Clearly, production and consumption are two mutually dependent spheres of activity, but which drives which? Here we shall look at three important perspectives on this debate. The first, which can be characterized as the *consumer sovereignty* view, emphasizes the agency and free will of individual consumers. Deriving from *neoclassical economics*, consumption is seen here as an individual market-based economic transaction dependent on price-based decisions. Rational consumers, acting on full information about different products and prices, will make informed decisions about which products to buy. These are individual decisions taken by autonomous consumers unaffected by society and its various norms and expectations. In turn, these actions will have an impact on the production process as market information about which products are popular feeds back through retailers to manufacturers. In short, consumers can choose whether to buy the products on offer or not, and as such they have considerable sovereignty, or control, over the economic system as a whole. From this viewpoint, then, the customer *is* always right!

This can be contrasted with the *consumers-as-dupes* viewpoint, which emphasizes the primacy of *production* within the economic system as a whole. From such a perspective, consumption is read as the outcome of changes in the nature of the production process. It is a pleasure-seeking but ultimately regressive activity in which passive consumers are enticed into parting with their money, thereby delivering profits for producers and refuelling the basic capitalist model described in Chapter 3. As we shall see later in this chapter, this argument can be extended to explore how particular consumption spaces – for example, a shopping mall – are actively designed to induce consumers to spend as much money as possible during each visit. The rise of retail capital over the last few decades has complicated the picture in that consumption is increasingly shaped by *retail* rather than manufacturing corporations. Most importantly, however, the consumers-as-dupes perspective places capitalist corporations rather than consumers in control of consumption processes and spaces.

Third, what we might call a *culturalist* perspective emphasizes the way in which consumers actively construct their own identity through their consumption practices. Such an approach draws attention to the other intersecting facets of identity – such as gender, ethnicity, age and sexuality (see Chapters 12 and 13) – and looks at how consumers knowingly purchase certain commodities and use them in certain ways as part of an active process of identity construction. These processes of identity consumption, however, are not individualized as in the consumer sovereignty view, but rather are part of larger social forces and trends within society (e.g. teenage fashion). Here the interactions between

Table 10.1 Mass consumption and after-Fordist consumption compared

Characteristics of mass consumption	Characteristics of after-Fordist consumption
Collective consumption	Increased market segmentation
Demands for familiarity from consumers	Greater volatility of consumer preferences
Undifferentiated products/services	Highly differentiated products/services
Large-scale standardized production	Increased preference for non-mass produced commodities
Low prices	Price one of many purchasing considerations, alongside quality, design etc.
Stable products with long lives	Rapid turnover of new products with shorter lives
Large numbers of consumers	Multiple small niche markets
'Functional' consumption	Consumption less 'functional' and more about aesthetics
	Growth of consumer movements, alternative and ethical consumption

consumers and corporations are complex and two-way. Consumption processes will clearly be influenced by manufacturing and retail corporations and the commodities they offer, but consumers purchase such goods selectively and knowingly: firms, in turn, will respond to emerging consumption dynamics within society. As we shall see later, this approach also allows the geographies of consumption to be laid bare, as the processes of identity construction are very much place-specific. This perspective also has implications for how we might read particular consumption spaces, suggesting that consumers are actively aware of the way in which spaces are designed to induce consumption.

What is clear is that the *nature* of consumption has changed in the past few decades. In parallel with our discussion of after-Fordist production systems in Chapter 5, it is possible to identify the key attributes of what might be called *after-Fordist consumption* (Table 10.1). In general, the Fordist era was characterized by the large-scale, mass consumption of a relatively limited range of standardized commodities (originating, most famously, with the Model T car, available as Henry Ford declared, in 'any colour so long as it's black'). The system was driven by economies of scale and low prices that enabled the widespread consumption of a broad range of household and personal goods. The after-Fordist era, by comparison, seems to have engendered a much more fragmented consumption pattern in which many highly differentiated products are offered to a wide range of consumer groups or *niches*. This mode of consumption is driven less by the price and functionality of commodities, and more by the aesthetic and symbolic value they bring to consumers. While there is general agreement among commentators on the nature of these changes, how might we explain them? For some, these changes have been *caused* by the shift from Fordist to more flexible after-Fordist production systems. In other words, after-Fordist consumption simply reflects a more nimble and subtle form of capitalism, but a production-driven one nonetheless. Alternatively, however, after-Fordist consumption can be seen as a mode of consumption shaped by more strategic and knowledgeable end consumers that have thrown off the constraints of mass consumption.

One characteristic of after-Fordist societies of the developed industrial world that is undisputable, however, is the growing importance of consumption-related employment. Box 10.1 points to some of the features of this type of work, not all of which are positive.

This chapter seeks to combine the insights from both the consumer-as-dupe and culturalist approaches to consumption: as already explained in Chapter 1, we see limited utility in approaches grounded in the assumptions of neoclassical economics. Consumption, then, is quite clearly neither entirely determined by corporations or consumers, but is a complex process of interaction between the two groups. More specifically, in this chapter we want to advance the more nuanced argument that the extent to which producers/retailers or consumers hold sway over the consumption process will vary both according to the commodity in question, and geographically. For example, in the UK food retailing sector introduced above it is hard to escape the huge influence wielded by Tesco and other leading retailers such as Sainsburys and Asda. In other less concentrated food retailing markets, however, consumers will have more choice and hence greater sovereignty over their food consumption choices. In different kinds of activities – such as independent high quality restaurants in global cities or high fashion – consumer choices will be crucial to the ongoing viability of the enterprises involved, and as a result the range of options may fluctuate quite rapidly depending on changing reputations and customers' reviews. It is, however, hard to escape the conclusion that there is a *growing* concentration of power and influence in the hand of large retail corporations. We now turn to look at their efforts to shape the geographies of consumption. In so doing, we will see how retail geography has moved beyond the mathematical modelling of the intra- and inter-urban distribution of retail outlets, a preoccupation during economic geography's quantitative phase (Box 10.2).

Box 10.1 Consumption work

Another important characteristic of what we are calling the after-Fordist era in developed economies is that a growing proportion of the workforce is accounted for by consumption-related jobs in sectors such as retailing, restaurants, tourism and entertainment. Tesco's 250,000 employees in the UK, for example, make it the largest private sector employer in the country. At the same time, the rapid growth in these kinds of jobs has raised questions about their quality and desirability. These doubts have prompted economic geographers (and other social scientists) to look *within* workplaces and explore the nature of work in the consumptive sphere. This work has revealed that consumption-related jobs tend to be characterized by some, or all, of the following features:

- socially constructed as low status jobs within society (e.g. 'the waitress', 'the flight attendant', 'the checkout assistant') and receiving relatively low wage rates;
- low levels, or indeed the complete absence of, collective representation and unionization;
- dominated by part-time and temporary contracts rather than full-time positions. In some instances (e.g. tourism) there may be a marked seasonality to availability of employment;
- technology-based surveillance of the workforce both to monitor workers (e.g. listening in to telephone sales calls, measuring productivity of fast-food operatives) and deploy them efficiently (e.g. gauging the appropriate number of checkout staff in a supermarket);
- based on labour-intensive, repetitive tasks which cannot be replaced by technology (e.g. waiting on tables or cutting hair);
- having a *performative* component in which workers are required to take on a particular role or personality. This can occur in both face-to-face (e.g. the scripts given to fast food workers) and technologically-mediated situations (e.g. Indian call centre workers taking on English names as part of scripted encounters with customers). These performances may challenge workers' true senses of identity and lead to substantial stress levels;
- the social construction of many of these jobs as essentially 'female' (e.g. flight attendants), reinforced through recruitment practices;
- the social construction of many of these jobs as being for young people (e.g. fast food workers, bar and nightclub staff), reinforced through recruitment practices;
- selective recruitment on the basis of a wide range of other personal attributes – ethnicity, bodily appearance, weight, bodily hygiene, dress and style, interpersonal skills – to fulfil the requirements of the performative encounter with customers.

Box 10.2 Central place theory

The development of central place theory was indicative of the quantitative modelling approach to economic geography that predominated during the 1950s and 1960s in the Anglo-American context (see Box 1.2). A central place is simply a place that offers a good, or goods, for sale. The theory concerned developing models of the size and distribution of urban settlements as determined by retail activity. Perhaps the best-known model is that developed – initially in the 1930s, but translated to English in the mid-1960s – by the German economic geographer Walter Christaller. His model was based on two key variables: the distance a consumer will travel to purchase a particular good (the *range* of a good), and the minimum level of business required for a retail business to be viable (the *threshold* of a good). These variables were used to group retail establishments into categories or *orders*. In deriving the geography of location for each order, he argued that retailers would establish their shops as close to their customers as possible in order to minimize travel costs and maximize turnover. His analysis revealed that this requirement of centrality would lead to an orderly spaced hexagonal network of shop locations in central places, reflecting the fact that hexagons are the most efficient geometrical figures for serving territory without overlap (Figure 10.1). When the distributions of all seven orders were intersected, a hypothetical urban hierarchy was created in which the lowest order shops were available in all locations, and higher order shops would be available further up the hierarchy. While such models have proved influential and have been applied historically in real-world settings, they are simplifications of a very complex reality. Their contemporary applicability to a world shaped by cheap and efficient transport systems, powerful retailers, variable regulatory and planning frameworks, and geographically uneven distributions of wealth is severely limited.

10.3 The Changing Geographies of Retailing

The development of large, powerful, highly profitable retailers has tremendous downstream implications for consumption processes. Such retailers have considerable power to shape the geography and nature of contemporary consumption patterns through influence over which products are sold, where they are sold, how they are sold, and at what price. In this section we will consider the changing geographies of contemporary retailing in three stages. First, we will profile the ongoing globalization of retail activity and its implication for consumption patterns. Second, we explore the shifting intra-national geographies

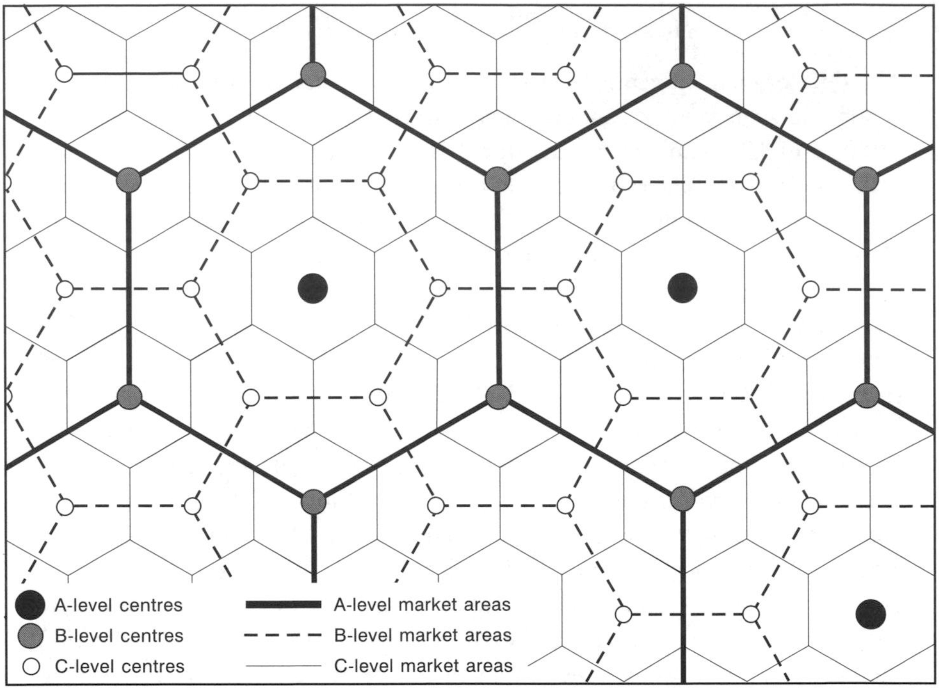

Figure 10.1 Christaller's hexagonal central place theory pattern
Source: Stutz and Warf (2005), Figure 10.10.

of retailing, and in particular, dynamics at the urban scale. Third, we move beyond the formal spaces of retailing to consider the significance of various *informal* modes of retailing.

The globalization of retailing

Perhaps the most important geographical outcome of the rise of retail capital has been a tremendous globalization of retailing over the past 10–15 years as retailers have sought new opportunities in which to invest the profits secured from their home markets (Coe, 2004). The period since the early 1990s has seen the emergence of a select group of transnational retailers that has used aggressive merger and acquisition activities, backed up by subsequent rapid organic growth, to assume dominant market positions across a range of countries. The leading transnational retailers, ranked in terms of their foreign sales, are detailed in Table 10.2. A number of important observations can be made from this data. First, it gives a sense of the *scale* of international retail operations, with all 15 retailers deriving over US$10bn of sales from foreign markets in 2005. Second,

Table 10.2 Leading transnational retailers, ranked by sales outside home market, 2005

Rank	Name of company	Country of origin	International sales (US$m)	International sales (% of total)	No. of countries of operation
1	Wal-Mart	US	62,700	20	15
2	Carrefour	France	50,050	52	29
3	Ahold	Netherlands	45,352	82	5
4	Metro	Germany	38,502	54	30
5	Aldi	Germany	20,119	45	12
6	Lidl & Schwarz	Germany	19,832	43	19
7	Tesco	UK	19,640	24	12
8	Auchan	France	19,535	45	12
9	Delhaize	Belgium	18,893	79	9
10	IKEA	Sweden	18,868	96	33
11	Tengelmann	Germany	16,706	51	16
12	Rewe	Germany	15,207	31	14
13	Ito-Yokado	Japan	12,010	34	4
14	Casino	France	11,849	42	19
15	Pinault	France	11,775	46	29

Source: Derived from Deloitte/Stores, *2006 Global Powers of Retailing*.

it gives a sense of the *scope* of international retailing, with many of the leading players having store operations in 20–30 countries, a level of international expansion comparable with many manufacturing sectors. Third, it reveals that the leading retail transnationals – with the notable exception of the world's largest retailer by far, Wal-Mart – are Western European, or more accurately, British, French, Dutch, German, Belgian and Swedish. While many of the world's very largest retailers hail from the US (e.g. Home Depot, Kroger, Target, Sears and Safeway), they can achieve that size without straying far beyond the borders of their home country. Fourth, it shows that the leading transnational retailers tend to be food retailers or general merchandisers, rather than specialty providers (e.g. toys or computer goods).

What, though, is the geography of this recent wave of expansion? The globalization of retailing is not a new process, dating back as far as the late 1800s. Foreign expansion really took off in the 1960s, however, and until the 1990s was largely dominated by investments between the leading economies of North America, Western Europe and Japan. Since the mid-1990s the expansion has taken on an entirely new geographical configuration. Figure 10.2, showing Tesco's global store distribution in 2004, is indicative. Tesco had stores in five markets in Eastern Europe and six in East Asia, in addition to the UK and

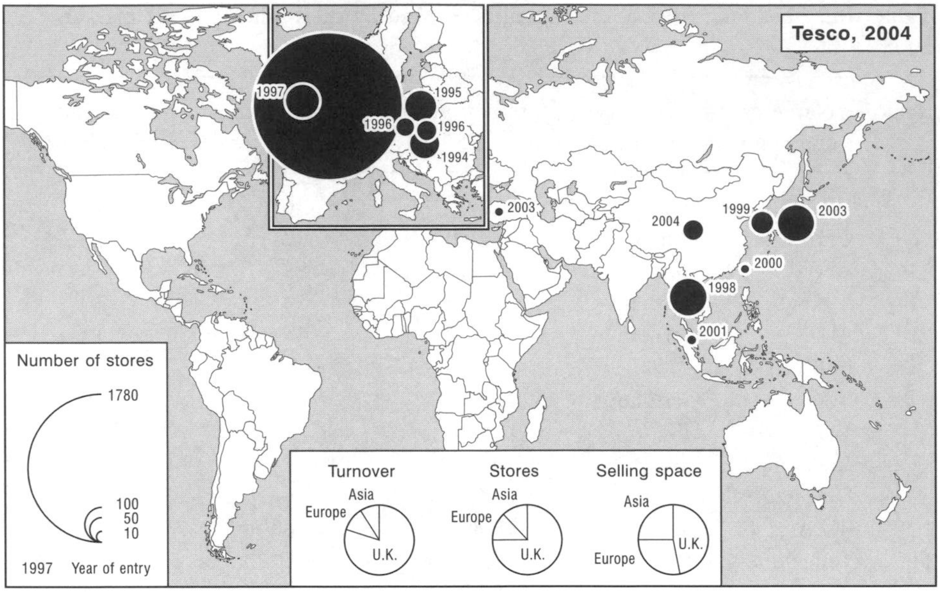

Figure 10.2 The global distribution of Tesco stores, 2004
Source: Company reports.

Ireland. In general, international retail investment by the leading retail transnationals is now targeted towards countries in so-called *emerging* regions, namely Latin and Central America, East Asia and Eastern Europe.

Three further characteristics distinguish this latest phase of retail globalization. First, the sheer *rapidity* of the international expansion is startling. As one measure, in 1990, Ahold was present in the Netherlands and USA, Carrefour was to be found in five countries outside of France, and Wal-Mart and Tesco were only present in their home market. By 2004, Ahold was present in 21 countries in total (operating brands such as Stop and Shop, and Tops, in the USA, for example), Carrefour in 30, Wal-Mart in 10, and Tesco in 13. Second, the *scale* of investment currently being undertaken is unprecedented. Static 'snapshots' such as Table 10.2 and Figure 10.2 may not give a real sense of the size of the phenomenon. For example, over the period 1994–2004, Tesco went from having no foreign floor space to having 53 per cent of its global total in its overseas territories. These huge expansion processes are creating retail giants with foreign operations of unprecedented size. Third, the *impacts* of this expansion on the retail structures of the host countries are significant. Table 10.3 profiles Poland's retail structure in 2004, and reveals that six of the top ten (and four of the top five) retailers are foreign, or more specifically Western European, owned. In the hypermarket segment, foreign dominance is particularly pronounced, with Tesco, Auchan, Metro, Carrefour and Casino together accounting for 81 per cent of

Table 10.3 Top ten retailers in Poland, 2004

Company	Ownership	Formats	No. of outlets (Sep 2004)	2004 sales (€bn)
Metro	Germany	Hypermarkets, cash & carry, specialty	85	2.64
Spolem	Poland	Grocery stores, discount, supermarkets	ca. 4,800	1.61
Tesco	UK	Hypermarkets, supermarkets	70	0.95
Jeronimo Martins	Portugal	Discount	700	0.91
Auchan	France	Hypermarkets, supermarkets	31	0.89
Ruch	Poland	Wholesale, kiosks, grocery stores	ca. 37,000	0.84
Carrefour	France	Hypermarkets, supermarkets	85	0.82
Eurocash	Poland	Cash & carry, grocery stores	1,930	0.80
Casino	France	Hypermarkets, discount	160	0.75
PSH Lewiatan	Poland	Grocery stores	1,610	0.66

Source: www.retailpoland.com, accessed 25 October 2005.

the total national market. A similar story can be told for many leading economies across Eastern Europe, Latin America and East Asia.

Leading transnational retailers, then, are coordinating increasingly extensive networks of stores at the global scale. However, we need to be cautious about inferring from these trends that the nature of retail spaces is becoming identical around the world. Clearly, there are tendencies in that direction, with elements of store formats and signage, for example, being common across all markets. A British shopper entering a Tesco shop in Prague (Czech Republic) or Bangkok (Thailand) would quickly know they were in a Tesco store and see aspects of the store design that are very similar to stores in the UK. Equally, however, the same shopper would also notice significant variations in terms of the product range on offer, how goods are displayed to shoppers, and the physical structure of the store. In Prague, for example, Tesco has a city centre department store stretching over several storeys, and in Bangkok, Tesco hypermarkets are often part of large, multi-function developments including a range of other food, retail and leisure establishments (Figure 10.3). Both are quite different to the free-standing single storey supermarket that predominates in the UK. Importantly, transnational retailers vary in the extent to which they try and offer a standardized retail experience across all the countries in which they operate. While some companies, such as Wal-Mart and IKEA, operate in a highly *centralized* and uniform manner, others such as Ahold pursue a more *federal* strategy, operating as a collection of relatively autonomous national subsidiaries.

Figure 10.3 A Tesco hypermarket in Bangkok
Source: The authors.

From centre to suburbs, and back again?

Having established the contemporary importance of processes of retail globalization, we can now move on to consider the changing geography of retailing at the urban scale. In particular, we want to profile the shifting of retail investment from the inner city and downtown to the suburban periphery and back again, and in so doing, reveal how retail capital plays a central role in the constant re-making of the urban built environment and its consumption landscapes.

Up until the 1950s, retailing was essentially a *central* urban activity. The post-war suburbanization of retail capital and related decline of city-centre retailing were arguably pioneered – and indeed most evident – in the US. The scale and significance of these dynamics can be illustrated through briefly looking at developments in and around Chicago, Illinois, over the period from 1950 to the mid-1970s (drawing on Wrigley and Lowe, 2002). By the end of the 1950s, four large open-air shopping centres with ample parking provision had appeared on the periphery of Chicago as department stores began to realize the potential of shifting their focus to the rapidly expanding middle-class suburbs (Figure 10.4). The 1960s saw the building of a series of enclosed shopping centres or *malls* in a ring around Chicago, fuelled in part by the expanding urban expressway network. By the end of the decade, a total of 11 suburban shopping centres had combined retail sales to rival central Chicago. By 1974, the total was up to

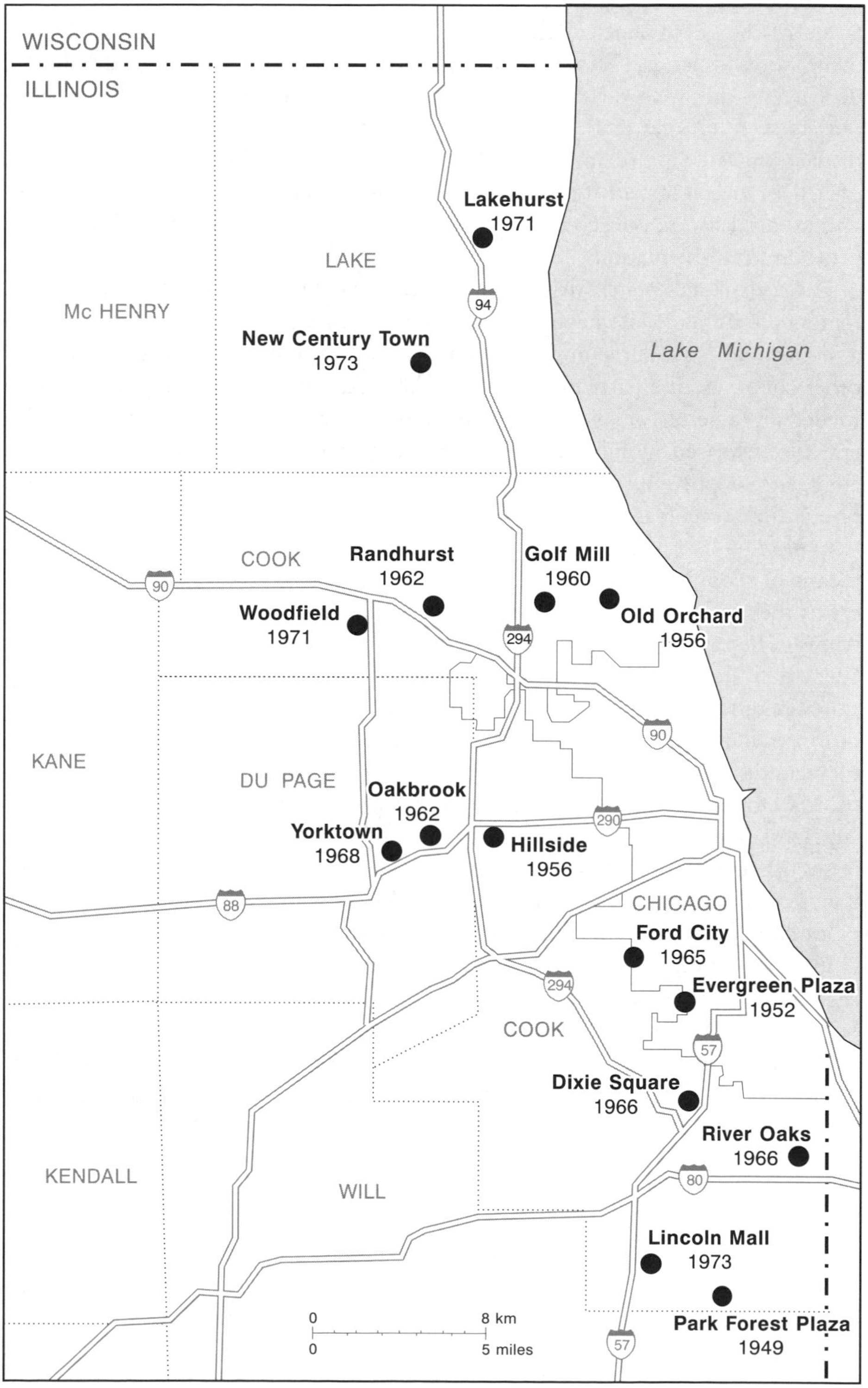

Figure 10.4 The development of Chicago's suburban shopping centres, 1949–74
Source: Adapted from Berry et al. (1976), Figure 22.

15 shopping centres, and a second 'ring' around Chicago was starting to emerge. Some of the key tenants of these new suburban centres were the very department stores – such as Sears Roebuck and J.C. Penney – that had once dominated downtown shopping districts. At the same time, central and inner city retail outlets were closing in their hundreds. Chicago's leading inner city shopping district, known as 63rd and Halsted, was severely hit, finding itself in the middle of a decaying area with falling income levels by the early 1970s. In just two-and-a-half decades, the geography of Chicago's retail provision had undergone a profound transformation.

These dynamics, often driven by powerful alliances of property developers, retailers and financiers, have been repeated in hundreds of cities across the US and Canada. A similar suburbanization of retail provision has occurred in many other countries, and particularly those of Western Europe. Such trends have been fuelled by a series of overlapping dynamics characteristic of the after-Fordist era: the increased mobility of consumers and high levels of car ownership; the emergence of portions of the population with large amounts of disposable income; population decentralization from urban to suburban areas; higher female participation in the workforce driving demand for quicker and more efficient means of shopping (e.g. the 'one-stop-shop'); and the growth strategies of retailers as they seek enhanced economies of scale through larger and more accessible stores. The *extent* to which the North American trends described above have been replicated elsewhere has been dictated by the strength of planning regulations in different countries: most European countries, for example, have been more resistant to suburban greenfield developments than in the US case. Regulatory conditions also critically affect the *timing* of decentralization processes. In the UK, for example, there was a boom in the development of out-of-town food superstores in the 1980s and early 1990s when planning guidelines on building in suburban areas were favourable. By the mid-1990s, these trends were promoting concerns about the development of *food deserts* – low-income areas of British cities characterized by the lack, or poor quality, of food retail provision.

In most countries, suburban retail parks are now a well established part of the consumption landscape. In the UK, the initial wave of food superstores has been joined by three subsequent phases of out-of-town development (Fernie, 1995). First, there has been a wave of retail parks housing, for example, home maintenance, carpet, furniture and electrical stores whose bulky products suit such accessible locations. Second, in contrast to the large scale decentralization of city centre retailing seen in the US, planning constraints have restricted such relocations (beyond the food superstores and retail parks just mentioned) to a small group of massive *regional* shopping centres that aim to serve more than their immediate urban hinterland. As shown in Figure 10.5, these are quite evenly spread across the country and, pioneers such as Brent Cross and the MetroCentre apart, were largely built in the 1990s. The Trafford Centre, for example, on the western outskirts of Manchester, has approximately 120,000 square metres of

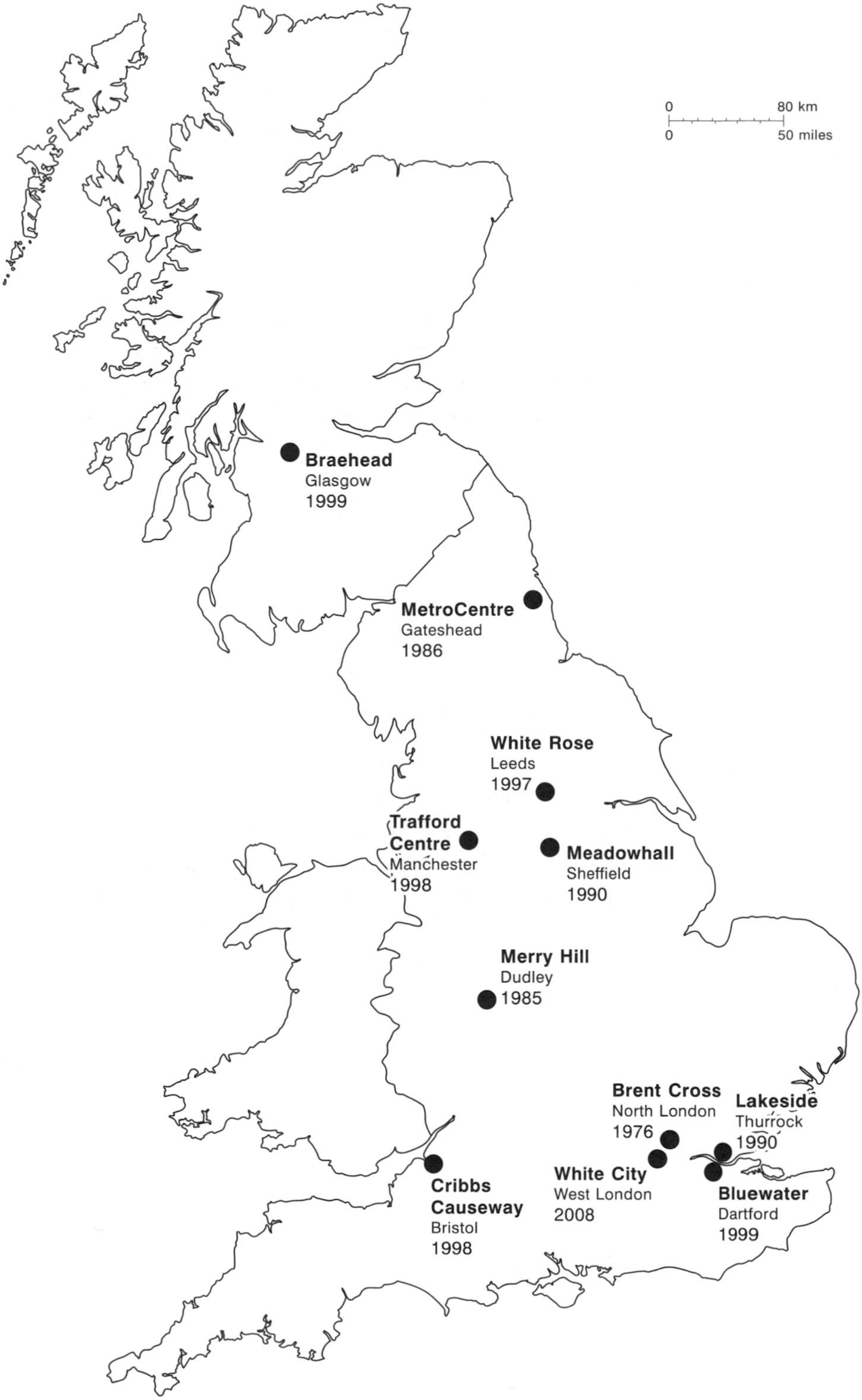

Figure 10.5 Britain's regional shopping centres
Source: Adapted from Lowe (2000), Figure 1.

retail space in addition to 30,000 square metres of catering and leisure space and 10,000 car parking spaces. The centre is home to 6000 employees working in 280 shops including six large 'anchor' tenants, 36 restaurants, a 1,600-seat food court, a 20-screen cinema and several other leisure attractions. The third, and latest, wave of suburban retail in the UK takes the form of outlet shopping malls that offer the excess stock of clothing manufacturers at reduced prices. Again, however, planning restrictions and property market conditions have meant that the expansion of such outlets has been much slower than predicted. Overall, then, it is important to recognize that the suburbanization of retailing now constitutes many different kinds and formats of retailing, and that its extensiveness varies significantly between different national regulatory contexts.

While suburbanization has had a massive and lasting impact on retail geographies, it is also important to note recent trends working in the opposite direction. While the nature and extent of the trend will again vary from context to context, three brief illustrations can be offered here to show the kinds of dynamics that are occurring. First, in the US context, since the mid-1970s when the level of retail decentralization was at its peak, considerable efforts have been made to *regenerate* the downtown districts of American cities. Central to such initiatives have been so-called *festival marketplaces*, stimulated by the success of the Faneuil Hall Marketplace which opened in Boston in 1976. The central idea here is to use combinations of architecture, cultural exhibits, concerts and ethnic festivals to attract people to speciality markets and shops, and restaurants. A different, but closely related, form of development has seen downtown retail revitalization as part of multiple use complexes comprising offices, hotels, leisure facilities and convention centres. Similar schemes have subsequently been tried in many cities around the world, including those where retail decentralization has been less pronounced. Second, in many of the post-industrial cities of developed countries (e.g. New York, Manchester) inner city retail has been boosted by *gentrification* processes which have seen the return of young middle-class professionals to live in new or renovated apartments in inner city areas. This has created high income areas that are attractive to shops, restaurants, nightclubs, and so on. These processes are also becoming apparent in certain developing country cities such as Shanghai. Third, and particularly in the UK context, large food retailers are starting to return to the inner city, prompted in part by the gentrification trends just mentioned, albeit in a patchy manner that does not mitigate the problems of food deserts identified above. Both Tesco and Sainsburys, for example, have established new small supermarket and convenience store formats to tap into these urban markets.

From formal to informal spaces of retailing

Thus far we have tended to focus on *formal* retail spaces as perhaps the most prevalent and important consumption spaces in capitalist society. It is important to recognize, though, that beyond the formal spaces of retail lies a wide variety of *informal* trading and exchange practices and spaces. Some of these activities are transient and temporary in nature (e.g. mobile street vendors and street markets) while others are a more permanent feature of the built environment (e.g. charity shops and retro-vintage shops). While some will be entirely legal operations, others may operate at the very limits of the law, or even illegally. The importance and nature of these informal consumption spaces vary tremendously geographically, as the following brief examples illustrate.

A fascinating example of informal retail spaces in the developed world is that of *car boot sales* in the UK (for more, see Crewe and Gregson, 1998). Car boot sales first appeared in the late 1970s but were booming by the 1990s, with an estimated 1500 taking place every week, attracting over one million people as either buyers or sellers. Car boot sales, as transitory and informal sites of second-hand exchange, have much in common with jumble sales, flea markets, yard sales, garage sales, and the like. Their distinctiveness comes from the fact that the household and personal goods offered for sale are transported, displayed and sold from, and around, the back of participants' cars or vehicles. Car boot sales usually take place in fields, car parks, or on other available spaces at the edge of urban areas, and are organized either by private sector promoters or by institutions such as schools and churches as part of fund-raising initiatives. The car boot sale environment can be contrasted with formal retail spaces on a number of levels. First, prices are agreed through negotiation and bargaining, rather than being fixed. In this sense, there is more consumer agency than in the standard shopping experience structured by retailers. Second, car boot sales are less predictable, as attendees do not know what goods will be available, and at what prices. Instead, the shopping experience unfolds as an exciting and highly tactile (i.e. hands-on) search for bargains. Third, car boot sales are a more sociable retail environment in which ongoing discussion with sellers and other buyers is an integral part of the process. Such informal retail spaces can, then, thrive in developed market contexts, but tend to operate around the margins of the mainstream, formal retail sector.

Informal retail spaces take on far greater significance in the context of developing world cities, where such exchanges may be more important than formal sites of retailing. While there are always problems of definition and measurement when it comes to informal economic activities, it is nonetheless possible to gauge their overall significance. Estimates suggest, for example, that in the developing world, some 37 per cent of employment is provided by informal activity (ranging from 33 per cent in Latin America to 54 per cent in Africa) compared to 3 per cent in high income countries (Daniels, 2004). Such employment is particularly concentrated in urban areas, where formal employment growth often lags well behind population increases fuelled by high rates of rural to urban migration. In African cities, for example, informal work is estimated to provide two-thirds of urban employment. While some of this activity is constituted by small scale manufacturing activity, a considerable proportion is derived from informal retailing activities.

Evidence from the Latin American city of Quito, Ecuador, gives an indication of the scale and varied nature of this informal retail activity (drawing on Bromley, 1998). Informal retailing has grown over a period of decades to occupy a variety of spaces in the central city. Here, several different types of informal retailing can be distinguished on the basis of their location, timing and legality. There are three main types of location: the market building (*mercado*), the open-air market (*plataforma*), and the street. The market buildings are permanent structures utilized daily, while the open-air markets vary in their frequency: both are generally legalized forms of retail. Street vending can be split into those who have a stall/site, and those who operate on the move, often illegally. In the mid-1990s, the historic centre of Quito city had 4,500 trading sites in 15 daily markets (up from 2,400 sites in 1976). In addition, there were 3,400 street traders operating daily, of which 2,800 had established some kind of temporary stall or site, and 600 carried their goods and moved continually. On one particular day, two key central streets (Chile and Cuenca) were found to have 666 and 411 traders operating on them. Recently, the scale and intensity of informal retail activity in central Quito have generated concern about the extent to which associated problems of hygiene and congestion are hampering attempts to re-model the area as a destination for wealthy tourists. The municipal authorities have responded by decentralizing market activities to the margins of the central area and clearing the key historic central plazas of all informal activity. Thus, even the geography of informal activity can be shaped by local regulatory conditions.

To summarize the arguments of this section, retail geographies are restless and continually changing. In the *formal* sector, these changes reflect both the growing size and power of retailers, and wider societal dynamics such as patterns of urban decay/renewal and the movement of high income segments of the population. Due to the acute sensitivity of retailing to planning and regulatory conditions, the timing, nature and extent of change will vary substantially from place to place. A range of *informal* retail activities and spaces exist alongside these formal retail operations. The importance of such activities varies geographically: in some contexts they will be relatively insignificant, filling particular market niches in a largely formal retail sector, while in others they form the mainstay of the urban economy, representing the only way for millions of poor city inhabitants to make a living.

10.4 The Changing Spaces of Consumption

In the previous section, we outlined some of the key changes to the geographies of consumption in contemporary societies. We now move to looking *within* different kinds of consumption spaces to examine the way in which firms, designers and planners try to manipulate the layout and design of those spaces in order to try and stimulate consumption. Geography is important in two ways here: first, in terms of the management of space within these different consumption arenas, and second, in the way in which different places are invoked as part of the design strategy (reinforcing the importance of representations of place outlined in Chapter 1). A further theme running through the analysis is the tension between *public* and *private* space – in other words, the extent to which different consumption spaces are truly open access, freely usable spaces, or if they are closely monitored and policed by corporate interests. In order to do this we will look in turn at examples of four kinds of managed consumption space: the store, the street, the mall and the theme park (Wrigley and Lowe, 2002).

The store

As we have already seen in this chapter, the retail store is an absolutely critical space of consumption in contemporary capitalist society. Retailers actively use geography in stores to try and promote consumption in two ways. First, the interior layout and use of space within stores are designed to enhance sales levels. Many of the techniques used in a standard supermarket are now well known: placing attractive fresh produce near the store entrance, using the smells of fresh bread or roasting chicken to entice customers in, putting important products at eye level, stacking promotional goods at the end of aisles, the use of signage to attract attention to particular products, grouping functionally related products (e.g. stir fry vegetables and sauces), placing 'impulse' products (e.g. confectionery, magazines) near the checkouts and so on. Similar techniques are used in all kinds of shops: window displays, for example, are critical in defining the target consumer for a clothing store. What these practices reveal is a conscious micro-management of the store environment on the part of the retailer to try and maximize sales. Furthermore, as we saw earlier, the precise range of products in a store may have already been shaped in line with the social and demographic profile of the area in which it is located. More broadly, retailers use the exterior and interior design of their stores as a strategic resource designed to distinguish themselves from their competitors and develop a brand image aimed at certain portions of the population.

Second, particular places are evoked within the store environment. We can illustrate this idea by looking at the coffee chain Starbucks (here we draw on Smith, 1996: see also Box 4.1). Starbucks outlets conform to a broadly similar design that seeks to evoke a particular sense of place. The design combines associations with European coffee houses and bars, and a North American 'Pacific Northwest' lifestyle of relaxed living and outdoor activities (the latter reflecting the chain's roots in Seattle, Washington). These associations are manifest in a simple décor of green and brown colour schemes with metal trims, the visibility of the machinery and experience of coffee-making, the prominence of educational information about coffee and its origins, and the use of large glass windows and bar-style seats that allow drinkers to observe and 'consume' the streetscape in front of them. Creating a particular notion of place – albeit a hybrid one in this instance – is central to Starbucks' strategy of bringing coffee to a mass market. Even more intriguingly, BritishIndia is a Southeast Asian retail chain that started in Malaysia in 1994, selling a range of clothes and furnishings evoking the British rule of India in the 1800s. With an emphasis on comfortable, elegant, high-quality products, the stores are explicitly aiming at middle class and expatriate portions of the population. The stores are decorated with photos and other memorabilia of 'British' India creating a colonial feel. Here, then, is a fascinating example of the contemporary mobilization of notions of place (and indeed time), with a retailer in postcolonial Southeast Asia using images of a colonial past in another part of Asia to sell products and develop a distinct brand identity.

The street

For centuries, streets have played a crucial role in the public and economic life of cities. As Jane Jacobs (1961: 29) once famously observed, 'Streets and their sidewalks, the most public places of a city, are its most vital organs. Think of a city and what comes to mind? Its streets. If a city's streets look interesting, the city looks interesting; if they look dull, the city looks dull.' What is of particular interest to us here is the relationship between the street and retailing. It is no coincidence that the main street (or streets) of most, if not all, cities is devoted to the retail function: think of Oxford and Regent Streets in London, Madison and 5th Avenues in New York, and Orchard Road in Singapore to name but a few iconic examples. This remains the case despite the processes of retail decentralization we have described above, and even though property prices and rental rates in these districts are some of the highest anywhere on the globe. This association between key streets and leading retailers has developed over time, reflecting the essentially urban nature of retailing for most of human history, and is reinforced through planning processes that seek to maintain important retail districts in a bid to attract both national and international visitors. The importance of these 'main streets', however, goes beyond the collection of stores they contain, as they become nationally and indeed internationally recognizable symbols of consumptive activity. In turn, retailers seek to feed off this reputation and benefit from the large numbers of visitors by maintaining a presence in these relatively expensive locations.

Beyond the prime importance of particular central urban streets for retailing in general, there is also a tendency for specific kinds of shops and restaurants to agglomerate in certain streets and districts of towns and cities. In part, they cluster together to benefit from agglomeration economies (see Chapter 5).

However, as with the 'main streets' just described, they also gather in particular streets for less tangible reasons to do with the image, reputation and mix of attractions in certain localities. In this way, certain streets become associated with specific forms of consumption, for example, foreign fashion designer outlets (Ralph Lauren, Calvin Klein, etc.) in Bond Street (Mayfair) and Sloane Street (Knightsbridge) in London. It is not just the shops themselves that are consumption spaces, but the street as a whole, leading some to suggest that the street itself is effectively *branded*. The street is also important on a more mundane level as an everyday retail and consumption environment found in almost every urban area. The street is largely shaped by two groups: urban planners, who determine what different zones of cities may or may not be used for, and property and retail corporations which invest in the urban built environment.

Retailers, for example, are highly *strategic* about where they choose to locate their stores as part of their efforts to distinguish themselves from competitors. This can be seen on two levels, using the example of high-quality women's clothing retailers such as Jigsaw, Monsoon and Oasis in the UK (Crewe and Lowe, 1995). First, during the early phase of their expansion in the 1990s, these retailers were highly selective about the kinds of urban areas they chose to locate in, tending to select locations in medium or large towns in the wealthy Southeast of the country, in addition to key sites in London. Second, during the same period, they were also very specific about the sites they chose *within* the urban area, tending to avoid both high street and mall locations in favour of less obvious locations towards the margins of central areas and in gentrifying districts, as part of building an image of exclusivity and difference. Building on this important early phase of development, all three of these retailers have now expanded into fully national operations with a mixture of location types including high street and mall sites. What this brief example serves to illustrate, however, is the strategic way in which retailers actively think about and shape their position within the urban streetscape.

The mall

In some instances, the notion of the street can blur into our next category of the mall. In Singapore, for example, Bugis Junction features an interior, air-conditioned street as part of the mall, and many other malls evoke notions of the street in their design. Importantly, however, malls are enclosed, privately owned and more controlled spaces than streets. As such, to understand them, we need to look beyond the surface 'magic of the mall' (drawing on Goss, 1993, 1999). Put another way, by exploring the form, function and meaning of shopping malls, it is possible to better understand how developers, retailers and designers actively encourage the purchase of goods and services. First, there are a wide range of seemingly mundane attributes of mall design that seek to promote consumer spending, including:

- the use of attractive central features to draw shoppers in particular directions and then direct them towards further options;
- the configuration of escalators, and the strategic positioning of toilet facilities and exits, in order to make consumers walk past as many shop fronts as possible;
- limiting long straight stretches so consumers are unaware how far they are walking;
- the use of signs and displays to encourage shoppers to keep walking;
- supplying rest points and seats for the weary consumer;
- the use of plants and vendor carts to encourage circulation within the walkways;
- the use of soft lighting and the absence of natural light to help suspend a sense of time;
- the use of mirrors and reflective glass to create an illusion of space;
- the manipulation of the ambient temperature;
- the use of 'soothing' background music to induce spending;
- the delineation of relatively segregated zones of shops designed to appeal to different types/classes of consumer;
- the presence of highly visible security staff to reinforce the safety of the environment;
- ongoing, regular cleaning to reinforce the cleanliness of the environment.

In sum, what at first might seem to be a rather bland shopping mall is in reality a highly designed and strategic space, reflective of the power of its owners, designers and tenants.

Second, and linking back to arguments made in Chapter 4, strenuous efforts are made to disconnect commodities from their real-world production systems and reposition them in a world of pleasure, fantasy and magic promoted through architecture, interior design, and theming that evokes other times and places. The huge Mall of America in Minneapolis, Minnesota, provides a powerful example. The Mall, which opened in 1992, covers some 4 million square feet, and contains over 520 specialty stores, four department stores, 30 fast food restaurants, 20 sit-down restaurants, a 14-screen cinema and many other leisure attractions including the 7-acre indoor theme park, Camp Snoopy. The Mall of America endeavours to create a sense of public, market and festival spaces through its design, and repeatedly evokes notions of travel, nature, primitiveness (through Native American references), childhood and heritage in its retail and leisure attractions. The Rainforest Café, for example, offers a 'safari' meal experience complete with passport, while there are many other 'destination' restaurants such as the California Café, Col. Muzzy's Texas Barbeque, Little Tokyo, Maui Wowi Smoothies and the Napa Valley Grill. The staging of events (e.g. fashion shows, cultural displays, children's entertainment, etc.) and the provision of a range of services to facilitate consumption (e.g. food courts, crèche facilities for small children) are also an integral part of the general strategy of encouraging 'guiltless' spending. Overall, as Goss (1993: 40) suggests: 'The shopping centre appears to be everything it is not. It contrives to be a public, civic place even though it is private and run for profit; it offers a place to commune and recreate, while it seeks retail dollars; and it borrows signs of other places and times to obscure its rootedness in contemporary capitalism.'

The theme park

Theme parks offer a particular, and increasingly ubiquitous form, of consumption experience (here we draw on Shaw and Williams, 2004). Visitors pay not just for a particular ride or show, but for admission to an entire complex. Hence they are paying to experience the whole built environment, which places particular emphasis on the architecture and its symbolism. Accordingly, theme parks are extremely carefully configured, and the use of space within them is heavily regulated and choreographed. In many cases they are explicitly designed to offer a contrast to the more messy spaces of everyday life, providing a safe, car-free environment in which entertainment is interactive rather than passive. At the heart of the theme park model is the consumption of multi-sensory experiences which may combine simulated environments (e.g. natural, cultural, historical or technological); the humanizing of those environments by live interpretations, performances and commentaries (e.g. re-enactments of historical events); state-of-the-art technological devices (e.g. films, rides and games); and themed exhibits and eating places. These experiences tend to be supported by high quality visitor services (shops, toilets, disabled provision etc.), a structuring of the experience to prevent information overload (route maps, suggested programmes etc.), and a constant overhauling and upgrading of facilities to attract repeat visitors. The net result is that – perhaps even more than in the cases of store and mall design – the theme park is a highly managed space designed to deliver a particular kind of structured consumption experience. In short, like the mall, it is a 'landscape of power' (Zukin, 1993: 221).

Within these spaces the use of geographical references and themes is rife. Consider Disneyland Paris, for example. This is a particularly interesting example, as it takes Disney's fantasy worlds, often derived from European-folklore, back to a European context (Warren, 1999). As in all Disney parks, upon entering, visitors find themselves on Main Street USA, an environment designed to replicate early twentieth-century urban America. In the Fantasyland zone, one can find Sleeping Beauty Castle, a French version of Disney's German-inspired original in Anaheim, California. Adventureland has a Pirates of the Caribbean attraction and the Frontierland zone offers the 'wild west'-style Fort Comstock and the Cowboy Cookout Barbeque. Discoveryland is the only area unique to Disneyland Paris, with a range of attractions – such as the Space Mountain rollercoaster – notionally inspired by the work of the French writer Jules Verne. The Hong Kong Disneyland which opened in 2005 has a very similar structure and layout with the exceptions that it has no Frontierland, and Discoveryland is replaced by the standard Tomorrowland found in the US parks. What these examples show is that the mobilization of place representations – whether real or fantasy, historical or futuristic – is central to the theme park experience.

In sum, this section has shown the ways in which corporate interests – most particularly retailers and property developers – seek to manipulate the physical built environment in order to stimulate consumption spending and thereby enhance their profitability. The rise in significance of large retailers and mall and theme park environments over the last few decades means that such practices are arguably an increasingly powerful shaper of contemporary consumption dynamics. In certain places that rely heavily on these kinds of economic activities – such as Las Vegas (Box 10.3) – it is almost possible to conceive of the entire city as a themed consumption environment. We must stop short, however, of uncritically accepting the consumer-as-dupes scenario described earlier. Hence, in the next and final section of the chapter, we move to exploring the relationships between place, consumption and identity, and the ways in which consumers – either individually or in groups – may subvert the intended uses of consumption spaces.

Box 10.3 Viva Las Vegas!

It is hard to think of a more iconic example of a consumption space than the city of Las Vegas, in Nevada, US. The city, famously founded on the site of a small desert railroad and agricultural town in 1905, was by 2004 a city of 1.87 million people. As is well known, the city has now grown into a massive centre of gambling, leisure, entertainment and retailing, attracting some 37.4 million visitors in 2004 – 11% from outside the US – who spent some US$8.7bn on gambling alone. Indeed, Las Vegas now claims to be the second most popular tourist destination in the world after Disney World (Florida). The employment structure of the city clearly reflects its reliance on consumption activities: in September 2005, 267,000 workers (31 per cent of the total) were employed directly in leisure and tourism activities, with many others employed in related and support industries such as government, health, education, business services and construction. Manufacturing, on the other hand, employed only 25,000 workers (3 per cent of the total). This, clearly, is a massive urban economy founded on, and sustained by, consumption. Over the period from the mid-1970s until the mid-1990s, Las Vegas transformed itself from an American casino resort to a tourist attraction with global appeal, a transformation supported by national and local regulatory conditions which facilitated quick decision-making on planning and development requests. The leading theme park, retail and casino attractions now available on the city's famous 'Strip' are shown in Figure 10.6 (see also Figure 10.7),

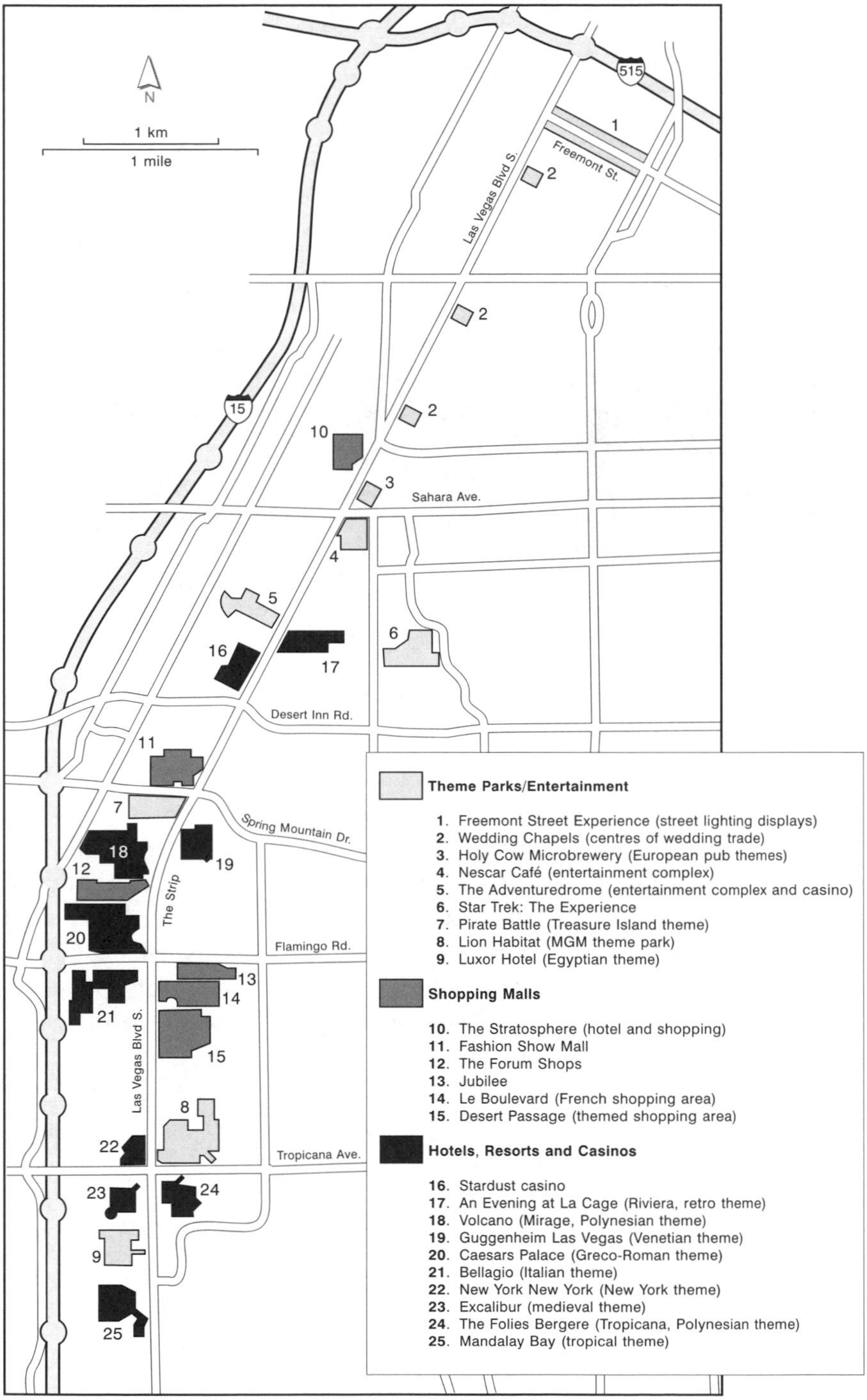

Figure 10.6 Major themes and attractions in Las Vegas
Source: Adapted from Shaw and Williams (2004), Figure 10.1.

Figure 10.7 A themed landscape: Las Vegas, Nevada
Source: The authors.

which also reveals the way in which many of the sites have been *themed* in order to increase their attractiveness to visitors by 'transporting' them to alternative times and places. Places as far away as France, Venice, Egypt, and Polynesia are evoked, in addition to fantasy (e.g. Treasure Island) and historical (e.g. Greco-Roman) references. In the case of the MGM Grand, a 5,000-room hotel, casino and entertainment complex that includes a 17,000 seat auditorium, the theme is actually MGM's own Hollywood studio complex! Overall, the example of Las Vegas speaks both to the importance of consumption for the economy of particular places, and also the way in which attractions are themed as a strategy for inducing visitors to spend money, and in this instance, gamble.

Drawn from http://www.lasvegasnevada.gov/, accessed 27 October 2005, and Shaw and Williams (2004).

10.5 Consumption, Place and Identity

It is wrong to suggest that all consumers are simply at the mercy of corporate interests, consuming what firms want to sell them. Consumers are active agents making informed decisions about which commodities to buy in order to construct particular *identities*, or senses of self. This is particularly true of certain kinds of highly visible *symbolic goods* – such as clothing, jewellery and cars – that are often taken to signify a certain standing or position within society. Other consumptive activities including the choice of restaurants, bars, and holiday destinations may be part of the same process. Equally, purchases may be used as markers of difference, to indicate a position outside, or towards the margins of, the mainstream of society (e.g. buying a car that runs on vegetable oil or wearing distinctive forms of clothing). It is important to recognize, however, that the ability of consumers to engage in these forms of consumption varies both socially and spatially. Socially, it may only be certain groups within the total population, and particularly those with the necessary disposable income, who choose to take part in symbolic consumption. Spatially, such consumption will be more widespread in wealthy countries than in poorer countries, where simply meeting basic needs requirements (water, food, shelter, etc.) may be the only priority. Even within wealthier countries, poor urban or rural dwellers with low income levels, limited local retail provision, and poor access to transportation, may effectively be excluded from the consumption of symbolic goods. In short, we need to recognize the profound social and spatial variation in the degree of consumer agency.

Moreover, the *nature* of the relationship between consumption and identity works out differently in different places – in other words, it is *place-specific*. This is because of the social and cultural specificity of consumption practices such as shopping and eating (Figure 10.8 and Box 10.4). Even where the spaces of consumption are apparently being homogenized by the globalization of retailers, property developers, hotel companies and so on, how those spaces are used will vary from place to place. This variation may reflect a consumer's unique individuality, their membership of a particular group in society (e.g. youth or ethnic cultures), or their position within a particular regional or national culture. Take the cartoon in Figure 10.9, for example, which offers a tongue-in-cheek look at two national stereotypes in the Singapore context – 'the cosmopolitan' and 'the heartlander' – based on their consumption preferences and what this suggests about their position within both Singaporean society and the wider world. The clothes worn, bodily appearance, accessories carried and newspaper preference are all taken as signifiers of particular, place-specific social roles and categories. The same consumption practices may signify something different in another geographic context. There may be still more geographies at work in individual consumption practices. Commodities can be purchased and

Figure 10.8 The same the world over? Coca-Cola in St. Lucia, the Caribbean, and Starbucks in the Forbidden City, China
Source: The authors.

combined to show pride in, and attachment to, particular places (e.g. elements of national dress), and/or places where one has visited (e.g. the tourist T-shirt or a souvenir displayed in the home). In short, the purchases and consumption practices we pursue as individuals are always, and inherently, geographical.

Box 10.4 Shopping

Shopping is clearly much more than a robotic response to the stimuli provided by retailers and shopping mall designers. Instead, it can be thought of as a culturally and societally specific *set of practices*. Shopping is a social activity as well as a simple exchange of commodities, entailing, for example, interaction with shop workers, and discussions with friends about the relative merits of products both before, and after, the act of purchase. Social relations within the family are particularly important, as many purchases are made for other members of the family group. Equally, shopping is an *everyday* activity, meaning that consumers do not constantly reflect on why and how they are shopping. However, it is based on certain skills (e.g. how to navigate round a supermarket, how to select a product) that are learnt within particular socio-cultural contexts. Shopping is not entirely unreflective, however. Consumers take decisions in relation to society-wide debates and dilemmas concerning, for example, what might constitute 'ethical' or 'environmentally responsible' consumption. Shopping is also about place and identity, with different facets of identity – such as family, class, ethnicity and gender – being constructed by participating in particular retail or shopping spaces.

With this nuanced understanding of shopping practices, we can revisit our discussion of shopping malls. Three observations can be made. First, while it is important to recognize the general ways in which shopping malls are designed, this does not mean that they are all the same. Shopping malls will each have their own distinctive histories, mixture of attractions, and senses of place, and consumers will make choices between different malls accordingly. Second, shopping malls are designed for an average, universal customer. In reality, different kinds of shoppers will use the mall in a wide range of ways. Not all shoppers, for example, will browse the space in a relaxed way as a form of leisure. Many will not be enjoying the experience, and will want to get in and out again as quickly as possible once a series of pre-determined purchases have been completed. Third, shopping malls have to be seen in the context of the other retail formats on offer in a particular society (e.g. home, second hand markets, city centre retailing etc.) Again, different shoppers will use a mall as part of different combinations of retail options.

Drawn from Miller et al. (1998).

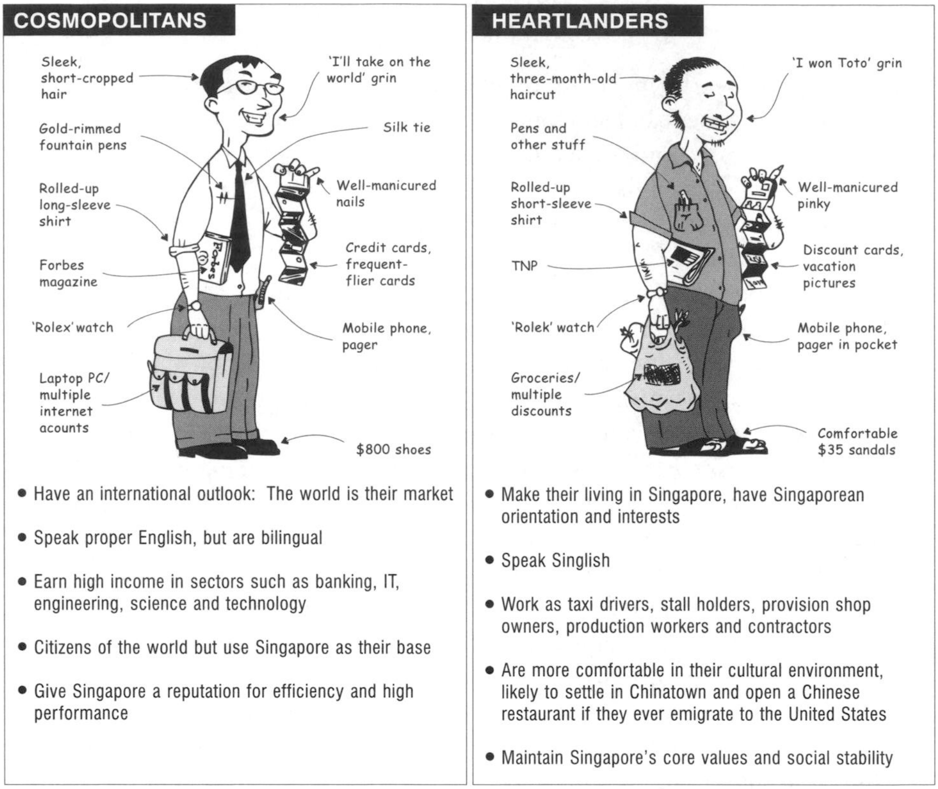

Figure 10.9 'Cosmopolitans' and 'Heartlanders' in Singapore
Source: Adapted from *The Straits Times*, 1999, p. 29.

We can illustrate these arguments with some specific cases. Coca-Cola, for example, although undeniably a 'global' commodity, is always consumed and experienced within particular localities. In Trinidad, in the Caribbean, it is possible to argue that Coca-Cola essentially becomes a 'Trinidadian' product in the way it is consumed. The national drink is Trinidad is rum, which is always consumed with a mixer drink, and that mixer is nearly always Coca-Cola. In this way, Coke is part of the drink which is most directly associated with what it is to be Trinidadian. Equally, Coke is seen less as an American brand, and more as one of a number of 'black' sweet drinks that have urban and sophisticated connotations in comparison to 'red' sweet drinks. This distinction in turn relates to differences in consumption patterns between white and black Trinidadians ('black') and the East Indian population ('red') (see Miller, 1998).

The expansion of McDonald's fast-food restaurants in East Asia since the early 1990s provides another illustration (here we draw on Watson, 1997). Far from being a simple imposition of the American fast food model onto the Asian context, what has in fact happened can be characterized as a two-way *localization* process. On the one hand, McDonald's has had some effect on local diets, particularly among younger people. What was initially seen as an exotic food choice – a taste of Americana – has gradually gained acceptance as an 'ordinary' food for certain sections of the population. Once broadly accepted socially, the retail chain can then expand more rapidly. In South Korea, for example, the first McDonald's restaurant opened in 1988: by 1990, there were 22 restaurants, by 1995 there were 58, and by 2005 the number had grown to 243. To facilitate this process, the company has altered its offerings and developed particular products for different national contexts (e.g. the McSpaghetti product which originated in the Philippines, or Halal inspections in Singapore and Malaysia). On the other hand, consumers have transformed McDonald's into local institutions through the way in which they consume McDonald's products and restaurants. While some aspects of the McDonald's system have become accepted – queuing and self-service for example – the notion of 'fast' food has been adapted and altered. While food delivery is fast, its consumption may not be, as young consumers have turned their local McDonalds into a kind of leisure centre or after-school club, a place for socialization and hanging-out rather than a brief visit. Again, a global icon is given local meaning through place-specific consumption practices.

Finally, we return to the essentially individual nature of these practices by considering the rapidly expanding consumption of home furnishings in markets such as the UK, Canada and the US (Leslie and Reimer, 1999). While most commodities that we buy end up in the home, one group of commodities, home furnishings (e.g. furniture, curtains, bed sheets, paint, flooring, etc.), are explicitly deployed to create particular environments and spaces within the home. This material process of home decoration and maintenance is very closely related to identity formation. The increased emphasis on home-making in contemporary society is reflected in the emergence of dedicated retail spaces such as IKEA, Habitat, and Do-It-Yourself (DIY) chains such as B&Q and Home Depot, magazines such as *Elle Decoration*, *Grand Designs*, *Renovation Style*, *Wallpaper* and *World of Interiors*, and television makeover shows such as *Changing Rooms*, *Property Ladder*, *DIY SOS* and *Groundforce*.

In this way, the home becomes an arena of creativity and personal expression, fuelled by frequent visits to furniture stores. Even in this instance, however, we must not totally lose sight of the corporate interests that are lurking in the background. Consumers are still, in part, responding to the various 'prompts' provided by retailers and the media, even if they are using and combining commodities in highly unique and personal ways within their homes. Furniture retailers, for example, promote frequent changes in 'fashionable' colour schemes and actively seek to reduce the durability of their products. Strip away the playfulness and individuality of home renovation, and the profit-seeking imperative of capitalism is lurking not far beneath the surface.

10.6 Summary

This chapter has investigated the various geographies inherent in the consumption process. First, we have explored the changing patterns of where consumption takes place, geographies that are increasingly shaped by large retail corporations. Most importantly, the suburb has risen to challenge the city centre as the pre-eminent site of consumption in contemporary society. Second, we have looked within particular consumption spaces such as the store and the mall to explore how those spaces are designed and manipulated in a bid to induce consumption. These practices involve not only the shaping of physical space, but also the use of a wide range of geographical references and images within the built environment. Third, we have profiled the way in which consumption and the practices of identity formation are unavoidably place-specific, with commodities assuming different meanings in different socio-cultural contexts. Running through our account has been a concern for whether consumption dynamics are being shaped by corporate interests such as retailers and property developers, or by the actions of consumers. While it is hard to escape the power and influence of retailers in the modern world, the extent to which consumers are free to assert their agency is highly socially and spatially variable.

Our analysis of consumption has clearly revealed the need to take ideas of social and cultural variation seriously in order to understand how capitalist practices 'play out' on the ground. It is to this topic that we turn in the next part of the book, which offers discussions of corporate and national economic cultures, and the ways in which gender and ethnicity work to shape and influence economic relations.

Further reading

- Wrigley and Lowe (2002), Mansvelt (2005) and Jayne (2006) provide student-friendly reviews of the best geographical work on retailing and consumption.
- Coe (2004) provides an overview of the globalization of retailing from an economic-geographical perspective.
- Shaw and Williams (2004) provide a comprehensive account of the changing nature of tourism and leisure spaces.
- Bromley (1998), Crewe and Gregson (1998) and Daniels (2004) offer more conceptual and empirical material on informal consumption spaces.
- For two challenging but rewarding readings on the modern shopping mall, see Goss (1993, 1999): on theme parks, see Zukin (1993: Chapter 8) and Warren (1999).

- For more on the case studies of Coca-Cola in Trinidad, McDonald's in Asia and home furnishing in the West, see Miller (1998), Watson (1997) and Leslie and Reimer (1999), respectively.

Sample essay questions

- Why is consumption such a geographically variable process?
- Who controls consumption: corporations or consumers?
- How, and why, have the geographies of consumption been transformed in the past three decades?
- How are different consumption spaces manipulated in order to try and induce consumer purchases?
- In what ways can consumers actively contest and subvert the intended meanings of consumption spaces?

Resources for further learning

- The websites of leading global retailers such as Tesco (http://www.tesco.com), Wal-Mart (http://www.walmartstores.com) and Starbucks (http://www.starbucks.com) provide a range of information on their global store operations.
- Similarly, the sites for regional or 'mega' malls such as Manchester's Trafford Centre (http://www.traffordcentre.co.uk/) and Minneapolis' Mall of America (http://www.mallofamerica.com/) or theme parks such as Tokyo's Disney Resort (http://www.tokyodisneyresort.co.jp/index_e.html) give some sense of what is on offer at such attractions.
- Leading shopping streets, such as London's Oxford Street (http://www.oxfordstreet.co.uk/home.html) and New York's Madison Avenue (http://www.madisonavenuenyc.com/) also seek to promote their own brand of shopping experience.
- http://www.consume.bbk.ac.uk/index.html: the Cultures of Consumption site hosted by Birkbeck College, University of London, offers a range of cutting-edge research materials on consumption dynamics worldwide.

References

Berry, B.J.L., Cutler, I., Draine, E.H., Kiang,Y-C., Tocalis, T.R. and de Vise, P. (1976) *Chicago: Transformations of an Urban System*, Cambridge, MA: Ballinger.

Bromley, R.D.F. (1998) Informal commerce: expansion and exclusion in the historic centre of the Latin American city, *International Journal of Urban and Regional Research*, 22: 245–63.

Coe, N.M. (2004) The internationalization/globalization of retailing: towards an economic-geographical research agenda, *Environment and Planning A*, 36: 1571–94.
Crewe, L. and Gregson, N. (1998) Tales of the unexpected: exploring car boot sales as marginal spaces of contemporary consumption, *Transactions of the Institute of British Geographers*, 23: 39–53.
Crewe, L. and Lowe, M. (1995) Gap on the map? Towards a geography of consumption and identity, *Environment and Planning A*, 27: 1877–98.
Daniels, P.W. (2004) Urban challenges: the formal and informal economies in mega-cities, *Cities*, 21: 501–11.
Deloitte (2006) 2006 Global powers of retailing (downloaded from www.deloitte.com, 20 June 2006)
Economist, The (2005) This sceptered aisle, 6 August, p. 23.
Fernie, J. (1995) The coming of the fourth wave: new forms of retail out-of-town development, *International Journal of Retail and Distribution Management*, 23: 4–11.
Goss, J. (1993) The 'magic of the mall': an analysis of the form, function, and meaning in the contemporary retail built environments, *Annals of the Association of American Geographers*, 83: 18–47.
Goss, J. (1999) Once-upon-a-time in the commodity world: an unofficial guide to the Mall of America, *Annals of the Association of American Geographers*, 89: 45–75.
Guardian, The (2005) Tesco stocks up on inside knowledge of shoppers' lives, 20 September, p. 23.
Jacobs, J. (1961) *The Death and Life of the Great American Cities*, New York: Vintage Books.
Jayne, M. (2006) *Cities and Consumption*, London: Routledge.
Leslie, D. and Reimer, S. (1999) Spatializing commodity chains, *Progress in Human Geography*, 23: 401–20.
Lowe, M. (2000) Britain's regional shopping centres: new urban forms? *Urban Studies*, 37: 261–74.
Mansvelt, J. (2005) *Geographies of Consumption*, London: Sage.
Miller, D. (1998) Coca-cola: a black sweet drink from Trinidad, in D. Miller (ed.) *Material Cultures: Why Some Things Matter*, London: UCL Press, pp. 169–87.
Miller, D., Jackson, P., Thrift, N., Holbrook, B. and Rowlands, M. (1998) *Shopping, Place and Identity*, London: Routledge.
Shaw, G. and Williams, A.M. (2004) *Tourism and Tourist Spaces*, London: Sage.
Smith, M.D. (1996) The empire filters back: consumption, production, and the politics of Starbucks coffee, *Urban Geography*, 17: 502–24.
Straits Times, The (1999) Cosmopolitans and heartlanders, 23 August, p. 29.
Stutz, F.P. and Warf, B. (2005) *The World Economy*, 4th edn, New Jersey: Pearson.
Warren, S. (1999) Cultural contestation at Disneyland Paris, in D. Crouch (ed.) *Leisure/tourism Geographies*, London: Routledge, pp. 109–25.
Watson, J.L. (ed.) (1997) *Golden Arches East: McDonald's in East Asia*, Stanford, CA: Stanford University Press.
Wrigley, N. and Lowe, M. (2002) *Reading Retail: A Geographical Perspective on Retailing and Consumption Spaces*, London: Arnold.
Zukin, S. (1993) *Landscapes of Power: From Detroit to Disney World*, Berkeley, CA: University of California Press.

PART IV

SOCIALIZING ECONOMIC LIFE

CHAPTER 11

CULTURE AND THE FIRM

Do countries and companies have economic cultures?

Aims

- To understand how and why firms are not the same everywhere
- To appreciate how place matters in shaping the culture and behaviour of firms
- To demonstrate why corporate culture matters in terms of influencing firm performance
- To explore whether nations and regions develop their own economic cultures.

11.1 Introduction

Ford, BMW, and Toyota are all transnational automobile manufacturers and among the best-known brand names in the world. In 2004, their sales and revenues exceeded the GDP figures of most national economies. Ford, the largest of the three of them, achieved a massive US$171.6 billion turnover in 2004. The figures for Toyota and BMW were respectively US$154.2 billion (net revenues) and US$53.5 billion (sales). For these multi-billion dollar automobile giants, everything seems to be *global* – from their product presence and production locations, to their workforce and shareholders. We would therefore expect these 'global' firms to be similar in their corporate practices and business cultures. Upon a closer look, however, they are far from being generic global corporations – rather, they remain firmly rooted in the cultures of their countries of origin.

All three corporations started off as family businesses – Henry Ford and his family in the US, the Quandt family in Bavaria, Germany, and the Toyoda family in Japan. Today, family members continue to be involved in all three companies. In 2005, two members of the Ford family sat on Ford's board of directors (Edsel B. Ford II and William Clay Ford, Jr.), one of the Quandt family members was the deputy chairman of BMW's supervisory board (Stefan Quandt), and two of the Toyoda family members were involved in Toyota's board of directors and senior management (Shoichiro Toyoda and Akio Toyoda). Most of Ford's board of directors were Americans; virtually all of BMW's supervisory board and board of management were Germans; and all of Toyota's board of directors and senior management were Japanese men. Clearly, in terms of ownership and management at least, the three companies are very much rooted in their national home territories.

But do Ford, BMW, and Toyota have different economic *cultures*? Embodying the American culture of market-based price competition, Ford's adversarial approach to its suppliers differs sharply from its Japanese competitor, Toyota, which seeks cooperative relationships with its business network – a reflection of the Japanese approach to cooperative group dynamics (Womack et al., 1990). Within intra-firm relationships, their approaches to transnational management also vary significantly. While Ford is much more amenable to hiring international executives to manage its worldwide operations, Toyota remains emphatically Japanese in its international operations, and even more so in its domestic business in Japan. Its presidents for sales and manufacturing in North America are 'company men' sent from Japan. BMW, on the other hand, embraces both of these two corporate cultures of adversarial competition and cooperative partnership. Its distinctive culture of engineering excellence, however, originates from Germany's *national* quest for engineering precision and power, and is even embodied in its corporate headquarters in Munich, modelled after its famous four-cylinder engines (Figure 11.1). BMW built its new corporate headquarters just in time for the 1972 Munich Olympics to symbolize 'prosperity, autonomy and technical perfection. Plus a hint of utopia.' This nationally-specific and firm-specific quest for engineering excellence is much less evident in Ford or Toyota's production and marketing strategies. It is no coincidence, however, that other German auto giants, such as Daimler-Benz and Volkswagen, are also well known for their engineering capabilities.

Clearly, these firms are really only 'global' in terms of the market reach of their products, and perhaps their component sourcing. By virtue of their different origins, historical development, and management structures, they remain *national* firms that carry with them distinctive traits manifested in their corporate practices and management processes. Geography matters here, as the country of origin – through the so-called 'home country effect' – shapes their various approaches to inter-firm and intra-firm relationships. These firm- and national-level differences can be summarized in the concept of *economic culture*. While culture is a notoriously complicated concept, and difficult to define with any precision, we use it here to refer to a wide range of shared practices and norms, as explained in Box 11.1.

This chapter marks the beginning of the final Part of the book, entitled 'Socializing economic life'. In these concluding three chapters, we seek to

Figure 11.1 The BMW headquarter office in Munich, Germany
Source: The authors.

demonstrate that the 'economic' processes we have explored in the book thus far are always, and unavoidably, shaped by social and cultural forces. In this chapter, we are concerned primarily with those aspects of culture that directly influence economic processes. Although the examples we have noted so far relate to corporate and national differences, these cultural forms may take shape at the level of departments or divisions within a company; *or* the company as a whole; *or* regional territories comprising many individual corporations; *or*, at the national level. There is, therefore, a complex geography to be unravelled when examining economic cultures. In the subsequent two chapters we look more

Box 11.1 What is culture? A simplified approach

Cultures are systems of shared meanings (e.g. language, religion, custom, tradition, or ideas about 'place'). They are shared by people who belong to the same community, group or nation and help them to interpret or make sense of the world, and to reproduce themselves through everyday living. They are embodied in material artefacts (e.g. architecture) and social worlds (e.g. myths and stories), and are dynamic rather than static. In other words, culture is a *process* rather than a thing.

Cultures are socially determined and cannot be divorced from power relations: dominant groups often attempt to impose their culture and are challenged by other groups or sub-cultures. Cultures give us a sense of 'who we are' and 'where we belong', and are one of the principal means through which our individual identity is constructed.

To sum up, there are three distinctive elements in the concept 'culture':

- the ways of life or practices of the members of a society, or a group within a society (how they dress, marriage customs, family life, patterns of work, religious ceremonies, leisure pursuits, and so on);
- the values the members of a society/group hold (abstract ideals);
- the norms people are expected to observe (principles or rules).

closely at the role of gender and ethnicity, respectively, and the ways in they are intertwined with economic relationships. In this final part of the book, then, we explicitly tackle the challenge laid down in Part I – namely to look beyond the role of 'economic' factors in our understandings of economic processes.

The current chapter is structured as follows. In the next section, we highlight a common myth concerning the basis of firms' behaviour – the idea that firms operate according to a common logic and set of practices the world over and economic processes are therefore 'un-cultural'. We then tackle this myth in four interrelated sections. First, we zoom in on *the firm* as a scale of inquiry in Section 11.3. By unpacking the typical firm, we show how corporate cultures and subcultures shape firm behaviour in different places and contexts. Second, we move the geographical scale of inquiry up from the firm to its *national business system*. Section 11.4 shows how firms embedded in different national business systems exhibit distinctive practices and management structures. Third, in some national contexts, there are also important regional differences in economic culture, which can have significant implications for regional prosperity and the geography of development (Section 11.5). Finally, in section 11.6 we suggest the need to incorporate *multiple cultures* and *multiple scales* in understanding firms as capitalist institutions in today's global economy. Here, we also show how economic cultures get mixed, and change and adapt, as they travel across space through processes of internationalization and globalization.

11.2 Firms Are the Same Everywhere, or Are They?

To most economists, the firm is viewed simply as a set of production units responding to competitive initiatives and changes in costs. In other words, the firm is a 'black box' that converts inputs into outputs according to specified production functions and market demand. Few microeconomics textbooks devote even a single chapter to explaining what constitutes the firm. The reason is simple. Once the theoretical construct known as the firm is stripped of its geographical and social contexts, the economic analysis of the firm becomes a relatively straightforward technical exercise. The firm thus exists in most standard economic thinking as a simple intermediary between demand and supply.

We argue, however, that precisely because of this supposed 'straightforwardness', a host of common myths are produced and reproduced in popular understandings of the firm. First, the firm is widely held to embody the capitalist logic of *profit maximization*. This notion has become so powerful and pervasive that it is universally assumed to be the driving force behind all decisions made by a firm's managers. Most would take for granted the notion that firms act *unconsciously*, at least, according to this capitalist logic. A part of this logic is the idea that firms must maximize profits in order to remain competitive and therefore to survive. As Box 11.2 shows, however, this logic has significant shortcomings in reality.

Second, if the firm is merely a 'black box', there is no compelling reason why it should be different in different *places* and geographical contexts. After all, firms are supposed to respond merely to market signals and cost changes. The notion that firms are *the same everywhere* has become popular with the rise of what we might call the 'global corporation discourse' (see also Section 8.2). In this perspective, business gurus and consultants have denounced any possible influence of particular places on the formation and behaviour of firms. In his best seller *The Borderless World*, for example, Kenichi Ohmae proclaims that 'country of origin does not matter. Location of headquarters does not matter. The products for which you are responsible and the company you serve have become denationalized' (1990: 116). Ohmae's caricature is rather misleading, despite coming from a business guru who is very familiar with the distinctive Japanese culture of business organization (see more below).

Third, the media often celebrates a *global convergence* in corporate practices as a natural phenomenon. The central tenet of this popular myth is fairly simple:

Box 11.2 Competitiveness questioned

Erica Schoenberger (1998) points out that in a capitalist economy it is difficult to contradict the idea that competitiveness is important to the prosperity of a person, a firm, or a region. Indeed the concept has such a pervasive place in discussions of economic development and strategy that its implications are left immune from examination. But at the same time, it can be invoked to justify actions that are only marginally related to the realities of competitiveness. Competitiveness assumes that human well-being is best served through a process of selection in the market place. As Schoenberger argues, this is a process in which the weak, or the less competitive, 'deserve to die'.

Schoenberger suggests that the power of competitiveness is in its ability to silence any contradiction. To invoke competitiveness is to appeal to the inevitable, the natural, and the common-sensical. She shows, however, that thinking outside of this box can yield some surprising results. It is generally accepted that to remain competitive a company like Nike must pay minimal wages in the developing world to those who stitch together its running shoes. Thus, an Indonesian factory worker might be paid just a few cents to assemble a pair of shoes that will retail for over US$100. Schoenberger argues that while increasing wages significantly to Nike's Asian workforce would reduce the company's profitability, it *would* still be profitable and it could be done without affecting the price of its products. What, then, are we to make of the idea that Nike must pay minimal wages in order to be competitive? In reality, it is a rather hollow justification for the distribution of value along the commodity chain for sports shoes.

global competition drives global firms to become more and more alike in their business practices. For firms in different regions and countries, there is supposedly no holding back the tide of globalization – all 'irregular' and 'idiosyncratic' business and organizational practices are flushed out. In short, only a standardized and efficient form of business organization will survive global competition.

In the remainder of this chapter, we will develop a critical perspective on such ideas by exploring the diversity and importance of corporate cultures. Far from operating according to the cold logic of profit maximization the world over, firms are in fact distinctive cultural entities with fissures, fragments and contradictions. And the places where firms operate also have economic cultures which can, in some instances, prove instrumental in their economic success or failure.

11.3 Fragmenting the Firm: Corporate Cultures and Discourses

As noted in the Introduction to this chapter, one of the principal differences between firms in general occurs in the distinctive, firm-specific practices and behavioural norms that can be broadly conceptualized as *corporate cultures*. To go beyond the 'black box' conception of the firm in neoclassical economics, we need to understand corporate cultures in two ways. First, we can consider how firms differ from each other (*inter*-firm) because of their corporate cultures. For example, even among what most consumers consider to be similar retail giants, such as Wal-Mart and Costco in the US, there are significant differences in the ways they treat their employees and customers (Herbst, 2005). Second, we can open up the firm by viewing it not as a black box, but as an organization that is internally heterogeneous and contested by different *intra*-firm interest groups. These groups and actors enjoy varying degrees of power and access to resources. They represent different fragments and thus different subcultures of a firm.

It is worth noting that firms are becoming increasingly interested in understanding themselves in terms of cultural and sub-cultural practices. Indeed, as self-knowledge becomes ever more critical in managing firms – a phenomenon described in Box 11.3 as *soft capitalism* – we can expect greater importance to be attached to the understanding of how corporate cultures and subcultures influence learning and competitive outcomes in today's firms.

Corporate cultures

Business firms from around the world may vary in their corporate cultures and conventions. Sometimes, corporate culture can lead firms into irrational or misguided decisions that are ultimately harmful to their best interests – a failure that has been described as 'the cultural crisis of the firm' (Schoenberger, 1997). The stakes involved in understanding corporate cultures and conventions are therefore very high.

What, then, exactly *is* corporate culture? How do we go about identifying the distinctive cultural traits of an organization? We can see corporate culture operating in four ways (here we follow Schoenberger, 1997):

1 *Ways of thinking*: This refers to the ways in which thinking at a particular company is constrained, directed, or focused. Inevitably, over time, organizations develop elements of 'group think' in which certain ideas and practices are accepted uncritically. It is often easier for an outsider or newcomer to spot these elements than it is for someone who has been immersed in a particular environment for a long time. There might, for example, be a strong

Box 11.3 The rise of 'soft capitalism' since the late twentieth century

Does contemporary economic organization remain the same as it was when Karl Marx used the term *industrial capitalism* to describe what he saw in the factories of nineteenth-century England? The answer, of course, is 'no'. Contemporary capitalism has become far more knowledge-driven. This is true not just in terms of knowledge or information as commodities to be sold, but also self-knowledge or *reflexive knowledge* among managers who wish to know themselves, and the system in which they work, a little better. Reflexive knowledge has emerged to be a central competitive tool for firms and businesspeople to rival their competitors in an era that Thrift (1998) labels 'soft capitalism'. Corporate culture has become an essential weapon in competitive battles. In part, this is because economic processes have become much more *uncertain* and *complex* in today's global economy, and so firms and their managers are facing a much more challenging operating environment (Thrift, 2000a). They require some kind of 'stabilizers' in the form of organizational conventions and corporate cultures to make sense of this uncertainty and complexity, and to maintain some degree of coherence in the corporate structure in the face of constant change.

But the importance of corporate culture to managers goes even further. In the high velocity world of electronic trading (in shares, commodities, bonds, etc.), prices can move based on the instincts and sentiments of traders. A good example is how the sky-high crude oil price of US$75 per barrel in May 2006 reflected the sentiment of commodity traders more than the reality of crude oil demand and supply. As most traders were expecting the price to push higher by the winter of 2006, many began to accept higher bids for future delivery and, in doing so, made the rising price a reality. The same 'herd behaviour' can occur in stock markets throughout the world. Investors' concerns over rising interest rates in the US, for example, lead to a month-long stock market decline throughout the world in May/June 2006, resulting in at least US$2 trillion in stock value being wiped out. Again, it was worries, sentiments, expectations and instincts that played a large part in causing this decline. The shared meanings and norms that exist within the trading community – its culture – are therefore of central importance.

imperative at certain car manufacturers to emphasize innovative approaches to design and engineering, while others may emphasize affordability and market share. In the financial sector, some banks may see themselves as 'safe' institutions making conservative investment decisions and loans, while others may take on higher risks for greater returns. To take a rather different example, some employers may see their relationship with employees as purely functional and contractual, while others might value highly the loyalty, career development, and even family life of their workers. While these differences are hard to pinpoint precisely, comparing firms either as an employee, client or consumer does reveal many intangible differences in the ways of thinking that prevail within them.

2 *Material practices*: When ways of thinking are converted into *action*, corporate culture is manifested in the everyday practices of business. In short, this relates to what a firm should be doing, and how it should be doing it. In particular firms and workplaces this encompasses a wide range of activities, including conversations, meetings, work tasks, divisions of labour, production processes, service delivery, and so on. Over time, practices become routinized and unthinkingly accepted as a 'natural' part of how things are done, but different organizations can have very different practices. A new proposal, for example, might be circulated for comment and revision in some companies, while in others it will be passed down the chain of command as an order. In some workplaces, the working day is strictly observed with well-defined arrival and departure times and fixed break times; in others, working hours are treated as flexible. This may even extend to the types of products that are made – a company that strictly defined itself as a typewriter manufacturer, for example, and took pride in the quality of its product, would no longer be in business. But a typewriter manufacturer that was flexible enough in how it viewed its product to shift into computer keyboards or inkjet printers might still be successful.

3 *Social relationships*: Implicit in ways of thinking and material practices are the relationships that exist within a firm between its various employees. How do people work together, get along, and respond to orders from superiors or suggestions from subordinates? These relationships form the social 'glue' that holds a company together and enable a complex division of labour to work successfully. There are, however, different models that might be adopted, ranging from informality and friendship, team spirit and collegiality, through to competitiveness, rigid hierarchy, and mutual animosity. These different cultures of relationships within the firm affect the way in which work is distributed, obligations are met, orders are followed through, and strategy is executed. Such differences may have tangible effects on a firm's business – they influence the commitment of employees to doing a good job; they affect the communication of good ideas and the adoption of innovations; and they may determine employee turnover. Once institutionalized

and shared, the social relationships in a firm become *conventions* that are taken for granted and are perpetuated over time.

4 *Power relations*: The final dimension of corporate culture emerges from internal social relationships. It concerns the ways in which power is distributed and wielded within a company. Obviously any firm, even the smallest enterprise with just a few employees, has a hierarchy comprising those who make decisions and those who mostly just follow orders or prescribed work patterns. In a large firm, these power relations are complex and comprise a long chain of command. But the way in which power is allocated and exercised will vary across different firms. In some settings, the hierarchy will be clear and a strong vertical chain of command will exist. In others, a more flexible and flattened reporting system is implemented. The type of structure adopted can have important effects on communications within a firm – if, in a flattened hierarchy, a sales assistant can interact with a senior executive, then good ideas are more likely to emerge from the 'grassroots'. On the other hand, a firm with a strong hierarchy might be able to move quickly in response to new opportunities. Beyond the structuring of power relations in this way, corporate culture will also dictate the appropriate response to orders – are they followed without question, or are employees empowered to discuss and negotiate? This too will affect the degree of 'grassroots' innovation that is possible in a firm.

Taken together, these four dimensions provide a way of characterizing a corporate culture and highlight the ways in which everyday business practices become *stabilized* within a set of conventions that may vary from firm to firm. This is where interesting economic-geographical questions might be asked: how do corporate cultures evolve in specific places and how do they exhibit tendencies of path dependency – a process whereby past behaviour guides present and future decisions (see also Section 5.2)?

To answer these two questions, we now turn to look in more detail at how some large and powerful American corporations, once dominant leaders in their respective industries and markets for decades, have been progressively outcompeted by foreign rivals (drawing upon Schoenberger, 1997). One example involves Xerox, the American giant in the copying and imaging business, and Canon, its Japanese challenger. Xerox, the first company to introduce and market the xerographic machine in 1949 (Figure 11.2) and a billion-dollar American firm by the 1960s, was on the verge of collapse by the early 1980s. Its market share in copier machines dwindled from over 90 per cent in the early 1970s to just below 15 per cent by the end of the decade. The main beneficiaries were Japanese entrants such as Canon, Ricoh, and Sharp. As early as 1970, Canon developed its own plain-paper copier, the NP-1100, by avoiding Xerox patents in copier technology (Figure 11.3). In 1971, Xerox did approach Canon with the idea of licensing Canon's copier technology, but Xerox's technologists,

Figure 11.2 Xerox's first copier in 1949
Source: Courtesy of Xerox Corporation.

Figure 11.3 Canon's first copying machine in 1970
Source: Courtesy of Canon Inc.

presumed to be the most informed people in the organization, condemned the possibility of collaboration as selling out to an inferior competitor and the proposal was dropped.

While Japanese entrants were focusing on developing small, low-volume, simple, cheap and reliable copiers, 'this segment of the market had been written off [by Xerox] as uninteresting. The dominant culture valued large, fast, technically elegant, and high-margin [machines] and, in the absence of any viable alternatives in the market, these commitments were unchallengeable' (Schoenberger, 1997: 195). The central command in Xerox was so focused on its American competitors (Kodak and IBM), and their familiar style of competition, that it failed to recognize or respond to a new kind of competition from 'outsiders'. To Xerox, the Japanese competition simply did not matter, not because it did not exist, but because it was invisible and it came from the 'wrong' place. The corporate culture at Xerox was characterized by a blinkered 'way of thinking', in which the important competitors were American counterparts. In addition, as noted, Xerox's conception of appropriate 'material practices' involved the production of large and technologically-advanced copiers for the corporate market. This contrasted with Canon's approach, which was to focus on a simple machine for a wider consumer market. Finally, the nature of social relationships and power at Xerox evidently gave the upper hand to technologists who dismissed Canon and its product as inferior. In a different corporate culture, in which marketing specialists had a stronger voice, a different decision might perhaps have prevailed. In short, then, various elements of Xerox's corporate culture immobilized the firm and prevented it from seeing what was objectively apparent – a serious competitive challenge from a new product and a new producer. The results were costly: Xerox was almost forced out of the very industry it helped to create.

Corporate subcultures

While the Xerox case study shows how corporate culture varies *between* different firms, corporate culture can also be internally contested by different segments *within* the same firm. Indeed, the firm itself is not a coherent and homogeneous 'black box' that behaves according to some pre-determined logic. Rather, its competitive market behaviour can represent outcomes negotiated by different groups within the firm. Often, what is at stake is the definition of corporate culture – for example, will the view of research scientists prevail or will the engineering division set the agenda? Will the finance people or the marketers call the shots? Or, will one regional division assert its priorities over another?

In all these cases, the purpose of internal negotiations is to assert the *discourse* that will define a firm's internal culture. We noted in Chapter 2 that a discourse comprises the language, concepts and practices that make the world understandable and help us to interpret what is going on around us. A corporate discourse is thus the way in which a company understands itself and the reality in which it is operating. It provides the worldview that fosters a particular corporate culture.

At the *intra-firm* scale, corporate discourses such as 'fierce global competition', 'lack of competitiveness', and 'cost pressures' are commonly produced as mantras to guide thinking and action. The board of directors of a firm may call for organizational re-engineering based on their perception of the worsening corporate 'bottom line' or a new 'business model'. But the issue of 'competitiveness' is sometimes far from life-threatening to the survival, let alone profitability, of the firm (see the case of Nike in Box 11.2). The discursive construction put forward by one group may not, however, be accepted by others. A labour union may pose a counter-discourse of 'exploitation' that demystifies the 'competitiveness discourse' (see also Chapter 9). Even within the management structure of a firm, alternatives may also be articulated, each vying for the power to define the company's purpose and self-image. Thus, far from being the homogeneous and unitary entity that is assumed in economic theory, the firm may in fact be a fundamentally incoherent and unstable amalgamation of competing interests and viewpoints.

This characterization of the firm can be illustrated with the example of the mining and steel conglomerate BHP in Australia and its threat in the 1980s to close a steel mill in the town of Newcastle, north of Sydney (drawing on O'Neill and Gibson-Graham, 1999). The critique of this action by those on the political Left in Australia viewed it as another example of a heartless corporate calculation to maximize profits and ignore the needs of workers and communities. But from an insider's perspective the situation was a long way from this stereotype of corporate rationality. Various individuals within BHP viewed job cuts or plant closures very differently. One figure was particularly critical in this story. Robert Chenery joined the Newcastle steelworks as assistant general manager in 1981 when the BHP treasury decided to rationalize its steel-manufacturing division in the face of declining profitability in Australia. He fought very hard against his direct boss, general manager Jack Risby, to keep the plant open, despite not knowing how to keep the plant running with half of the workforce after the rationalization exercise. In the end, there was an unexpected injection of public money into BHP under the National Steel Industry Plan in 1983 and the Newcastle plant was saved from closure. As Chenery recalled:

> Melbourne [BHP's HQ] were going to close it. And it would be better for BHP to have less steelworks. No doubt about it. Why have two? But . . . suddenly, Newcastle was profitable. Suddenly, Newcastle's not going to close any more. Suddenly, they are approving a blast furnace reline and I think it was about a $26-million development, which actually modernized one of our little blast furnaces.
>
> (Quoted in O'Neill and Gibson-Graham, 1999: 19)

Why is this story important? In some respects the various debates that go on within a firm could be seen as just 'background noise' – inevitable 'chatter' that ultimately resolves itself in the direction of the capitalist imperative for profit. But the example here suggests that internal discussions within the firm are a little more than that – they may actually change the course of the company's plans to close a steel mill. They seem to matter in ways that go beyond simply understanding how firms work. They also matter because they allow us to see the firm as rather more vulnerable, divided, and susceptible to influence than the conventional understanding in economics would permit. It actually opens a window on possible political strategies in relation to corporate activities, e.g. the possibility of new stakeholders having a legitimate voice – environmental groups, indigenous people, unemployed workers, and so on.

To sum up, this discussion of corporate culture has raised critical issues for the notion of global convergence of business practices described in Section 11.2. It demonstrates that even if globalization and global standards are clearly laid out in the corporate vision of a particular firm, it can be undone by the sheer *lack of coherence*. Rallying the cooperation of different actors in a firm is typically difficult because diverse groups tend to have different aims and objectives, some of which may be in conflict with a well-defined corporate vision. As firms grow larger, and spread themselves further, this internal diversity and contestation is likely to increase. Just as Chapter 9 provided several examples of how workers in different parts of a firm can unite and contest organizational change, managers too may be pulling in different directions.

11.4 National Business Systems

So far we have focused on the firm as a geographical scale of inquiry, but we have said little about its country of origin. We have shown how firms in the same industry have different economic cultures and in many cases, the dominant corporate culture in these firms is challenged internally by a variety of subcultures. But when we move up in scale, how do national and regional cultures shape firm strategies and behaviour? What can we learn if we place the firm in its distinctive national and regional context? In the introduction to this chapter, we highlighted the enduring *national* differences in the *modus operandi* of three automobile giants – Ford (American), BMW (German), and Toyota (Japanese). How do we explain this 'embedded' relationship between firms and their country of origin? To answer this question, we turn to a concept developed by organizational sociologists and apply it to geographical contexts at the national scale. This concept is known as the *business system* approach (Whitley, 1999). Put simply, the approach argues that in different national economies, there are distinctive ways of doing business and that over time, these systems of business practices become *institutionalized* as ways of organizing economic activities in those national economies. The business system comprises three important elements: (1) ownership patterns and corporate governance; (2) business formation and management processes; and (3) work and employment relations.

Ownership patterns and corporate governance

One of the most visible ways to distinguish different firms in various national business systems is to examine their patterns of corporate governance – defined as ownership and control mechanisms. A diversity of corporate governance structures exist, ranging from public companies traded on stock exchanges, to state-owned enterprises, family businesses, cooperatives, and private companies. In some countries (e.g. the US and the UK), the shareholders of most firms do not exercise direct control and prefer to leave their businesses in the hands of professional managers. In other countries (e.g. in Scandinavia and South America), key shareholders, and sometimes their family members, get directly involved in management control. Elsewhere, the state may step in to develop dominant state-owned or government-linked enterprises (e.g. Venezuela, Saudi Arabia, and Singapore).

Two contrasting examples illustrate these different approaches to ownership and control. Figures 11.4 and 11.5 illustrate the ownership structure of companies associated with two of the world's richest billionaires – Microsoft's Bill Gates and Cheung Kong's Li Ka-shing. In 2005, Bill Gates topped the *Forbes* list of the world's billionaires with US$46.5 billion net worth, while Li Ka-shing's US$13 billion placed him 22nd on the list. Although neither individual owns more than 40 per cent of all shares in their respective flagship companies, they have very different approaches to corporate governance.

In 1996, stakeholders other than Bill Gates were already involved in governing Microsoft, including co-founder Paul Allen and Steven Ballmer (CEO since January 2000). By 2005, Microsoft had brought together a much more diverse group of top executives. In June 2006, Bill Gates announced publicly that he

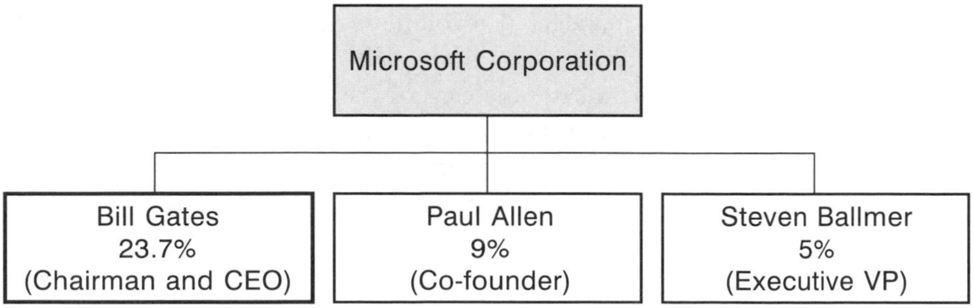

Figure 11.4 Bill Gates and Microsoft Corporation, 1996
Source: Redrawn from La Porta et al. (1999), Figure 1.

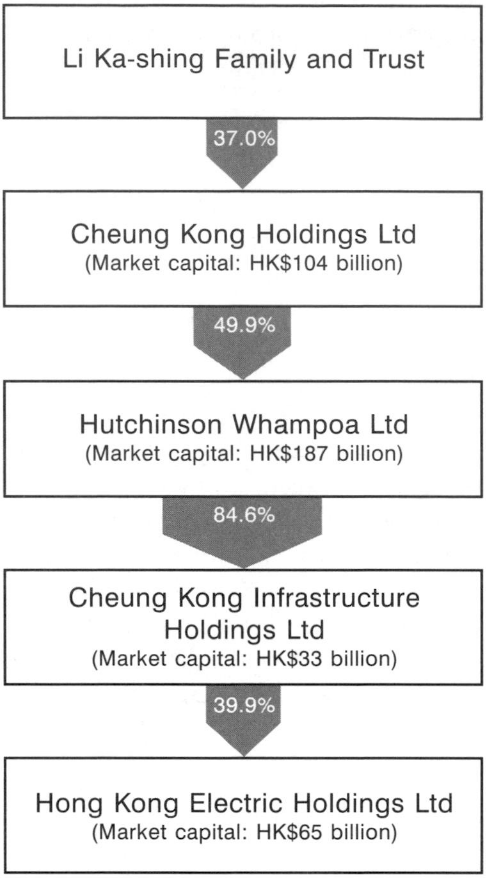

Figure 11.5 Li Ka-shing and the Cheung Kong Group as of 25 March 2003
Source: Yeung (2004), Figure 5.2.

would pass on Microsoft's executive baton to his successor and focus on his own philanthropy work. This separation between ownership and control is the hallmark of the American 'managerial revolution' that contributes enormously to the competitive strength of American firms such as Microsoft. This is because capable professional managers can take care of the firm on behalf of its shareholders, motivated by incentives in the form of promotion and financial rewards. As explained further below, this dominant form of corporate governance in the US is facilitated by unique combinations of capital markets and employment relations that define contemporary American capitalism.

Hong Kong's Li Ka-shing, on the other hand, has continued to exercise very tight control of his family-owned corporate empire. Figure 11.5 shows that he has adopted a pyramid ownership structure to control four very large companies publicly traded on the Hong Kong Stock Exchange. The pyramid structure allows the controlling shareholder of the ultimate parent firm to exercise effective management control of subsidiaries without having to incur the substantial financial investment required for a majority shareholding. As the controlling shareholder of Cheung Kong (Holdings) Ltd, Li Ka-shing and his family members can and *do* exercise effective *management control* over its major subsidiary Hutchison Whampoa Ltd (the world's largest port operator) and the other two listed companies down the ownership chain. Unlike Bill Gates' delegation of management control to Steven Ballmer when he relinquished the position of CEO in January 2000, Li Ka-shing remains firmly in control as the Chairman of Cheung Kong (Holdings). Victor Li, his elder son, serves as the Deputy Chairman and Managing Director, while his brother-in-law is the Deputy Managing Director. These three members of the Li family also hold all the top posts in the four listed companies. This pattern of strong involvement by family members in the management of publicly listed firms tends to prevail in most Asian economies, where corporate governance remains largely family-based and capital markets are relatively underdeveloped (Yeung, 2004). Some have even argued that there is an innate cultural preference for family ownership and control of businesses among ethnic Chinese – 'the spirit of Chinese capitalism' as one analyst has called it (Redding, 1990). In reality, however, such traditions of ownership emerge from the particular circumstances of doing business in different institutional contexts, as we will now show.

Business formation and management processes

In countries with highly competitive markets (e.g. the US and the UK), firms tend to be established by private entrepreneurs. Firms in these contexts are also more likely to engage in arm's-length transactions and to hold adversarial relationships with customers and suppliers. The development and maturity of financial markets in these countries put tremendous pressure on top management to seek short-term returns and to practise financially-focused corporate strategies. In contrast, in countries practising 'cooperative capitalism' (e.g. Germany and Japan), stable and enduring relationships between banks, firms, and the state tend to encourage strategic and long-term investments by corporate managers. More resources can be devoted to those activities that promise to generate *future* returns. Management is less focused on financial returns *per se* and more managerial autonomy is exercised.

For a specific case, we can turn to the global financial sector and consider how German pension institutions differ from the Anglo-American model. The German model of collective decision-making can be thought of as a *social partnership* that permeates from nation to region, from industry to firm, and is constituted by a system of interdependent management systems in which all the key stakeholders in the firm have a say. For example, in the case of BMW, joint decision-making works through both its supervisory board and its board of management. This two-tiered management structure allows the stakeholders and partners of the German firm (e.g. banks and pension institutions) to be involved via the supervisory board, while giving the firm's top executives an opportunity to seek cooperation and support, via the board of management, from its stakeholders and social partners. There is thus coherence in the relationships between the systems of finance, labour, and retirement income provision in Germany.

This relatively stable model of management and pension systems in Germany contrasts sharply with the system in the US (Clark et al., 2002). Many American firms, particularly in the manufacturing sector, provide pension benefit plans in both union and non-union environments. Unlike German firms, however, these US firms do not involve these pension institutions through their boards of directors. In the US, the private pension and retirement benefit system is underpinned by the US financial services industry – not banks or manufacturers as in the case of Germany. Although they are stakeholders in these manufacturing firms, US pension institutions do not enjoy the kind of social partnership that prevails among their German counterparts. Moreover, American manufacturing firms have utilized much more aggressive investment strategies – such as engaging the services of investment banks – to manage their pension funds in an effort to reduce their pension fund contributions. In the US, where the financial industry is highly innovative and well developed, there are numerous financial instruments and intermediaries that can help firms in their quest to *maximize* the financial returns from their pension-related assets. There are, in short, very significant national variations in the way in which pension systems are controlled and operate.

Work and employment relations

The delegation of responsibility to skilled workers, and the demarcation of workplace functions, can vary significantly between national business systems. In the UK, for example, the main divisions in the workplace exist between generalist top managers and managerial specialists, as well as between skilled workers and unskilled operatives. In the US, the principal division exists between college-educated managers and manual workers. In Japan, it occurs between male, core employees and female, temporary workers (Table 9.1 laid out some of these national variations in labour practices). Diverse national business systems thus tend to foster different relationships between employers and their employees, and between different employees in the same workplace.

An intriguing study of workplace practices in relation to production machinery illustrates these national differences in work culture (Gertler, 2004). The example involves the use of advanced machinery technologies from Germany in workplaces in Canada. The typical firm in Canada tends to assume that technology is something embodied entirely within the physical properties and design of the machine. Machines are understood to be operated by an employee, but the maintenance and reconfiguring of a complex piece of equipment are assumed to be the work of a specialized technician. This industrial culture is largely a legacy of the Fordist mode of production in which workers were deskilled due to production fragmentation and deepened divisions of labour (see Chapter 5 on Fordism and Chapter 9 on deskilling).

Anglo-American users of advanced machinery thus tend to undervalue the importance of on-site training and learning and instead emphasize the clear demarcation of specific roles in the production process. The culture of industrial practices in both Germany and Japan, on the other hand, tends to take a more cooperative and consensual approach. Workers and technicians in these two industrial cultures possess high skill levels, work in fairly stable employment environments, tend to be involved in joint decision making, and will place a strong emphasis on training. As a result, machinery designed and built in Germany and Japan requires considerable workforce expertise to be implemented successfully (for full list of contrasts, see Table 11.1).

From a database containing 407 cases of distinct technology implementation in Ontario, Canada, it was found that Canadian-owned plants have much greater difficulty in adopting foreign-produced machinery than Canadian-produced machinery (Gertler, 2004). In short, Canadian-produced machinery conforms to the local culture of workplace practices, in which the producers of machinery provide the servicing and adaptation of their products, rather than relying on the workforce using them. Foreign-produced machinery, meanwhile, is designed with a different industrial culture in mind. Interestingly, this results in a greater demand for close proximity to machinery producers among Canadian firms than among foreign-owned plants in the same province.

In a variety of ways, then, distinctive national business systems can be identified. Different norms of ownership and control can be seen to exist across national boundaries – often resulting from the institutional contexts (or the historical legacy of past contexts) in which business is being conducted. Moreover, firms in different countries have contrasting approaches to involving their stakeholders, resulting in significant differences in the management practices and supervision within firms. Finally, broad cultures of workplace practice differ across national boundaries, such that the expectations of the workforce can vary significantly.

11.5 Regional Cultures

The previous section assumed that business systems are fairly coherent and homogeneous at the national scale. At the *regional* level, however, different cultures of business practice may exist, even within the same country. This has attracted particular attention because it appears to have implications for varying

Table 11.1 Contrasting cultures? German and North American use of German advanced machinery

	German users	North American users
Purchasing practices	Production managers and shopfloor workers participate in vendor and machinery selection, customization, and design process	Production managers buy machinery, set specifications
Contracting procedures	Usually informal; based on handshake, reputation, trust	Always formal, legalistic; full, explicit contracts
Payback periods	Lengthy; 10–20 years the norm	Very short; 1–3 years increasingly common
Training and labour practices	Require minimal training from vendor at time of purchase	Require extensive training from vendor at time of purchase, and repeatedly afterward
Operating manuals	Satisfied with general descriptions	Demand detailed, exhaustive descriptions of machine operation and trouble-shooting
Service expectations	Undemanding	Expect rapid response on 24/7 basis
Maintenance practices	Spend lavishly on upkeep of machinery; operators responsible for upkeep	Assign low value to regular, preventive maintenance; reluctant to spend money
Operational success	Few problems; high reliability and quality of output	Frequent, lengthy breakdowns; complain that machinery is 'too complex', 'over-engineered'

Source: Gertler (2004), Table 4.1. Reprinted with permission of Oxford University Press.

levels of innovation and entrepreneurship. The social and cultural factors that lie at the heart of regional economic success are best encapsulated through the notion of *institutional thickness* (Amin and Thrift, 1994). By institutions, we mean enduring social structures and processes that shape our everyday life. They range from educational and healthcare institutions (e.g. schools and hospitals) to political and economic institutions (e.g. governments and central banks). Some regions appear to be endowed with a favourable combination of certain institutions that engender pro-growth economic cultures and a strong regional capacity for innovation (Storper, 1997). Whether this institutional combination is 'thick' depends on the following four elements:

- the strength and breadth (i.e. quality and quantity) of pro-growth institutions in a particular region;
- the degree of cooperation and cohesiveness among local institutions, creating shared rules, conventions and knowledge;
- the emergence of progressive local power structures, or forms of collective representation, that allow the region to bargain with authorities at the national and supra-national scales;
- the development of a consciousness among the participants in institutional networks that they are involved in a common enterprise.

In the most favourable circumstances, the outcome of institutional thickness will be a regional economy characterized by dynamic, flexible institutions, innovation, high levels of trust and effective knowledge circulation. Some geographers have used the term *learning region* to describe these successful territorial ensembles of innovation and production. Important regional institutions may emerge in the form of research centres and universities, other educational and training institutions, local associations of producers, chambers of commerce, business associations, and technology-promotion agencies. These institutional forces often engender a regionally-specific set of industrial practices and corporate culture.

This regional culture is best illustrated in a comparative study of Silicon Valley in California and Route 128 near Boston (drawing upon Saxenian, 1994: see also Figure 11.6). The key question is why, during the 1980s and the early 1990s, certain industrial sectors (software and electronics, in particular) flourished in California's Silicon Valley, while along Route 128 in Massachusetts they declined. Despite similar histories and technologies, Silicon Valley developed a decentralized but cooperative industrial system while Route 128 came to be dominated by independent, self-sufficient corporations. The two regions in the same country apparently have different *regional cultures*, characterized by more cooperative *inter*-firm relationships in Silicon Valley, and greater reliance on *intra*-firm vertical integration in Route 128.

As early as the 1940s, the Boston area in Massachusetts hosted a sizeable group of electronics manufacturers. At around the same time, the Santa Clara Valley (today's Silicon Valley) in California was mostly an agricultural region famous for its apricot and walnut orchards. Much was to change in the ensuing four decades. How did Massachusetts, home to world-class educational institutions such as Harvard and MIT, lose out to Silicon Valley by the 1980s as the world's leading region in the electronics industry? In short, engineers and other

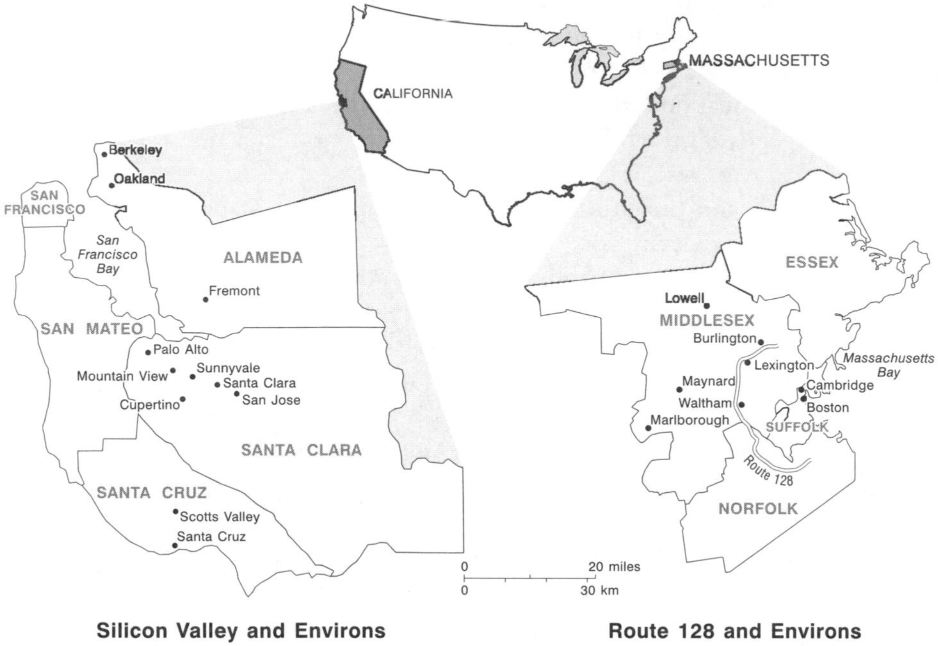

Figure 11.6 Location map of Silicon Valley and Route 128, USA

professionals in Silicon Valley were much more entrepreneurial in their business culture. Frequently moving to a new job, or quitting a job to start a new firm, were both seen as perfectly acceptable in Silicon Valley's cultural environment. In Massachusetts, however, the industrial culture was very much big firm-centred in the sense that engineers and professionals were expected to remain working for the large firms that hired and nurtured them. Hopping from one firm to another or leaving to set up one's own firm was generally seen as disloyal and ungrateful.

As a consequence of this difference in regional industrial cultures, there was much more cooperation and collaboration among Silicon Valley firms than their counterparts in Massachusetts. Silicon Valley firms collaborated with one another in formal and informal ways, developing alliances, contracting for components and services, or simply sharing information. Employees of different firms mixed frequently at local business and social gatherings. In contrast, Massachusetts firms were highly secretive and self-contained, and employees had fewer inter-firm contacts. The industrial culture in Silicon Valley was much more relaxed about sharing of information and skills, whereas firms in Massachusetts anxiously sought to protect their intellectual property and prohibited inter-firm sharing of information and knowledge. Venture capitalists in Silicon Valley also played a critical role in transferring skills and knowledge among firms, while in Boston, where more traditional sources of finance were sourced, the banks supplying such capital simply did not have the technical expertise to advise and promote start-ups.

In sum, these contrasts between Silicon Valley and Route 128 illustrate the differing regional industrial cultures that can emerge, and the implications of these differences for entrepreneurship and innovation. The contrasts emerged gradually over time but eventually became very clear distinctions in the networks of knowledge circulation and support for small enterprises that existed in each place. Once these distinctive practices are entrenched, they tend to remain, and success stories such as Silicon Valley become examples that other localities try to emulate. Indeed, fostering a local culture of innovation and learning has become a celebrated cause for many local governments.

11.6 Multiple Cultures, Multiple Scales

There is sometimes a tendency to see culture as something fixed – something that defines how we behave because of some innate characteristics, hence the notions of 'youth culture', or 'Chinese culture', or 'scientific culture'. In each case, the behaviour of an individual is assumed based on their identity as a youth, Chinese, or a scientist. In this chapter, we have steered away from this notion of culture and have instead used the concept to refer to shared practices, norms, and conventions, rather than innate characteristics. Culture does not determine practices; instead, practices are where culture is to be found. This is a subtle distinction, but an important one. It allows us to distinguish an argument that 'Germans do business in a certain way', from one that says 'we find certain kinds of business practices in Germany'. While the first statement implies that things are being done a certain way *because* they are being done by Germans, the second statement implies only that we find a certain commonality of practices in Germany and leaves open the reasons why this might be the case. This liberates us from dead-end discussions that imply a person with a certain identity *must* belong to a certain cultural group and therefore behave in a certain manner.

Taking this view of culture allows us to analyse economic practices in a more sophisticated manner. First, it allows us to identify culture in a wide range of everyday behaviour and practices. We can consider management culture, work culture, risk culture, ownership culture, innovation culture, teamwork culture – any aspect of business practice can be considered an element of corporate culture.

Second, as noted, viewing culture as a range of practices allows us to avoid implying that all members of a certain group will share the same economic culture. Clearly, generalizations about how things are done in Germany, or Japan, or the United States must be tempered with the possibility that many firms and individuals in those places *do not* conform to the general patterns described. There may be as many differences in economic practices *within* these countries as there are *between* them. By understanding culture as comprising a range of practices, we leave open the possibility that there will be both commonalities and differences in the practices of a given culture.

Third, by looking for culture in the practices of firms, we have avoided 'fixing' the scale at which economic culture must be defined. Hence we have been able to talk about economic culture as emerging within corporations, within regions, and within nations. Any given set of practices is thus the outcome of a complex mingling of influences at multiple scales. There are subcultures within the firm that 'do things' in a particular way. Then there is a dominant worldview within a specific corporation that fosters a distinctive corporate culture. But some practices are also the product of regional or national business cultures. To characterize economic cultures, then, requires that we keep multiple scales in view.

A fourth dimension of economic cultures complicates the picture still further. Over time, cultures are not fixed or predictable, but will inevitably change and adapt. New leadership, a period of crisis, or simply the passing years, can all bring about changes in the way firms conduct themselves. Perhaps the most potent force for change in corporate culture, however, is the geographical juxtaposition of different sets of practices. As firms spread their operations around the globe, or employ individuals who have migrated from other places, then there will inevitably be a process of cross-cultural learning and adaptation.

Ethnic Chinese businesses in East and Southeast Asian societies provide an example of the changes that can occur in economic cultures (Yeung, 2004). As firms from these regions have expanded elsewhere around the globe, and as foreign firms increase their presence in these areas, changes have occurred in the practices of ethnic Chinese firms. In particular, we see the greater inclusion of non-family members and professional managers, the broadening of 'Chinese' business networks to include non-Chinese actors, and the increasing participation of Chinese elites in non-Chinese business networks. These processes represent significant steps towards the hybridization of Chinese capitalism in a context of globalization. The resultant form of Chinese business system resembles neither the earlier form of family-centred business organization nor the Anglo-American form of economic culture driven by professional managers. Instead, a new hybrid culture of business practice is becoming apparent. This new culture comprises some elements of earlier Chinese business practices (continual role of family members in ownership and management) *and* new elements of professionalism and knowledge-based decision-making processes developed through cross-border learning in different international business environments.

Similar processes of cultural hybridization also occur between German firms and Anglo-American pension systems (see Section 11.4). As German firms active in international markets learn about the financial management practices that are popular in the Anglo-American system, they begin to transfer *some* of them back to their home country, and in so doing, play an active role in transforming the German pension system itself. Such practices may include the establishment of new forms of pension institutions, the adoption of international accounting standards relevant to the reporting of corporate pension schemes, the increased importance of wealth creation through stock market valuations, and the spread of Anglo-American-style pay packages with bonuses and stock options for managers. This does not necessarily lead to a global convergence in how pension funds are managed, however. Instead, these new practices will *co-exist* with and, over time, transform the dominant German system into a mixed and hybrid system (Clark et al., 2002). The key point in this and the Chinese business network example is that when different cultures interact, it is not simply a case of one overriding the other, but rather that new, hybrid economic cultures are created through the mixing process.

11.7 Summary

Clearly, our answer to the question in this chapter's title is that countries and companies do indeed have economic cultures. Firms do not work the same way all over the world, despite a common capitalist logic. Instead, when their economic cultures are understood as a range of practices and norms, we see that they are highly differentiated, often reflecting influence of *where* they come from. But the picture is complicated by the multiple scales at which corporate culture is constituted – as we have seen, even the corporation itself cannot be said to have a unitary and uncontested culture. Thus, while firms, regions and nations can be seen to have commonalities in their business practices, there is also a great deal of diversity within them. The task of identifying economic cultures is therefore a complicated one.

If this complexity leads us to emphasize the diversity of economic cultures that exist, it also implies with some certainty that we are a long way from a global convergence of economic cultures. While it is true that firms must increasingly compete on a global scale, this does not mean that they necessarily converge around a single model of business practices. At most, we may see a hybridization of economic cultures as processes of globalization lead to mixing, juxtaposition and cross-cultural learning.

Further reading

- Schoenberger (1997) provides a classic study of corporate culture in economic geography.
- See Dicken (2003) for general arguments against the 'global corporation' as a homogeneous phenomenon.
- Gertler (2004) and Saxenian (1994) offer fascinating accounts of the importance of regional/national industrial cultures.

- Thrift and Olds (1996), Lee and Wills (1997), and Thrift (2000b) are pioneering works in the so-called 'cultural turn' in economic geography.
- Yeung (2005) offers a concise review of the economic-geographical perspective on the firm as social networks.

Sample essay questions

- How do economic cultures vary across different firms and nations?
- What are the geographical foundations of economic cultures?
- What are the differences between corporate cultures and national business systems?
- How might different regional cultures emerge within the same national business system?
- Does globalization mean that economic cultures will inevitably become more alike?

Resources for further learning

- http://www.thinkingmanagers.com: this website by Edward de Bono and Robert Heller contains many insightful articles and blogs on corporate cultures and subcultures.
- http://www.xerox.com and http://www.canon.com/about/history: the Xerox corporate website provides a fascinating history of the origin and development of the Xerox machine. The Canon website offers a history of the Japanese challenge to Xerox's dominance in the copying machine business.
- http://www.ecgi.org: this official website of the European Corporate Governance Institute hosts some very useful information and publications on corporate governance throughout the world.
- http://www.microsoft.com and http://www.ckh.com.hk: both corporate websites contain further information on corporate governance in Microsoft and Cheung Kong Holdings.

References

Amin, A. and Thrift, N. (eds) (1994) *Globalization, Institutions, and Regional Development in Europe*, Oxford: Oxford University Press.

Clark, G.L., Mansfield, D. and Tickell, A. (2002) Global finance and the German model: German corporations, market incentives, and the management of employer-sponsored pension institutions, *Transactions of the Institute of British Geographers*, 27: 91–110.

Clark, G.L., Thrift, N. and Tickell, A. (2004) Performing finance: the industry, the media and its image, *Review of International Political Economy*, 11: 289–310.

Dicken, P. (2003) 'Placing' firms: grounding the debate on the 'global' corporation', in J. Peck and H.W.-C. Yeung (eds) *Remaking the Global Economy: Economic-geographical Perspectives*, London: Sage, pp. 27–44.
Gertler, M.S. (2004) *Manufacturing Culture: The Institutional Geography of Industrial Practice*, Oxford: Oxford University Press.
Herbst, M. (2005) The Costco challenge: an alternative to Wal-Martization?, http://www.laborresearch.org/print.php?id=391, accessed 15 August 2005.
La Porta, R., Lopez-de-Silanes, F. and Shleifer, A. (1999) Corporate ownership around the world, *Journal of Finance*, 54: 471–517.
Lee, R. and Wills, J. (eds) (1997) *Geographies of Economies*, London: Arnold.
Ohmae, K. (1990) *The Borderless World: Power and Strategy in the Interlinked Economy*, London: Collins.
O'Neill, P. and Gibson-Graham, J.K. (1999) Enterprise discourse and executive talk: stories that destabilize the company, *Transactions of the Institute of British Geographers*, 24(1): 11–22.
Redding, S.G. (1990) *The Spirit of Chinese Capitalism*, Berlin: De Gruyter.
Saxenian, A. (1994) *Regional Advantage: Culture and Competition in Silicon Valley and Route 128*, Cambridge, MA: Harvard University Press.
Schoenberger, E. (1997) *The Cultural Crisis of the Firm*, Oxford: Basil Blackwell.
Schoenberger, E. (1998) Discourse and practice in human geography, *Progress in Human Geography*, 22: 1–14.
Storper, M. (1997) *The Regional World: Territorial Development in a Global Economy*, New York: Guilford Press.
Thrift, N. (1998) The rise of soft capitalism, in A. Herod, G.Ó. Tuathail and S.M. Roberts (eds) *An Unruly World: Globalization, Governance and Geography*, London: Routledge, pp. 25–71.
Thrift, N. (2000a) Performing cultures in the new economy, *Annals of the Association of American Geographers*, 90: 674–92.
Thrift, N. (2000b) Pandora's box? Cultural geographies of economies, in G.L. Clark, M.A. Feldman and M.S. Gertler (eds) *The Oxford Handbook of Economic Geography*, Oxford: Oxford University Press, pp. 689–704.
Thrift, N. and Olds, K. (1996) Refiguring the economic in economic geography, *Progress in Human Geography*, 20, pp. 311–37.
Whitley, R. (1999) *Divergent Capitalisms: The Social Structuring and Change of Business Systems*, New York: Oxford University Press.
Womack, J.P., Jones, D.T. and Roos, D. (1990) *The Machines that Changed the World*, New York: Rawson Associates.
Yeung, H.W.-C. (2004) *Chinese Capitalism in a Global Era: Towards Hybrid Capitalism*, London: Routledge.
Yeung, H.W.-C. (2005) The firm as social networks: an organizational perspective, *Growth and Change*, 36, pp. 307–28.

CHAPTER 12

GENDERED ECONOMIC GEOGRAPHIES

Does gender shape economic lives?

Aims

- To show how certain kinds of work are excluded in conventional economic analyses
- To demonstrate the gendered nature of the workforce
- To explore how certain jobs and workplaces are constructed as feminine or masculine
- To reveal the complex connections between workplaces and homeplaces.

12.1 Introduction

In 2004, Stephanie Villalba, a senior manager with a major investment bank in London sued her employer for US$20 million, claiming she was humiliated, underpaid relative to equivalent male colleagues, and sacked as a result of her gender. Ms. Villalba, who had worked for Merrill Lynch for 17 years, recounted to an employment tribunal one incident in which she was asked to serve drinks to male colleagues during a flight on the company's corporate jet, and made to sit in the stewardess' seat. The case arose just as the bank settled a ten-year battle with 1,000 female workers in the US who alleged that the bank had a culture of bias against women. Merrill Lynch was not, however, the only bank caught up in allegations of sexual discrimination, harassment, and unequal pay. Between 1999 and 2004, a series of cases erupted in both London and New York in what became known as the 'Sexism in the City' scandal. Some were successful and some were not, but all revealed the very masculine, aggressive and sexist culture that prevailed in the high-powered financial and legal sectors in both cities. In another example, a female banker was sacked while on maternity leave. In yet another, this time in New York, a 44-year-old director at the

Union Bank of Switzerland (Europe's largest bank) was told by a male executive that she was too 'old and ugly' to do the job. After giving evidence that she had been invited by a male superior to a strip club and had been excluded from all-male golf outings with clients, she was awarded almost £15.5 million in damages.

These cases have made it to the headlines, and in some instances made it to court, because they have involved highly paid elites – people who could afford lawyers (and, in some cases, *were* lawyers). But they also raise some broader issues related to how gender affects a person's experiences of economic life, and the workplace in particular. First, in several cases it was the incompatibility of work schedules with the rest of an employee's life – particularly child-raising and other family responsibilities – that led to conflict. This was especially true when maternity leaves and flexible schedules were at issue. That it was exclusively women bringing forward these complaints reminds us how much domestic responsibilities still tend to fall to wives and mothers. Importantly, it also indicates how invisible many of these responsibilities are, and the way in which they are treated as distractions from real and productive activities at 'work'. Conventional economic analysis draws a very clear boundary between the *productive*, value-added work that goes on in the *public* space of the productive economy at large, and the *reproductive* activities that are undertaken in the non-economic spaces of the *private* sphere.

A second issue illuminated by the 'Sexism in the City' cases is the gendered allocation of jobs to men and women in the labour market. It remains the case that some jobs are selectively 'coded' for men and others for women, and so their presence in particular occupations is very uneven. The women whose cases are noted above were unusual in that they had achieved professional and managerial positions in a highly male-dominated world. In short, banking, and especially investment banking, is still very much for 'the boys'. Furthermore, the financial industry illustrates not just the over-representation of men in a particular occupation and sector, but also the ways in which this can create exclusionary masculine environments in the workplace. At a micro-scale, these processes can exacerbate the identification of a particular job with a particular gender. But they do more than this. Gendered workplaces also call for specific identities to be played out: particular kinds of masculinity and femininity need to be performed in order to conform to the culture of a working environment.

The chapter is divided into five sections. We start by exploring how certain kinds of work are excluded from the analysis in both human capital theory, which underpins mainstream economic approaches to labour markets, and in conventional economic accounting of productive activities (Section 12.2). We then examine the increasing participation of women worldwide in the waged labour force (Section 12.3): a recurring pattern is the devaluation of this work in much the same way as domestic labour is devalued. Next we move on to explore the ways in which gender is played out in the labour market, determining the type of work that is available to women, and how it is performed in the workplace (Section 12.4). Building on these ideas, we then consider how women in the workforce tend to find their professional lives in tension with the responsibilities that they take on in the domestic sphere, and we explain the fundamentally geographical nature of this home–work linkage (Section 12.5). Most of the chapter involves expanding our idea of the spaces in which economic activity is conducted (e.g. bringing the 'lived' spaces of the home and workplace into view, see Figure 1.4), and exploring how gender differences are fundamental to understanding how the productive economy is linked to the supposedly non-economic sphere. In concluding, however, we consider how a feminist economic geography might go further than simply taking notice of a gender-differentiated economic world, and might actually see that world differently (Section 12.6).

12.2 Seeing Gender in the Economy

Later in this chapter we will look in greater detail at the ways in which the economy is experienced differently by men and women. But it is first worth asking how conventional economic analysis views gender differences. We will do this in two ways: first, by exploring the ways in which economic actors (men and women, as workers) are understood in mainstream economics, and second, by exploring how conventional understandings of the economy as a whole exclude much of the work that women actually do.

Gendered economic actors

In mainstream versions of how the labour market works, people, male or female, are simply labour inputs to the production process. Workers may have certain attributes that derive from their gender (such as physical strength), which make some jobs more appropriate for men than women, but this is taken to be a 'strength issue' not a 'gender issue'.

In most cases, however, it is difficult to use biological differences as a basis for explaining why women or men in particular do certain jobs. The prevalence of women among elementary school teachers, or men among electricians, can hardly be explained on the basis of fundamental biological differences. In these cases, mainstream economic analysis, known as human capital theory (Becker, 1993), assumes that gender concentrations are driven by the *supply side* – meaning that, for one reason or another, more men (or women) are qualified for, and apply for, particular types of jobs. If more men than women present themselves for a particular kind of work, then this sorting process is seen as happening outside of the labour market, i.e. beyond the realm of the economic, and in the sphere of the social or the cultural. If more boys choose to be engineers or architects, and more girls choose to be secretaries or nurses, then, it is argued, the rational and objective mechanism of the labour market is not to blame.

If the market mechanism is assumed to be impartial and objective in relation to gender, then women and men are likewise assumed to act as rational economic decision-makers (see Chapter 1). When looking for a job, individuals are assumed in the neoclassical economic tradition to be divorced from their social contexts and to be searching for a job with purely economic motives in mind. They are treated as rational actors with full access to information on job opportunities and wages, and an ability to analyze that information effectively. The job seeker is assumed to accept a job offer rationally when it provides an acceptable level of monetary return, or at least when it no longer seems worthwhile continuing to look. In other words, finding a job is a purely economic calculation.

Preferences, prejudices, discriminations, expectations, hopes, insecurities – all these things are seen as being exogenous to the labour market. They may affect the ways in which individuals see themselves in relation to the labour market, but mainstream economics would not see them as being created or reproduced by the labour market. For example, when responsibilities for child-raising fall primarily to women and cause interruptions in labour force participation, a conventional economic analysis would argue that this means women will tend to look for jobs that permit these periods of absence with a minimum of long-term financial cost. The notion that there might be systemic biases against women would not be entertained.

There is a great deal missing from this picture. First, people do not have access to perfect information about opportunities in the labour market. Their options are constrained by the information networks available to them. As we will see later, these information networks are often configured very differently (and with different geographies) for men and women. Second, women and men might be understood as interchangeable in economic theory, but in the practice of everyday life there are different cultural expectations bearing down upon them. In many societies, the domestic sphere is seen as a 'woman's place' – the site where women primarily work, and where the work to be done is women's work. As we will again see later, this has profound implications for how women experience the labour market.

The gendered economy

We have seen that economic analyses tend to make clear distinctions between the economic (the site of rational, objective calculations and impartial mechanisms) and the non-economic (the site of the social and cultural dimensions of everyday life). These distinctions between economic and non-economic realms will remind us of the discussion in Chapter 2, with the economy seen as a self-contained 'machine' that operates outside of the social and cultural lives of the people who participate in it. But viewing the economy in this way ignores the differential experiences of economic life for men and women. We will now look a little more closely at the way in which the economy is bounded and defined in conventional economic approaches.

An important basis for the neglect of gender differentiation is found in the ways in which the economy is counted and analysed. The health of an economy is universally measured in terms of the monetary value of total output – Gross Domestic Product (GDP). As we saw in Chapter 2, however, this leaves a lot out of the picture. In short, those aspects of our lives that cannot be counted, do not count. Work that involves cleaning our homes, or renovating them, maintaining a car, raising children, caring for the elderly, preparing meals and so on, simply does not count unless money changes hands.

If an entire country decided to use professional home cleaners rather than scrubbing and polishing for themselves, or went to a laundromat rather than washing and ironing clothes at home, it could have a remarkable effect on economic growth, regardless of whether or not they did anything productive with the extra time available. Where more and more unwaged work is being shifted into the waged economy, as has happened a great deal in recent decades (in childcare, for example), then a rather false impression is created of greater overall output. Even though no additional production may be taking place, GDP inaccurately registers shifts from the household economy to the market economy as growth. It is estimated that such shifts from unpaid to paid work overstate GDP growth in industrialized countries by up to 0.8 percentage points a year (and this is a significant error, when one considers that many industrialized countries grow by only 2–4 per cent per year) (http://www.gpiatlantic.org/ab_housework.shtml, accessed 17 March 2006).

There is, it should be clear, a distinct geography to the inclusions and exclusions of national income accounting. The spaces of the private home are neglected, while the spaces of the public, non-domestic sphere are carefully accounted. This in itself would not be a particularly gendered exclusion, were it not for the fact that the prevailing gender regime in most societies has the domestic sphere as a primarily feminine space, and the public sphere as the domain of masculinity. Thus, the uncounted portion is largely the work done by women in the home.

This phenomenon has been highlighted in detailed research undertaken by the National Statistics Office in the UK. In 2000, statisticians conducted a household survey in order to discover who does household work and how much it might be worth. They first asked people to account carefully for how long they spent performing different tasks around the house and for their families. The categories of tasks are listed across the top of Table 12.1. The study's authors were careful in distinguishing the work of a household from purely leisure activities – hence just going out for a walk is left uncounted, but walking to get somewhere when one could alternatively catch a bus or drive, counts as household production.

For all these categories of unpaid work, the first row of Table 12.1 shows the number of hours that were estimated to have been spent in the UK on each of these tasks. Clearly, at over 60 billion hours in a single year, childcare is by far the largest category of work and represents almost half of total 'household

Table 12.1 Net value added, hourly effective return to labour and contributions of males and females to household production in the UK, 2000

	Child Care	Nutrition	Transport	Housing	Tenant services	Laundry	Adult Care	Clothing	Total
1 Time (millions of hours)	61,884	25,030	20,374	13,276	4,101	3,998	3,264	1,071	132,999
2 Net Value Added (£millions)	220,494	64,936	101,444	148,593	45,904	35,720	10,566	709	628,366
3 Contribution of males	42%	32%	54%	43%	39%	15%	43%	5%	41%
4 Contribution of females	58%	68%	46%	57%	61%	85%	57%	95%	59%

Source: UK National Statistics Office, Household Satellite Account, http://www.statistics.gov.uk/hhsa/hhsa/Index.html, accessed 21 July 2006.

production'. The next step in the analysis was to attempt to put a monetary value on the work being done. The study did this by calculating what these services would have cost if they had been delivered by a paid service provider, for example, the cost of a live-in nanny indicates the value of looking after children. By carefully deriving values in this way, the worth of all unpaid activities can be calculated, as shown in row 2 of Table 12.1. The final result is a total value of just over £628 billion. Given that the total *formal* output of the UK economy in the same year was £892 billion, the point is well made that the invisible economy is far from insignificant – indeed it represents a shadow economy of unpaid work that is of the same order of magnitude as the formal economy itself.

Of perhaps even more interest to us here, however, was the gendered distribution of unpaid work (see rows 3 and 4 in Table 12.1). Overall, women provide 59 per cent of the value of unpaid services, men only 41 per cent. Only in the case of transport services do men contribute more than women, while for some other functions (such as clothing upkeep, and laundry) the work is done almost entirely by women. In the case of food preparation, more than twice as much work is done by women as by men.

Gender divisions of labour are not the same the world over – gender roles are configured differently across various cultures (and, of course, there is also variation between different households in the same culture). But by looking inside the home, and by differentiating the people who make up a household in this way, we begin to see how conventional measures of the economy neglect a great deal of work that is being done, and how this work is unevenly allocated between men and women. The key issue, then, is the incorporation of private domestic space into the public, countable, space of the economy. Talk of 'spaces' here is not just metaphorical; the boundaries of the home are very much where the formal economy is assumed to stop. This has implications both for what we see as economic activity and for how we understand the relationships of power within the economy. Neglecting domestic space allows categories such as labour and capital to be viewed as the driving forces behind economic processes. But when we open up the household for examination, we can start to see that much work is done outside of 'the economy' and the power relations that determine who does this work, and how it is valued, are not captured in the capital-labour dynamic. Instead, another set of processes needs to be understood, namely *patriarchy* (Box 12.1).

Box 12.1 Patriarchy

Patriarchy refers to a system in which men are assumed to be superior to women and therefore to have authority over them. There has been substantial debate over whether patriarchy is an entirely separate system of domination and control from capitalism, or if the two are actually enabled by each other. The subordination of women is very useful to capitalist enterprises seeking controllable and compliant employees: around three-quarters of production line labour in the world's export processing zones are young women, for example. It is also debatable whether patriarchy is universal, or has culturally distinctive manifestations. The feminist sociologist Sylvia Walby (cited in McDowell 1999), identifies six structures of patriarchy:

- *household production*: where a division of labour is created in which women do large amounts of unpaid work in support of their family (as illustrated in Table 12.1);
- *waged work*: in which women are segregated into certain jobs, for example as secretaries, nurses, and elementary school teachers;
- *the state*: where men dominate government institutions and produce gender-biased legislation and policies;
- *violence against women*: both physical and emotional;
- *sexuality*: where male control is exerted over women's bodies, for example, in abortion legislation in many jurisdictions;
- *cultural institutions*: where various other forms of patriarchy are discursively reproduced, for example in the media and how it depicts society.

Patriarchy may thus be evident either in the public sphere of institutions and legal frameworks that differentiate in favour of men, or in the private sphere, in the ways in which relationships between men and women are conducted. Clearly, however, the two are connected.

Whichever way we conceptualize patriarchy and its relation to capitalism, we have established an important starting point by expanding our view of the spaces in which economic activity is seen to exist – beyond factories, fields, and shopping malls, to include gardens, kitchens and playrooms. This is critical because, as we will see later, the association of femininity with domestic spaces and domestic work shapes a great deal of what women experience when working *outside* the home. But it also points towards the possibility of a feminist interpretation of the economy that sees work – *skilled* work – happening in spaces that are not usually assumed to have productive activities going on. We will come back to this point in the final section of the chapter. For now, though, we will turn to the formal waged economy to see how women have been incorporated into the labour force.

12.3 From Private to Public Space: Women Entering the Workforce

The primary role that women play in the unpaid work of the domestic sphere has implications far beyond the home. The first, and most obvious, influence is the limitation that expectations of motherhood and marriage place on women – expectations that often keep them from entering the workforce at all. Such expectations vary across cultures, but even in societies that consider themselves to be the most emancipated, it is undoubtedly true (as we saw in the UK study) that the role of the 'primary parent' and domestic worker is still a gendered one. This is also evident in Figure 12.1, which shows consistently lower rates of

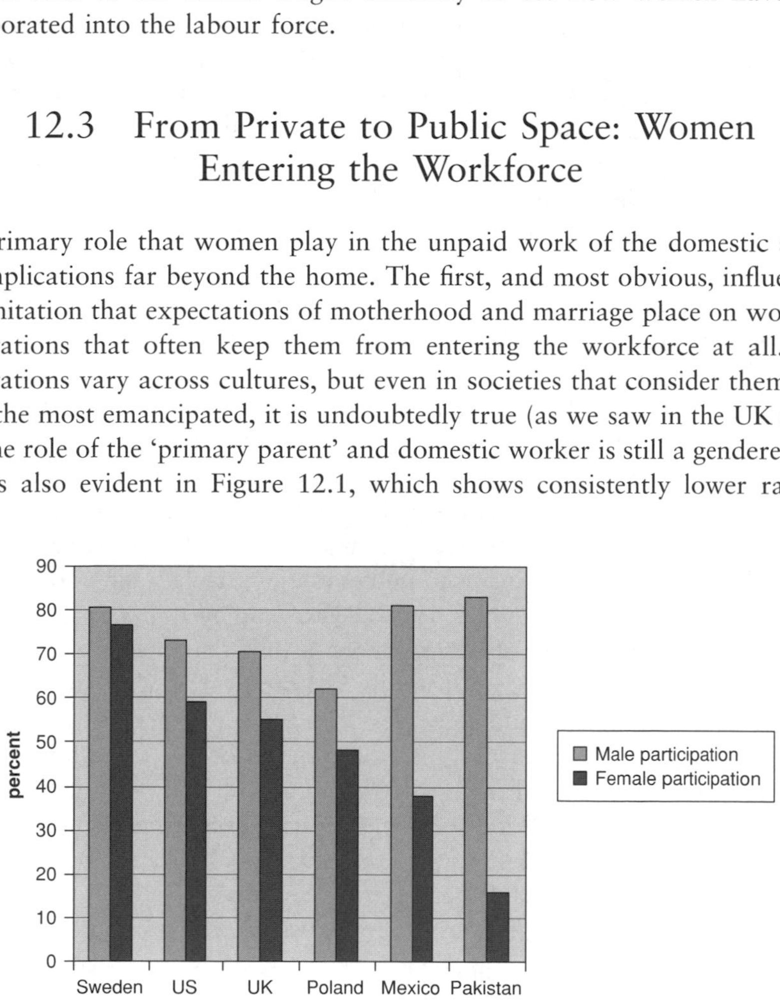

Figure 12.1 Labour force participation rates, selected countries
Source: Based on data from International Labour Office, LABORSTA database http://laborsta.ilo.org/, accessed 17 November 2005.

participation in the labour force worldwide for women than for men, although exactly how much lower varies greatly between different national contexts (e.g. Sweden vs. Pakistan). Participation is defined as the number of people in the waged workforce (either employed or looking for work) as a percentage of the total population of working age.

Although the overall global pattern is one of women participating less in the waged workforce than men, the trend in the past few decades has been towards greater participation. In the developed industrialized countries, women's participation in the labour force has increased steadily since the mid-twentieth century. Even more dramatically, female participation has increased greatly in newly industrializing countries. In Malaysia, for example, the female participation rate has grown from just over 30 per cent when the country achieved independence from British colonial rule in 1957, to around 47 per cent at the turn of the millennium. Moreover, this increased participation is not due to larger numbers of women entering traditionally female occupations, for example, running market stalls or working as elementary school teachers. Instead, much of this participation is due to the rapid development of export-oriented garments and electronics sectors in the 1970s and 1980s, which drew young rural women into assembly factories.

We see the same pattern, then, in both the Global North and the Global South. These parallel changes are different in nature but ultimately part of the same process. The key is the restructuring of the global economy in recent decades (as part of a restless capitalist landscape noted in Chapter 3). One component of this has been the deindustrialization of the global North and a structural shift towards service sector employment. With this change, the secure and well-paid industrial work of the Fordist era, which supported male-breadwinner families, and stay-at-home mums, has largely disappeared. As levels of pay and job security have declined, dual income families are increasingly the norm in developed countries and larger numbers of people are employed in part-time work or temporary jobs. This change also came about at the same time as women's roles in broader society were changing from the 1960s onwards, with greater participation in the public sphere, and higher levels of education increasingly available. Partly because of this, but partly causing it too, has been a trend towards later marriage and childbirth (and the greater social acceptability of doing neither).

At the same time, certain kinds of manufacturing jobs from the Global North have migrated to Mexico, China, Pakistan, India, Indonesia, and other developing countries. In these places, it has been women that have been the rank and file of the industrial workforce. In many industrial estates around the world, where clothing, toys, electrical appliances and computers are assembled, at least three-quarters of the workforce is female (Figure 12.2). Asking why this is so would elicit different answers depending upon who was asked. Factory managers usually note a series of personality traits that are assumed to be distinctively feminine, and which are useful to them given the specific labour process

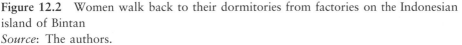

Figure 12.2 Women walk back to their dormitories from factories on the Indonesian island of Bintan
Source: The authors.

involved: a high tolerance for boring, repetitive work; manual dexterity and precision (the infamous 'nimble fingers'); and a submissiveness towards authority, usually in the form of older supervisors or male managerial staff. Critics of labour conditions in the developing world would frame the situation somewhat differently, arguing that patriarchal power structures are recreated as a disciplining mechanism in the workplace so that the workforce is kept under tight control. They would also point out that workers are usually paid a very low wage, often so little that it barely covers their cost of living. Employers are able to do this essentially because their workforce is often young and single, without a family to support. Indeed in many cases, support is provided by rural families to assist industrial workers, rather than the other way around. These contrasting perspectives have led to much debate about whether employment in the factories of the developing world is actually liberating for young women, or whether it just represents further exploitation and oppression (Table 12.2).

A final point that should be made about the increasing participation of women in the workforce is that their household work is not usually taken up by men, but by other women. Many countries that have seen growing female participation have also accommodated immigration flows to supply low cost domestic workers who will perform domestic cleaning and childcare functions for working

Table 12.2 Contrasting views on the emancipatory potential of industrial employment for women

Female industrial work as emancipation	Female industrial work as oppression
• Working hours and conditions are better than in agriculture or domestic work. • Wages are good in terms of local purchasing power. • Some employers are looking for longer term commitment and skills development for employees. • Employment provides participation and respect in the public sphere and a chance to associate with other women. • Income gives a greater social role and respect within the household, and greater control in terms of delaying marriage and childbirth.	• Low wages and benefits, sometimes below subsistence levels. • Often debilitating or dangerous working conditions and harsh factory discipline. • Insecure employment – jobs may disappear with changing international circumstances; women may be laid off if they marry or have children. • Few opportunities for upward mobility or skills development. • Women are confined to particular job categories, especially those involving repetitive, boring, manual work. • Patriarchal dominance and discipline is often reproduced inside the factory. • Waged work may not mean a reduction in household responsibilities; and wages may become the property of male household heads.

mothers. The work of these women is especially devalued. Overall, then, we see a process of women's increasing participation in the workforce worldwide. But women in the developing world find themselves incorporated in particularly subordinate roles. There is, therefore, an important intersection between issues of gender, ethnicity and nationality in the valuation of working people, as Box 12.2 illustrates. Thus, while women's access to the labour market was once considered the litmus test for gender equality, it is now recognized that femininity is actually being used in the work place as a form of devaluation, often in conjunction with race and nationality.

12.4 Gendering Jobs and Workplaces

We saw in section 12.2 that women tend to take primary responsibility for household work, but the assignment of particular kinds of work to men or women also happens outside the household. Sometimes, this is as an extension of household work. For example, childcare work, elementary school teaching,

Box 12.2 Devaluing the 'third world woman'

In the contemporary global economy, women in the developing world appear to be especially favoured as employees by international investors in the manufacturing, tourism and hospitality businesses. They are also favoured in the industrialized world as the nannies and maids who carry out the domestic chores of those women who have entered the waged workforce. For example, in various parts of the world, notably Singapore, Hong Kong, the Middle East, North America and several countries in Europe, Filipina women are employed as domestic helpers, maids and nannies. Indeed, the word 'Filipina' has, in some places, become almost synonymous with domestic servitude. But being 'favoured' does not necessarily mean they are 'valued' – their work is generally lowly paid and they are treated as unskilled. In fact, in some cases, they are viewed as being *incapable* of skilled work, and the intersection of ethnicity, gender and nationality creates strong stereotypes. These include the ability to do tedious and repetitive work without complaint, to work with close attention to detail, and to acquiesce to managerial control without the resistant pride of masculinity.

This devaluation has been especially and tragically evident in the factories of Mexico's Northern border, known as *maquiladoras* (see also Box 8.1). In the 1990s, the border town of Cuidad Juarez gained notoriety when almost 200 female corpses were discovered in the surrounding desert over a five-year period. It has been argued that these murders were indicative of the worthlessness accorded to women's bodies, lives, *and* work, in the factories of the region. Within the factories, they are seen as unskilled, untrainable, and docile, and yet women constitute about 70 per cent of labour-intensive assembly line workers. Managers and supervisors see themselves as the 'brains' behind the operation, with women simply acting as robotic manual workers. As a result, it is seen as pointless to try to train such workers, and in any case they are viewed as far too inclined to resign or switch jobs to make it worthwhile. In essence, female Mexican workers are seen as entirely disposable and replaceable – a piece of the factory that eventually will become worn out (with stiff fingers, tired eyes, aching backs) and discarded. Outside the factory, newly independent women are often seen as sexually promiscuous, dangerously liberated, and on the same level as the prostitutes for which the city has long been known. In this way, it is argued that it was the devaluation of Mexican womanhood in general that was happening in Cuidad Juarez, and not just of factory employees in particular. The low pay, poor conditions and lack of upward mobility for women factory workers were part of a more general devaluation of Mexican womanhood. Where industrial production called for a 'disposable' kind of worker, northern Mexico became an ideal location. For more, see Wright (1999).

nursing, and cleaning are all often characterized as 'women's work': their feminine associations in domestic space are transferred into the public domain of the labour market. In this section we will examine some of the evidence for this process and its consequences for women's wages. We will also explore the ways in which waged work that does occur in the home, such as some forms of childcare, becomes devalued. But it is not just domestic spaces, and the jobs associated with them, that become gendered. Workplaces can also become very masculinized or feminized environments, with exclusionary consequences. We will therefore examine how spaces of work become gendered.

The gendering of jobs

The UK, as we saw in Figure 12.1, has a comparatively high level of female participation in the workforce. And yet, if we start to look at some specific occupations we see that gender makes a significant difference to the jobs that are available to women and men. For example, London's first female bus driver was not employed until 1976, and by 2003, only 5.7 per cent of London bus drivers were women (http://www.metroline.co.uk/180804.html, accessed 17 March 2006). Clearly this is still very much a 'man's job', and it is perhaps not coincidental that in Table 12.1, 'transportation' was the only household chore that men carried out more than women. At the same time, only about 10 per cent of nurses in the UK are male, again reflecting the types of activities that are accepted as feminine (and which correspond to assumed nurturing and caring roles in the household).

Not all occupations are so clearly gendered as bus drivers and nurses, but across the labour market there is still a very clear sorting or *segmentation* of the workforce into those jobs that are coded as male jobs, and those that are coded as female (see Box 12.3 for more on segmentation). Some, as we have noted, are gendered because they are extensions of what are seen as 'naturally' feminine domestic work. But others are gendered because they demand the performance of a particular form of femininity or masculinity. This is especially so in jobs that require some form of social interaction with a customer or client (see Box 10.1). These are increasingly widespread in service-oriented economies and require not just technical skills, but an embodiment of certain characteristics. At the top of the list is gender and sexuality, but linked to this are height, weight, skin tone, grooming, dress, comportment, accent, elocution, emotional presence and so on; in other words, the full range of ways in which the body is presented in a social interaction.

In the introduction to this chapter, we noted that one of the 'Sexism in the City' cases involved a banking director being described as 'too old and ugly' to do the job. Clearly this implies that a particular kind of femininity was deemed acceptable, while others were not. There are many other cases where a particular form of gendered identity and sexuality are 'required' to do a job. A closer

Box 12.3 Theories of labour market segmentation

The process by which a labour market is divided up into different parts is known as segmentation. These segments might be geographically separated, but more commonly the term is used to describe different types of jobs, and usually to contrast better, higher paid, and more stable work from less desirable occupations. Once matched to different segments of the labour market, it generally becomes quite difficult for individuals to move outside of them. If people were randomly distributed across different segments, we would expect that the characteristics of people in particular occupations would reflect the characteristics of the population as a whole. Clearly, however, this is not the case. Gender, ethnicity and immigration status represent three primary characteristics by which the labour force is differentiated, with women, ethnic minorities and immigrants disproportionately occupying subordinate positions in the labour market. The purpose of labour market segmentation theory is to explain such patterns of over- or under-representation. It rejects the view that job allocation is the outcome of the skills, qualifications and experience (i.e. the human capital) of applicants. Instead it argues that each segment has its own distinctive set of rules that govern access to it, and mobility within it. The historical, institutional and geographical contexts for a labour market are hence more important than the human capital endowments of individuals. Early versions of segmentation theory identified primary and secondary labour markets, with the former characterized by stable, well-paid jobs generally occupied by white males, and the latter comprising unstable, low-paid jobs with few prospects for internal promotion. In recent years, segmentation theory has become more subtle, going beyond this dual labour market structure to consider multiple segments. In addition, the identities of employees themselves, as well as the types of jobs they occupy, have received attention. Thus the context provided by families, schools, neighbourhoods – all of the circumstances in which a workforce is socially reproduced – are important in shaping labour market outcomes. As a result, labour markets vary greatly from place to place, not just because social relationships tend to be localized, but also because particular intersections of state control – city, region, national, and even international – regulate the labour market in place-specific ways. For more, see Peck (1996) and Hanson and Pratt (1995).

examination of the merchant banking industry in the City of London reveals the ways in which dimensions of gender identity are performed in the workplace (McDowell, 1997). Some functions, such as bond trading, are seen as requiring a quick, aggressive, competitiveness embodied in the brash, virile masculinity of young male employees. Femininity is seen as anomalous, even where a women 'tries' to act like a man. In the sober world of corporate finance, however, a steady, confident, rational and intellectual image is projected, this time associated with the blue-blooded assuredness of a grey-haired and bespectacled senior executive. In other jobs, such as secretarial work, subservience, attention to detail, organization, and multi-tasking are required, and these are seen to be correlated with femininity. This is not to say that women cannot be aggressive and competitive, for example, but it is seen as a less normal, less appealing, and certainly less feminine trait. McDowell's key point is that sexual and gendered identities are performed in the workplace. These performances create certain roles for men and for women, which are difficult to transcend.

In most jobs, then, it is not just masculinity or femininity that is seen as embodying the requirements of the work, but particular *kinds* of masculinity or femininity. These are then enacted in selective recruitment, training strategies, evaluation and promotion criteria, as well as more subtle techniques for ensuring adherence to organizational norms. The demeanour, including the gender and sexuality of the provider of a service is thus crucially important, and indeed are a fundamental part of the 'product'. For example, since the early 1970s Singapore Airlines has used the marketing slogan 'Singapore Girl, you're a great way to fly' and has emphasized the attentive and caring demeanour of its flight attendants (Figure 12.3). And just as this is true in the merchant banking or airline industries, so it is also the case for real estate agents, bar tenders, lawyers, doctors or academics – any job that requires an interactive component.

Valuing gendered work

Having established some of the ways in which jobs become gendered, we can now turn to look at the patterns of employment for women in the labour force. What we find is that, just as women's work in the home tends to be invisible and ignored, classically feminine jobs in the workforce tend to be poorly paid and accorded a low status.

Table 12.3 shows the top ten jobs occupied by women in the United States. For some, such as secretaries, school teachers and registered nurses, the percentage of women in the job is over 90 per cent. For all of the jobs listed, women dominate, with over 80 per cent of total employment. This is a pattern repeated across much of the world: women tend to be employed in female-dominated occupations. They may work alongside men, but their peers who are actually doing the same job are predominantly women. Another pattern is the generally low pay in these jobs. In Table 12.3, only two of the ten occupations (nursing

Figure 12.3 The 'Singapore Girl': feminine work personified
Source: © Singapore Airlines, reprinted with permission.

and school-teaching) have a median weekly wage for women that exceeds the overall male median wage of US$713. Finally, by looking at the last two columns, we can see how, even in the same job, women's wages are significantly lower, on the whole, than men's. In this top-ten list, only a single occupation ('receptionists and information clerks') recorded a higher median wage for women than for men. Overall, then, even in low-paid, feminized occupations, men are getting paid more. How do we explain this pattern? First of all, as we have already observed, many of the jobs in Table 12.3 are extensions of conventionally feminine roles in the domestic sphere – for example, teaching, nursing and secretarial support. In the domestic sphere they are provided without any financial reward and this devaluation of such work carries over into the public sphere. The example of childcare provides a particularly compelling example of this argument, as discussed in Box 12.4.

As the childcare example suggests, a second cause of devaluation of women's work revolves around how the skill involved is assessed (and therefore how much it should be rewarded). The notion of 'skill' is highly problematic as it is actually quite difficult to identify the tasks involved in most jobs that would enable us to rank them one after the other in terms of skill. In general, however,

Table 12.3 Top ten occupations of full-time employed women in the US, 2004

Occupation	% Women	Women's Median Weekly Earnings (US$)	Men's Median Weekly Earnings (US$)
Preschool and kindergarten teachers	97.7	515	–
Secretaries and administrative assistants	96.7	550	598
Elementary and middle school teachers	96.7	776	917
Receptionists and information clerks	93.9	463	454
Registered nurses	91.8	895	1031
Teacher assistants	91.7	373	–
Bookkeeping, accounting, and auditing clerks	91.2	542	563
Maid and housekeeping cleaners	89.7	324	402
Nursing, psychiatric, and home health aides	88.3	383	420
Office clerks	83.8	499	523
Total waged population with full-time work, 16 years and over	43.7	573	713

Source: US Department of Labor, Bureau of Labor Statistics, Annual Averages 2004 http://www.dol.gov, accessed 24 November 2005.

tasks associated with the home (which 'everyone' does) are perceived to be low in skill requirements. The use of machine technology in particular appears to be a key basis for differentiation between skilled and unskilled work, but the relationship is far from consistent. The *place* of work is also significant. The gender division of labour in Ecuador's garment industry illustrates this point nicely (Lawson, 1999). In the country's capital city, Quito, there exists a gendered division of labour within that sector which labels the work done by men as 'tailoring' but categorizes women as 'seamstresses'. The two occupations are spatially separated with tailors usually working in workshops or stores, while seamstresses work in their homes. Partly as a result of this, tailors enjoy state recognition and representation through guilds. The outcome is that tailoring is understood as a skilled 'profession' while seamstressing is 'just work'. This has left women working in the garment sector disadvantaged in a variety of ways, including a lack of union membership, welfare benefits, minimum wages, and job protection. There are, however, two interesting twists to this story. One is that restructuring and new competitive pressures in the garment industry have favoured the use of low-paid, unprotected, home-based seamstresses. The other is that women often rationalize their work in a very different way from men. While the tailors emphasize their professional standing and the masculine nature

Box 12.4 What keeps childcare cheap?

In recent years a popular new global genre of television programming has been concerned with parenting practices. Shows such as *Super Nanny* and *Nanny 911* demonstrate the skills involved in the disciplining and education of young children in their homes. And yet, with 'naughty chairs' and 'time-outs', they barely scratch the surface of the skills that are actually needed – from hygiene, nutrition, and psychology, to negotiation, artistic creativity and boundless patience. But childcare is a paradoxical occupation. While we can easily identify the skills that are needed, when it becomes someone's waged employment it is almost universally treated as unskilled, low-status, and insecure – with a wage level to match. The treatment of waged childcare work strikes at the core of how work in the domestic sphere is understood. It is, first of all, assumed to be women's work. And this applies whether it is done by mothers or by paid caregivers. In fact, it is important to note that just because women are increasingly entering the workforce it does not mean that childcare, whether in the home or in a day care, is no longer being done by women. Second, because so little attention is paid to the kinds of work that go on within the walls of the home, childcare is generally conflated with other kinds of domestic chores – cleaning, ironing, cooking, etc. These tasks require varying levels of skill, but they are usually bundled together as 'housework'. One would be very surprised if the various tasks undertaken in an accountancy office or a factory were left undifferentiated in this way, but in the domestic sphere this is exactly what is done. Third, childcare is often construed as recreational – since it involves 'entertaining' or 'playing games' with a child, it is easily confused with 'having fun'. Fourth, because there is little career progression in the field of childcare, it is generally assumed to be a dead end, de-professionalized and therefore temporary kind of work. Fifth, because it often occurs inside the home, it is usually treated as beyond the reach of normal workplace regulations. Finally, in many parts of the developed world, childcare at home is being done by migrant women. Such women have precarious citizenship rights in their country of work, and the differential between incomes in the Global North and Global South makes them 'cheap'. This is not, however, universally true. British 'nannies' are treated quite differently in North America than Filipina caregivers, for example. The ethnicity of the caregiver can therefore be just as important as their gender in determining their 'value'. For more, see England (1996) and Pratt (2003).

and status of their work, the women view their work in relation to domestic responsibilities and respectability – being a good wife, mother, or sister. From this point of view, the domestic sphere is a source of pride and identity for women, and not just a source of entrapment and marginalization. In fact, women see themselves as exercising power in the domestic sphere.

There are, then, cases where women may do the same work as men, but it will be paid less. This brings us to a third reason for the devaluation of women's work: the assumption that women are working in order to earn 'extra' money for the household, that their earnings are secondary, and that they have less need than men to derive satisfaction, professional self-esteem or identity from their work. In some societies the assumption that the main breadwinner will be male is quite explicit. In Singapore, for example, men are officially regarded as the head of a household and the principal earner. Thus, in the country's civil service, health benefits for family members are only channelled through male employees on the assumption that female employees have husbands whose jobs bring such benefits. In other societies, the secondary and supplemental nature of women's earnings is implicit. The effect, however, is to justify lower wages for women even if the work might objectively be seen as quite skilful. Thus we have moved away from a rationale that says women are doing less skilful work, to one that sees work done by women as less important *because* it is being done by women.

We have seen in this section how particular types of work are devalued, either because of their association with domestic spaces, or because of the ways in which skills are assessed, or simply because women's work is viewed as being secondary in the household.

Gendered workplaces

As we saw in the introduction to this chapter, the coding of jobs as specifically masculine or feminine may be reinforced by the formation of masculinized or feminized workplaces. Research has focused on masculine workplaces in particular and the ways in which they become exclusionary for women. For one example of this process we can return to the City of London's banking sector (McDowell, 1997). The ways in which certain jobs in merchant banking are seen as masculine or feminine is inscribed upon the landscape of the financial district and the environment of the banking workplace. The marble floors and dark wood of director's offices, the portraits of old patriarchs staring down, even in newer buildings, are reminiscent of the old gentlemen's clubs in which business would once have been transacted – reminding women of their historical exclusion. In social encounters in the workplace, comments, gossip or jokes about attractiveness, body size, and clothing place specific expectations on women, defining the ways they can acceptably be a part of the 'team'. Young women, in particular, are constantly reminded that they are seen as sexual objects in a male environment.

When work issues are discussed, military ('taking some flak'), sporting ('being a team player'), or sexual metaphors are often used, thereby creating exclusionary language. Even something as apparently well-intentioned as apologizing for bad language when there are 'ladies present' is a form of cultural exclusion (indicating that normal, masculine, interaction can resume when they have gone). Beyond the workplace, bonds and networks are formed between men in football teams, pubs, and other male events that often exclude women. Women are, it could be argued, marked as unnatural and exceptional interlopers in a masculine world, except where they play 'feminine' roles of subservience, for example, as secretaries.

Another example comes from the high technology sector (Massey, 1994). In university science parks (e.g. the famous Cambridge Science Park in the UK) and other research-intensive workplaces, there is a masculinity that is quite different from those found in merchant banking, but it is no less exclusionary. Massey notes, for example, the 'monastic' atmosphere of high tech research establishments, even including a reverence towards technology as a secular god. In many high tech science parks, over 90 per cent of highly qualified technologists and scientists are men. Thus the language, humour, employee interaction and other everyday dimensions of office life are very masculine. Beyond this atmosphere, the work culture also demands that researchers put in long days. But the ability to do so is based on certain assumptions: 'The point is that the whole design of these jobs requires that such employees do not do the work of reproduction and of caring for other people; indeed it implies that, best of all, they have someone to look after *them*' (Massey, 1994: 190). There are, then, both masculine aspects to the workplace and a demanding labour process that assumes a wife at home to take care of the reproductive side of life.

This section has largely focused on the patriarchal structures of male dominance that exclude women or devalue their work. It is, however, important to note as well that economic life is no longer quite so rosy for men in the labour market. While secure employment might once have been the expectation of most young men, economic restructuring in the developed world is increasingly removing the kinds of jobs that they would once have occupied. Box 12.5 examines this phenomenon in the United Kingdom in particular.

12.5 Home, Work and Space in the Labour Market

We have seen how the association of domestic roles with femininity makes certain jobs seem to be 'women's work', and how certain workplaces become exclusionary masculine places. We also know that women do indeed shoulder most of the burden for household work. This means that there is a key link between home and work that must be examined in order to understand how women participate in the labour force. It is a link that is created in the practical

Box 12.5 Redundant masculinities

In the developed world, urban labour markets have become increasingly polarized. At one end there are high-status, well-paid and technology-intensive jobs in what is often termed the 'knowledge economy'. They usually require high levels of formal education and many are in the business services sector – law, accountancy, marketing and various kinds of consultancies, as well as creative employment in design, entertainment, advertising, and journalism. Senior bureaucratic jobs in the public sector would also enjoy some of the same privileges. Although there are barriers to women in some of these fields, as we have seen in this chapter, they are increasingly open (at the entry level, at least) to men and women equally. The other pole comprises low-paid, low-status, and insecure jobs, largely in the service sector. These jobs exist, for example, in the retail sector, clerical work, food services, cleaning, and hospitality sectors. In these jobs, contact with the public is often a requirement and many are socially constructed as feminine jobs (see also Box 10.1). The classic attributes of masculinity effectively disqualify young men from such jobs. Meanwhile, traditionally masculine jobs in the manufacturing and resource sectors are declining in number as global restructuring takes such employment abroad to locations with lower labour costs. Secure employment at a factory, in a mine, or at a pulp and paper mill is now difficult to find in Western economies. The result, in British society for example, is that the traditional sources of employment for young working-class men with low levels of formal education are now very scarce. And with those jobs the traditional sources of employment that affirmed masculine identities have also disappeared. Not only are young men finding themselves redundant (i.e. out of work), but the 'laddish' masculinity from which they develop their self-esteem and identity is also increasingly redundant (i.e. out of date). For some this will mean a 'portfolio career' in which they move from one short-term job to another. For many, masculine identities will no longer be derived from the work that they do, but rather from the consumption decisions that they make – the clothes they wear, the music they listen to, and the drinks they consume. In short, while economic restructuring has brought women into the workforce in subordinate roles, in some contexts it has also led to the exclusion of young men and the removal of traditional bases for masculine identity. For more, see McDowell (2003).

everyday lives of women as they try to do all the work, paid and unpaid, that is expected of them.

Geographical considerations play a significant part in this integration of home and waged work. For example, in the town of Worcester, Massachusetts, in the early 1990s, women were constrained in their geographical scope when looking for waged work by the need to be at home when children return home from school, to collect them from school, or to incorporate shopping trips or other activities into their days (here we draw on Hanson and Pratt, 1995). All of this meant that they could not sustain a long commute to work, which in turn greatly limited the type of work they could find. In addition, women tended to use personal networks to find employment, while men relied more on formal job search processes such as newspaper advertisements or employment agencies. It also appeared that women's information networks tended to consist overwhelmingly of other women, while men got job information from other men. But because women's social networks with other women were primarily neighbourhood-based, the job opportunities they provided were also highly localized and limited (Figure 12.4). In this way, the concentration of women in certain jobs, and in workplaces that were within a short radius of their homes, was continually generated by the gendered social networks used to find work. Importantly, women in Worcester tended to find their jobs *after* a household had decided where to live – a decision often made on the basis of the husband's employment. Thus, women tended to be looking for work locally without actually having selected their residential location with a view to the possibilities for employment.

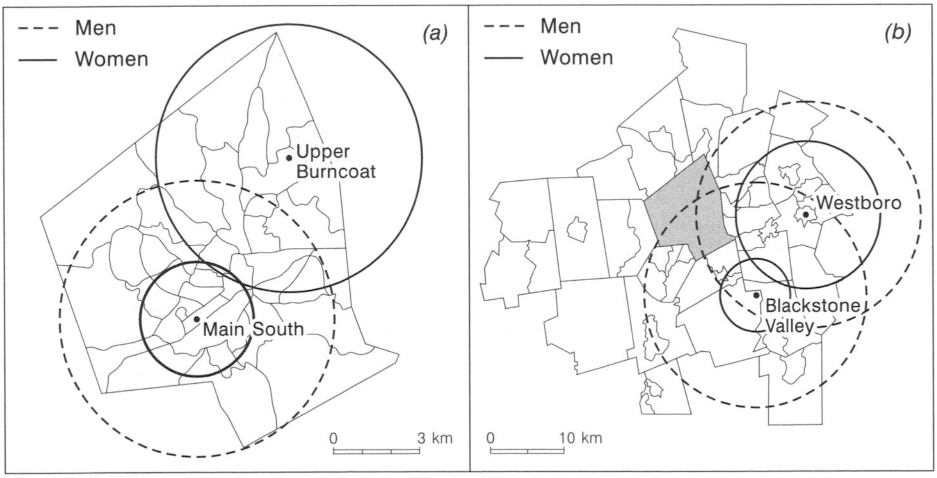

Figure 12.4 Median journey-to-work distances of men and women in four local areas in Worcester, Massachusetts (a) city of Worcester, (b) Worcester metropolitan area
Source: Hanson and Pratt (1995), Figure 4.2. Reprinted with permission of Routledge Publishing.

Another important aspect of Worcester's labour market was that women's concentration in certain types of work was not just because of their own decisions. Employers too were highly conscious of where different pools of labour could be accessed and would locate in particular neighbourhoods in order to do so. In fact, employers might even mould jobs to attract certain kinds of employees – such as married women with children. Employers might also consciously recruit through existing employees and through advertisements in neighbourhood newspapers in order to ensure a locally resident workforce. Both of these techniques will emphasize the clustering of employees' places of residence. This might in some cases be a tactic to ensure a stable, undemanding and non-disruptive workforce.

We can also think about the development of what have been called *pink-collar ghettoes* in the suburbs of large, especially American, cities. There, the relocation of service sector clerical jobs to the suburbs has been with the intention to tap into a married, white, female workforce which must juggle domestic responsibilities with employment. Hence a preference for shorter commutes leads to *entrapment* in certain locally available jobs. In Sydney, Australia, for example, suburban women are entrapped partly by the need for a car to access any kind of employment from areas poorly served by public transport, but also by the strong cultural expectation that 'good mothers' will spend their time ferrying children between after-school activities in the neighbourhood (Dowling et al., 1999). The evidence for this kind of spatial entrapment of women is, however, rather mixed. It would seem that a great many factors come into play when women's home–work interactions are negotiated. The presence of supportive neighbours or relatives in a particular place might enable a longer commute, for example; alternatively, loyalty to an employer, or seniority in a position, might encourage someone to face the costs of commuting rather than working locally (England, 1993).

The picture is also complicated when women are differentiated according to *race*. In the New York area, for example, on average, women tended to commute shorter distances than men, but black women (and immigrants in particular) commute much further than either white women or men (Preston et al., 1998). As we will explain in Chapter 13, this is largely due to the mismatch between the location of jobs and the location of people employed in those jobs, with ethnic minorities in urban centres left far from jobs because of the suburbanization of employment. But it is also worth adding that the picture of white and wealthy suburbs surrounding inner cities housing racialized minorities is a rather uniquely *American* image of a city's social geography, and even there it is increasingly out of date.

While the evidence for entrapment is therefore mixed, the larger point is that employment is embedded in the broader everyday activities, needs and responsibilities of men and women. Where those lives involve uneven responsibility for household work, then women will face barriers to participation even in gender neutral jobs. These limitations derive from the ways in which they integrate their 'economic' participation with their full, complex, and demanding lives. The truth, of course, as we have shown throughout this chapter, is that the two spaces or spheres are not separate, but are intimately connected. When women (in particular) make decisions about participating in the workforce and finding a job, they will be thinking about much more than purely rational economic calculations. Rather, they will be thinking about how flexibility of hours, proximity to home, and other aspects of the job will integrate with their lives. In this way, 'the economic realm is not confined to the workplace, nor the social realm to the home and community' (Hanson and Pratt, 1992: 398).

Up to this point, we have tended to see home as an anchoring point for women – a space where their responsibilities entrap them, ghettoize them, and marginalize them. There are, however, other ways of viewing the home–work link. The home can, in some instances, facilitate resistance to the tactics of labour control used by employers. This was, for example, the case for Jamaican women working as data-entry clerks and telemarketers in Kingston and Montego Bay, as noted in Chapter 9. Households, and especially extended households with relatives in the US or Canada, provided a critical source of support that allowed women to moderate the intensity of their work, and engage in the risky business of resistance to workplace discipline. The home, then, can also be seen as a potential site of resistance. This is an interesting twist on the entrapment thesis discussed earlier.

We can, however, go even further. If instead of viewing work in the home as undervalued and therefore in need of careful counting, we rather seek alternative value systems, then household work may take on a different meaning altogether. It is this re-scripting of value through a feminist approach to the economy that we turn to in the final section of this chapter.

12.6 Towards a Feminist Economic Geography?

So far, our approach in this chapter has been to see how gender might be recognized and included in an analysis of the economy, and of labour markets in particular. All of this might be summarized as an *adding on* or *counting in* approach, whereby we attempt to recognize the work done primarily by women in the household, and we include gender as a variable when trying to understand how people get allocated to different jobs (Cameron and Gibson-Graham, 2003).

The *adding on* approach involves recognizing the role of the domestic or reproductive sphere in the economy as a whole. Without understanding how the workforce of a capitalist economy is reproduced on a day-to-day basis and from one generation to the next, we cannot really comprehend the system as a whole. Once the domestic sphere is added in this way, then a whole range of processes can become a part of the economy: caring work, barter exchange, cooperative or mutual aid arrangements, subsistence agriculture, recycling, voluntary community work, gift-giving, and so on.

The *counting in* approach involves trying to quantify the work done by women that is currently left uncounted. This approach has made some progress but remains rather limited. The data from the UK that we saw in Table 12.1, for instance, is an example of attempting to count in the work done by women in domestic spaces, but even that exercise remains officially 'experimental' and does not feature in any way when official statistics about the UK economy are calculated and published. Thus, policies are still determined very much by their effects upon official indicators of progress and well-being.

Both the adding on and counting in approaches, however, leave intact the idea that the whole economy is definable and quantifiable (as discussed in Chapter 2). Rather than questioning the notion of 'the economy', they seek to modify it and expand it from a feminist perspective. Some, however, would go further and ask whether there is a distinctively different *feminist* economic geography – not just a feminist perspective on the economy, but a feminist approach that entirely re-orders what we are looking for in studying the economy, and what we are seeking to change. In other words, not just *analysing* the economy differently but actually imagining a *different* economy.

What might such an approach look like? Rather than acknowledging women's economic lives, we might start to develop an alternative economic 'ethic' that sees them differently. This would involve a re-visioning of how we conceive the economy, so that traditionally 'feminine' activities such as 'nurture, cooperation, sharing, giving, concern for the other, attentiveness to nature, and so on' are promoted and valued throughout the economy, and not just at the margins (Cameron and Gibson-Graham, 2003: 153).

This point sits within a larger argument that a feminist approach to the economy would look beyond the overwhelming power of capitalism. Rather than using a *capitalocentric view* of the economy, some propose instead a diversity of economic relations and a diversity of *subject positions* – meaning that women and men should be viewed as having more diverse dimensions to their lives (and identities) than the way in which they participate in capitalist production, labour and exchange relationships (Gibson-Graham, 1996). In advocating alternative/diverse forms of economy, this project has a great deal in common with some of the ideas about alternative economies that we discussed in Chapters 2 and 9.

One corollary of this approach might be that entirely different calculations are made about the worth of the domestic sphere. This can be illustrated by returning to the example discussed earlier concerning the tailors and seamstresses of Quito, Ecuador. There, it was noted that seamstresses saw the home not as a site of containment and repression, but as a place in which they constructed their self-esteem based on respectability and power within the household (Lawson, 1999). By re-visioning the economy, then, different ethical visions of what is good and bad, and what is valuable, might be enabled. Not everyone would see the home as a space of empowerment, but the broader point is that an alternative, feminist, approach to the economy opens up the possibility that there are multiple interpretations of 'worth' and 'value'.

12.7 Summary

In this chapter we started with the view from human capital theory that saw people as rational economic actors and treated the economy as entirely separate from social and cultural processes, and the domestic and community spheres in which they are played out. We then proceeded to look at how work in the home is highly gendered, and this provided an important starting point for much of what followed. In exploring how women are increasingly participating in the waged labour force around the world, we saw that the power of patriarchy was, in many instances, reproduced in the workplace. Furthermore, this patriarchal power is also integrated with devaluation based on ethnic difference.

We then turned to look at exactly what kinds of jobs women tend to be doing. In many instances, we see that jobs are gendered and that those jobs that women usual do, and which are considered most appropriately feminine, are also extensions of domestic tasks. Where women do work that is similar to men's, it is often seen as unskilled, de-professionalized, and secondary to the 'breadwinner' role of a male spouse. It is therefore paid less. But we also noted that workplaces can be gendered as well as jobs. The raucous masculinity of merchant banking perhaps provides an extreme example, but the increasingly 'interactive' nature of service sector employment means that employees have to *embody* the job and not just *do* the job. In this way, gender and sexuality are fundamental to whether someone is seen to 'fit' with an occupation.

We have also seen that the gendering of jobs is not just about *associating* women with certain kinds of work, i.e. representing femininity in certain ways. It is also related to the very practical integration of work with the continued domestic expectations of women. This integration of 'economic' and supposedly 'non-economic' sides of their lives means that job searches and employment possibilities can often be limited.

We have seen, then, how feminist approaches to economic geography pay attention to the way in which women are viewed differently from men in a patriarchal society – starting in the home, but in other spaces as well. We have also seen how feminist approaches tend to transcend a lot of the binary divisions that we use to understand the economic world: between economic and non-economic; structural change and lived experience; production and reproduction; and, finally, the study of social processes alongside a political project to change them. In the final section of the chapter, we therefore explored the possibility of an alternative feminist economic geography, in which the economic world is not just counted differently, but evaluated differently as well.

Further reading

- McDowell (1999) and Domosh and Seager (2001) provide excellent and accessible introductions to feminist geography.
- On the subject of childcare and the devaluation of caregivers in Canada, see Pratt (2004) and England and Stiell (1997); in Singapore, see Yeoh and Huang (1999), and in the UK, see Gregson and Lowe (1994). More broadly, England's (1996) edited collection examines how women negotiate work and childcare concerns.
- For more on the devaluation of 'Third World Women', see the work of Wright (1999) and Cravey (1998) on Mexico, Chant and McIlwaine (1995) on the Philippines, and Silvey (2003) on Indonesia.
- At the other end of the spectrum, McDowell's (1997) study of banking in London is full of insights, and recent research by Mullings (2005) in the Caribbean examines gender and race in banking in an entirely different context.
- At various points in this chapter we have noted the variation in gender regimes across space. This is an idea developed in some detail by Massey (1984) through her *Spatial Divisions of Labour* concept discussed in Chapter 3.

Sample essay questions

- Does increasing access to the waged labour force represent liberation from patriarchal powers structures for women?
- How might workplaces become masculine or feminine environments?
- How does the division of domestic labour within the household affect women's experiences of work in the waged labour force?
- Thinking about the Global North *and* the Global South, do contemporary processes of economic restructuring favour men or women?

Resources for further learning

- http://www.ex.ac.uk/~RDavies/arian/scandals/behaviour.html: a website entitled Bankers Behaving Badly contains an archive of press reports on the Sexism in the City scandals and related issues.
- http://www.statistics.gov.uk/hhsa/hhsa/Index.html: further data and information on household work in the UK are available at the National Statistics Office website for the Household Satellite Account.
- http://www.maquilasolidarity.org/: the Maquila Solidarity Network has campaigns and further information concerning women's work in export factories around the world.

- http://www.unifem.org/: the United Nations Development Fund for Women (UNIFEM) has information on women's work around the world.

References

Becker, G.S. (1993) *Human Capital: A Theoretical and Empirical Analysis*, 3rd edn, Chicago: University of Chicago Press.

Cameron, J. and Gibson-Graham, J.K. (2003) Feminising the economy: metaphors, strategies, politics, *Gender, Place and Culture*, 10: 145–57.

Chant, S. and McIlwaine, C. (1995) *Women of a Lesser Cost: Female Labour, Foreign Exchange and Philippine Development*, London: Pluto Press.

Cravey, A. (1998) *Women and Work in Mexico's Maquiladoras*, Oxford: Rowman and Littlefield.

Domosh, M. and Seager, J.K. (2001) *Putting Women in Place: Feminist Geographers Make Sense of the World*, New York: Guilford Press.

Dowling, R., Gollner, A. and O'Dwyer, B. (1999) A gender perspective on urban car use: a qualitative case study, *Urban Policy and Research*, 17: 101–10.

England, K. (1993) Suburban pink collar ghettos: the spatial entrapment of women? *Annals of the Association of American Geographers*, 83: 225–42.

England, K. (ed.) (1996) *Who Will Mind the Baby? Geographies of Child-care and Working Mothers*, New York: Routledge.

England, K. and Stiell, B. (1997) 'They think you're as stupid as your English is': Constructing foreign domestic workers in Toronto, *Environment and Planning A*, 29: 195–215.

Gibson-Graham, J.K. (1996) *The End of Capitalism (As We Knew It): A Feminist Critique of Political Economy*, Oxford: Blackwell.

Gregson, N. and Lowe, M. (1994) *Servicing the Middle Classes: Class, Gender and Waged Domestic Labour in Contemporary Britain*, London: Routledge.

Hanson, S. and Pratt, G. (1992) Dynamic dependencies: a geographic investigation of local labor markets, *Economic Geography*, 68: 373–405.

Hanson, S. and Pratt, G. (1995) *Gender, Work and Space*, London: Routledge.

Lawson, V. (1999) Tailoring is a profession; seamstressing is just work!, *Environment and Planning A*, 30: 209–27.

Massey, D. (1984) *Spatial Divisions of Labour*, Basingstoke: Macmillan.

Massey, D. (1994) *Space, Place and Gender*, Cambridge: Polity Press.

McDowell, L. (1997) *Capital Culture: Gender at Work in the City*, Oxford: Blackwell.

McDowell, L. (1999) *Gender, Identity and Place: Understanding Feminist Geographies*, Minneapolis: University of Minnesota Press.

McDowell, L. (2003) *Redundant Masculinities? Employment Change and White Working Class Youth*, Oxford: Blackwell.

Mullings, B. (2005) Women rule? Globalization and the feminization of managerial and professional workspaces in the Caribbean, *Gender, Place and Culture*, 12: 1–27.

Peck, J. (1996) *Workplace: The Social Regulation of Labor Markets*, New York: Guilford Press.

Pratt, G. (2003) Valuing childcare: troubles in suburbia, *Antipode*, 35(3): 581–602.

Pratt, G. (2004) *Working Feminism*, Edinburgh: Edinburgh University Press.

Preston, V., McLafferty, S. and Liu, X.F. (1998) Geographical barriers to employment for American-born and immigrant workers, *Urban Studies*, 35: 529–55.

Silvey, R. (2003) Spaces of protest; gendered migration, social networks, and labor protest in West Java, Indonesia, *Political Geography*, 22: 129–57.
Wright, M. (1999) The politics of relocation: gender, nationality and value in the Maquiladoras, *Environment and Planning A*, 31: 1601–17.
Yeoh, B.S.A. and Huang, S. (1999) Spaces at the margins: migrant domestic workers and the development of civil society in Singapore, *Environment and Planning A*, 31: 1149–67.

CHAPTER 13

ETHNIC ECONOMIES

Do cultures have economies?

Aims

- To explore how employment experiences can be shaped by ethnic identities
- To trace the emergence of ethnically-based economies in immigrant cities
- To examine the transnational dimensions of ethnic business practices
- To consider the implications of considering ethnic identity as a variable in economic life.

13.1 Introduction

Toronto is, according to one of its own slogans, 'the World in a City'. Despite being an exercise in civic boosterism, this image of global inclusiveness is one that the city can claim with some justification. With around 4.5 million people in the Greater Toronto Area, sprawling along the north shore of Lake Ontario and into the rolling farmland of southern Ontario, the city is Canada's financial, manufacturing, service, artistic, and academic hub. It is also home to immigrants and their descendants from almost every place on earth – few other cities combine the same proportion of foreign-born citizens (44 per cent in 2001) with such a diversity in their places of origins, and few can claim as large a proportion of 'visible minorities' (36.5 per cent) among their populations.

This is a feature of Toronto's social composition that a visitor would notice immediately upon landing at Pearson International Airport. Arriving passengers (and the relatives waiting for them) would be as diverse as could be found anywhere, as would the service sector workers who help the visitor during a stay in Toronto. But among that diversity of humanity, the observant visitor would start to notice some patterns emerging. All of the airport taxi drivers appear to be South Asians, and a brief conversation reveals that the driver has a Masters degree in engineering and comes from the Punjab in Northern India. Arriving at the hotel, the housekeeping staff appears to be almost exclusively Filipina or black Caribbean women. Stepping out to pick up a newspaper, the visitor might discover that the store around the corner is being run by a Korean couple, and so are the convenience stores a few blocks away in either direction. A shopping trip to a suburban mall reveals an environment where very little English is heard and all signage is in Chinese, reflecting the ethnicity of both retailers and the overwhelming majority of shoppers. It seems, then, that the distribution of people among different jobs, sectors, and neighbourhoods in this diverse metropolis is not entirely random. There is a visible pattern in which certain groups are either allocated to particular jobs in the labour force, or in which a particular entrepreneurial activity and neighbourhood is dominated by a specific ethnic group. Toronto is not unusual in this respect: in every large and diverse city around the world, ethnicity appears to play a part in determining people's experiences of the workplace, the commercial landscape, and more broadly, the economy.

These are also fundamentally *geographical* processes. The sorting of various ethnic groups into different labour market niches often reflects the ways in which the places where they live, or where they come from, are represented, or they reflect the geographical challenge of getting from ethnic residential neighbourhoods to places of employment. The creation of ethnic commercial clusters within a city – Chinatown, Little Italy, Little India, etc. – reflects distinct spatial patterns of economic activity that shape the economic lives of residents in both positive and negative ways. Finally, international migration flows into cities create social and economic ties with migrants' places of origin such that significant flows of capital now pass between households in different parts of the world, and in some cases transnational businesses are established with practices that appear to be distinctively ethnic forms of capitalism.

In all these ways, we will argue that ethnicity matters a great deal in the geographies of economic life. We will start, however, in Section 13.2, by examining the argument that economic processes and logics are blind to ethnic difference. We then start to refute that argument by looking at the role of ethnicity, both in the creation of a differentiated labour force (Section 13.3) and in the formation of distinctive ethnic economies (Section 13.4). In Section 13.5, we then look at a growing (but not necessarily new) phenomenon of transnationalism among migrant communities and explore some of its economic implications. Finally, we ask whether there is a specific form of business practice (even a distinctive form of capitalism) to be found in ethnic economies.

13.2 'Colour-blind' Economics

The above examples from Toronto allude to two ways in which ethnicity is important in shaping urban economies: first, the sorting of the workforce so that certain ethnic groups are disproportionately concentrated in particular kinds of jobs, and second, the patterning of urban landscapes so that businesses owned by certain ethnic groups tend to cluster in particular neighbourhoods and subsequently shape the economic lives of these ethnic populations. Before we elaborate upon a geographical interpretation of these processes, it is worth considering the notable *absence* of any role for ethnic identity in conventional economic analyses of these phenomena.

In Chapter 12, we noted the narrow horizons of human capital theory in relation to the experience of women in the labour market. Much the same argument could be made for ethnic minorities. Conventional economic analysis has very little space for human diversity. Instead, a person's place in the labour market, that is, the job and wage they obtain, are seen as a reflection of their human capital – the collection of skills, qualifications and experience that they bring to the market for human labour. The labour market is, in theory, blind to the visible or cultural differences that an individual might exhibit. Everyday experience suggests, however, that this is far from true. As we will show in this chapter, ethnicity is an essential variable in understanding who gets which jobs in the labour market.

Likewise, any social or cultural motivations for establishing or patronizing a business enterprise are obscured by conventional economic approaches, which focus on the incentive to profit from satisfying a market demand for particular goods and services that is driven by the price mechanism. Economic processes, in this form of technical analysis, become anonymous, universal, and rational. There is no room for personal ties, social relations, loyalty, discrimination, or culturally distinct practices. Indeed, it is often assumed that modern capitalism sweeps aside these sentimental bases for economic action, and they are taken to be vestiges of 'traditional' small-scale or agrarian societies. In reality, we will see that the enterprises established by ethnic minority groups are often a direct response to the marginalization they experience in relation to mainstream society.

The *location* of economic activities is also seen as the outcome of rational decision-making. In the economics of retailing, as well as in economic geography's past focus on location theory (see Box 1.2), the decision to locate an enterprise was rooted in an unerring rationality and logic. This logic was one based upon profit maximization and price minimization brought about, for example, by maximizing the consumer base available to a retailer, and minimizing the distance to be travelled by customers. This approach is captured by central place theory (see Box 10.2), which sought to model the ideal location for retailing and service activities in particular. In maximizing the available market area, a retailer would try to locate as far as possible from other similar retailers, as close as possible to customers, and consumers would minimize their travel costs by visiting the nearest supplier of a particular product. The urban landscape, however, tells another story that is much more complicated and 'irrational'. For any business, there may be advantages to be gained from clustering together with similar enterprises (as discussed in Chapter 5), but the social geography of many cities reveals a pattern in which it is not just similar enterprises that cluster together in space, but rather enterprises owned and patronized by people with similar ethnic backgrounds. Clearly, there are processes at work that go beyond either economic logic or the less tangible financial benefits of agglomeration.

Another issue that mainstream economics tends to neglect is the strong set of economic ties that bind immigrants in global cities with other parts of the world. Local labour markets are usually treated as just that – local. Individual and household decisions are analyzed on the assumption that they are made in a local context of opportunities, barriers, skills, and qualifications. In reality, however, lived experiences are more complicated. Many immigrants in world cities have responsibilities to family members back in China, the Philippines, India, Ghana, Mexico, and so on. This will sometimes force them into taking on several jobs in order to earn enough to satisfy these distant needs. When looking for jobs, immigrants will often depend upon their social networks of family and co-ethnics, usually forged long before they arrived, rather than 'objective' sources of information about opportunities and labour market conditions. Ethnic and family ties, transnational in their scale, are therefore an essential part of the economic lives of immigrants.

Clearly we need explanations for a variety of economic processes that go beyond economic logics to include cultural identities, belongings, and practices. While there are many dimensions to cultural identity and its intersection with economic processes (for example, one could explore the implications of youth culture or gay culture), here we will focus on ethnicity as the key variable. Box 13.1 explains the concept of ethnicity in more detail.

13.3 Ethnic Sorting in the Workforce

The market for labour is unlike any other market. Labour, as we saw in Chapter 9, cannot easily be bought and sold like other commodities. When employees are hired, fired, or promoted, the process of buying and selling is rather more complicated. 'Labour power' comes in the form of complete and complex human beings who have many more attributes than simply their ability to perform a particular function in a workplace. They have gendered, ethnic, class, regional and a variety of other identities that often lead to an unfortunate predictability to the type of work they will end up doing.

To illustrate this point, we can look closely at the labour market experiences of different groups in a major metropolis. Table 13.1 provides data on employment in the Los Angeles metropolitan area. It shows the distribution of four very broad ethno-racial groups (White, Asia-American, Black, and Hispanic), as defined by the United States Equal Employment Opportunity Commission.

Box 13.1 What is ethnicity?

Ethnicity is about difference, marked or coded in various ways. Thus when we talk about Chinese entrepreneurship, we are generally not talking about businesses in China itself, but about the business practices of the ethnic Chinese diaspora around the world. Chinese ethnicity only becomes significant when juxtaposed with different groups. In that sense, ethnicity is always a relational concept and as a collective identity it includes *and* excludes at the same time. Ethnicity is based upon a common (real or imagined) historical experience, ancestry or cultural commonality, for example, due to language or religion.

Ethnicity does not refer simply to minority groups, although it is often used in that way. To be ethnic is not to be somehow outside the mainstream or abnormal. In fact, everyone has an ethnicity. In that sense, a British pub in Bangkok owned by an expatriate is no less an 'ethnic' enterprise than a London restaurant owned by a Thai.

Ethnic identities are not necessarily stable and unchanging. Different aspects of ethnic identity may be brought to the fore in different circumstances, and the strength of an ethnic identity may only emerge over time. For example, Italian immigrants may have arrived in North America in the twentieth century with small-scale loyalties to villages and regions, but, once there, they developed an ethnic identity as Italians.

It is important to distinguish ethnicity from 'race'. 'Race' refers to the visible characteristics of human beings (hair, skin pigmentation, bone structure) but is a very unhelpful analytical category as most 'races' are poorly defined and show as much internal diversity as they do distinctiveness. The notion of 'race' tends to be used when groups with a particular appearance are racialized – that is, categorized and attributed certain characteristics by other groups. For this reason it is usual to recognize 'race' as a social construction and place the word in inverted commas.

Table 13.1 lists a variety of occupational types and then the number of employees in each category. For each occupation category, a 'concentration index' is also listed. This is simply the percentage of a particular group in a given occupation as a ratio of the percentage of all employees in a given occupation. Thus, where we see a concentration index of 1.6 for White employees in the 'Officials and Managers' category, we know there is an over-representation of white employees in that occupation – 1.6 times as many as one would expect if all occupations were shared evenly among different groups.

By examining the shaded boxes, showing distinct levels of over-concentration or under-concentration, we can see some patterns emerging. Most notably, White

Table 13.1 Distribution of ethno-racial groups in various occupations, Los Angeles, 2003

	All Employees	White	Asian-American	Black	Hispanic
Officials & Managers	136,783	90,509	15,326	9,292	21,047
Concentration Index		1.6	0.8	*0.7*	*0.4*
Professionals	207,824	117,670	50,261	14,428	24,594
Concentration Index		1.4	1.8	*0.7*	*0.3*
Technicians	64,094.0	25,400.0	14,424.0	6,718.0	17,166.0
Concentration Index		1.0	1.7	1.0	0.8
Sales Workers	157,166.0	70,286.0	14,072.0	17,958.0	53,844.0
Concentration Index		1.1	*0.7*	1.1	1.0
Office & Clerical Workers	199,603.0	73,039.0	28,816.0	29,799.0	66,914.0
Concentration Index		0.9	1.1	1.4	0.9
Craft Workers	68,203.0	26,336.0	5,804.0	5,161.0	30,446.0
Concentration Index		1.0	*0.6*	*0.7*	1.3
Operatives	111,102.0	22,480.0	8,909.0	11,381.0	67,741.0
Concentration Index		*0.5*	*0.6*	1.0	1.7
Labourers	78,377.0	11,471.0	3,824.0	7,113.0	55,664.0
Concentration Index		*0.4*	*0.4*	0.9	2.0
Service Workers	142,354.0	32,357.0	13,473.0	20,169.0	75,707.0
Concentration Index		*0.6*	*0.7*	1.4	1.5
Total Employment	1,165,506	469,548	154,909	122,019	413,123

Source: Calculated from U.S. Equal Employment Opportunity Commission data: http://www.eeoc.gov/stats/jobpat/2003/pmsa/4480.html, accessed July 20th 2006.
Note: This relates to private sector employment only. Public sector employees were not included. The ethno-racial categories used are those employed by the US Equal Employment Opportunity Commission.

employees are over-represented in managerial and professional occupations, and Asian-Americans are over-concentrated in professional and technical jobs. Both groups are under-represented in less desirable and less well-paid jobs, such as 'labourers' and 'service workers'. For Blacks and Hispanics, however, we see the reverse pattern, with under-representation in high-level managerial and professional jobs, and over-representation in these low-end occupations.

These figures could be analyzed in much more depth. In some cases the figures conceal the kinds of gender disparities discussed in Chapter 12, and certainly there is great diversity within the categories used. 'Asian-American', for example, bundles together a vast range of experiences from Taiwanese software engineers to Vietnamese refugees. But a broad pattern is clear – one in which Black and Hispanics disproportionately occupy subordinate positions in the labour market. A similar story, in which minority populations (in some, but not all cases, immigrants) are marginalized in the labour market, is replicated in many cosmopolitan cities around the world (see Box 12.3 on labour market segmentation). A key question is, why does the workforce get sorted in this way? Why are certain groups drawn into certain jobs? Why do employers take on disproportionate numbers of a certain ethnic group, thereby apparently excluding others? Why does an ethnic group continue to congregate in what is often an unappealing occupation? A wide range of processes can be identified, each with an important spatial dimension:

- *qualifications and skills*: the standard economic argument is that workers are allocated to jobs based on their qualifications and skills. Thus a subordinate position in the labour market would be expected if a group, as a whole, has relatively low levels of educational achievement. This assumes, however, that educational attainment is a reflection of innate abilities. In reality, it is often a reflection of unequal distribution of wealth and income within and between different ethnic groups that in turn affects the ability of parents to support their children's studies both financially and intellectually – there is, therefore, a strong process of inter-generational reproduction of low or high educational achievement. Different education attainment is thus a product of social relations. Furthermore, this is powerfully *place-based*: if you grow up in a working-class community where very few of the youth go off to university, the chances that you will do so are equally slim, even if another context might have made this a possibility. But even if we accept that education is a reflection of innate abilities and skills, the connection between this and employment opportunities is far from clear. Immigrants arriving in cities such as Vancouver, Los Angeles, London, and Sydney are often more educated than the native-born population. And yet they find themselves disproportionately concentrated in low-status, low-paid, and insecure jobs because their qualifications are not recognized in a new context, or because they have developed skills that are specific to a particular socio-cultural context. The skills of a teacher, journalist, architect, graphic designer, or marketer may not 'travel' well. In this way, the human capital associated with education or professional experience can be substantially *devalued* across international space. This points, in part, to cultural differences across space, but also to institutional barriers.
- *institutional barriers*: we have noted that many ethnic minorities who arrive as immigrants may have qualifications, but they are not treated as equivalent to the local credentials required by employers. In many labour markets, professions are regulated by institutions that decide who will be licensed to practise within their jurisdiction – colleges of physicians, bar associations, professional engineering societies, and so on. Where they do not recognize foreign qualifications, or demand costly upgrading courses, an immigrant may be destined for deprofessionalization. In other cases, certain immigration

programmes may place restrictions that limit an individual's ability to upgrade their qualifications or access their profession. In Canada, for example, many Filipina women have arrived under a programme that permits full immigration after two years of work as a domestic caregiver in the house of an employer. Its restrictive terms mean that while nurses, teachers, and accountants may enter the programme, those who graduate from the programme are usually destined for low paid work in healthcare or personal service occupations (Pratt, 2004). In the US, migrants from countries such as Mexico and El Salvador who have arrived without legal status often find themselves without any choice but to accept insecure, low-paid work and sometime dangerous work. In all of these instances, whether professional regulatory bodies, immigration programmes, or undocumented migrants, it is the institutions that govern labour markets in particular territories that create barriers to *upward* mobility for *geographically* mobile people. Understanding the place of immigrants in labour markets, then, requires a close examination of localized institutional structures.

- *discrimination and stereotypes*: the major immigrant-receiving countries around the world today (the US, the UK, Canada, and Australia, for example) present themselves as multicultural and tolerant societies and have laws that prevent discrimination or unfair treatment on the basis of ethnic identity. Nevertheless, when decisions related to hiring or promotion are made, employers and managers inevitably work within a set of cultural codes and understandings that are difficult to regulate. Over the past few decades, for example, hundreds of thousands of Filipina women have travelled overseas to work as domestic helpers, nannies, and nurses. In many countries, both in Asia and beyond, stereotypes have developed that equate being Filipino with aptitudes for nurturing, caring or domestic work. Furthermore, within workplaces such as hospitals or nursing homes it is the 'front-line' or 'bed-side' jobs that Filipinas tending to be doing, rather than managerial, supervisory, or instructional work. Rather than any innate ethnic aptitude, or lack of it, it would seem that this is due, at least in part, to the stereotypes of Filipina identity that exist in the labour market. Here, then, we see the stereotyped association between a particular *ethno-national space* and certain labour market outcomes. Such stereotyping may also take other more localized forms. In San Antonio, Texas, for example, particular neighbourhoods, especially those associated with Hispanic youth, have become stereotyped by employers and assumed to supply only certain kinds of (inferior and unreliable) labour (Bauder, 2001). There is, then, the possibility of a localized neighbourhood effect on labour market outcomes, and the more segregated and ghettoized an ethnic minority population becomes in urban social space, the greater the possibility of this occurring.
- *home–work linkages*: in some cities, residential segregation on the basis of ethnicity is quite pronounced. Certain neighbourhoods are strongly associated

with particular ethnic groups. Where this is the case, for example in many large American cities, there is often a geographical problem to be resolved in getting from home to work. It may mean that many visible minority immigrants have to travel much further to work than their white and non-immigrant counterparts – a phenomenon known as *spatial mismatch* (Preston and McLafferty, 1999). There are limits to this, however, and in large metropolitan areas ethnic residential enclaves may be sufficiently far from downtown or suburban concentrations of 'good' jobs, that for at least one adult in a family, with children to pick up from school and other errands to run, being close to home may be necessary (see Chapter 12 on the gendering of such roles). Hence ethnic residential *segregation* can translate into occupational *segmentation* as the geographical scope of possible jobs is limited by the need to minimize commuting time on public transport. Jobs may also themselves be on the move: where manufacturing or service industries have relocated to suburban areas, especially in American cities, they have often left behind ethnic minority neighbourhoods in downtown areas.

- *social capital and the search for a job*: the final feature of labour markets that needs to be incorporated into our explanation for ethnic segmentation is the role of social networks in finding work opportunities. The idea of a 'labour market' assumes that buyers and sellers of labour power (i.e. employers and their employees) will know about all possible alternatives in terms of jobs available, current market trends in wages and so forth. In reality, the labour market is far less easy to navigate – many people will find jobs because they had contacts who hired them, or actively lobbied for their hiring, or simply made them aware of an imminent job opening. These social networks are critical to labour market processes for new immigrants in particular who may otherwise have difficulty knowing where to start in finding a job. Such networks do, however, have distinctive geographies. They are frequently *neighbourhood-based*, and as we saw in Chapter 12, evidence suggests that this is more often the case for women than for men. Where ethnic minorities exhibit patterns of residential segregation, the likelihood of a neighbourhood-based and co-ethnic social network being used is very high. There are two important consequences of this geographical process. The first is that an ethnic group concentrated in a particular line of work will perpetuate its concentration in that line of work. Second, such co-ethnic network-based referrals tend to work when seeking low-level positions, such as jobs in kitchens, hotels, or factories, but a whole other set of contacts and networks are required to get managerial work. Thus a reliance on networks tends to keep ethnic minorities in certain sectors and in certain jobs in those sectors. Breaking out of particular sectors, and particular places in the city, can be very difficult.

There are, then, multiple processes operating to channel certain groups into particular occupations or parts of the labour market. In all cases, though, there is an important *spatial* dimension to these processes. First, place-based processes are often important in determining what kinds of qualifications a person will bring to the labour market, and the place where they come from may be stigmatized by employers. Second, this also extends to particular *ethno-national* identities, such as Filipinas, who may find themselves 'coded' or represented in certain ways in international labour markets. Third, the construction of institutional barriers to entering certain kinds of professions is based on administrative territories and jurisdictions, and the migrant populations that move between them are often deprofessionalized in the process. Fourth, the geographical 'balancing act' that workers have to engage in between work places and home places can also lead to limited employment opportunities, especially when ethnic groups are residentially concentrated in urban space. Finally, the social networks that are so important in finding jobs are usually local and place-based in nature and lead to certain kinds of work.

13.4 Ethnic Businesses and Clusters

We have now seen the ways in which ethnic groups (many of them immigrants) are differentially sorted into various kinds of jobs in urban labour markets. In some cases, ethnic minorities have responded to marginalization in the labour market by establishing their own businesses and, as we shall see, their own commercial neighbourhoods. This is an *entrepreneurial process* with a long history. For example, some Chinese and Jewish neighbourhoods in major cities around the world can be traced back several centuries. It is, however, a phenomenon that has become even more prevalent with increasing flows of migration since the last quarter of the twentieth century. Major immigrant cities may now have a distinctive Chinatown, Koreatown, Little India, Little Tokyo, Greektown, Little Italy, Portuguese Village, and so on (part of Singapore's Little India is shown in Figure 13.1). In this section we will address two issues: the reasons for the emergence of ethnic entrepreneurship, and the processes that sometimes lead to the spatial clustering of such enterprises in the urban landscape.

Ethnic entrepreneurship

Many ethnic minority groups, especially immigrants, have turned to self-employment as a means of livelihood. Indeed, immigrants from minority ethnic groups often show a higher tendency to establish their own businesses than members of the 'host' societies in which they have settled. The most visible, and identifiably 'ethnic', of these businesses would be restaurants, grocery stores, video rental shops, and travel agencies – those enterprises that are observable on the urban landscape. As Figure 13.2 illustrates, some businesses may combine multiple activities – in this case a Filipino enterprise in Toronto houses a grocery,

Figure 13.1 Singapore's Little India
Source: The authors.

general variety and video rental store, as well as a cash remittance agency and a prepared food take-out business. Beyond these activities, ethnic enterprises may also extend into a full range of less visible activities, including banking, real estate, construction, and manufacturing.

Different ethnic groups have varied in their tendency to establish businesses in their 'host' societies. Where immigrant groups have been quickly assimilated into the waged labour force (for example, British, Jamaican, or Filipino immigrants to Canada arriving with the necessary linguistic skills) then levels of self-employment are generally low. But where the cultural barriers to employment are difficult to surmount, then self-employment has been more common. This has been especially true where immigrant communities have also arrived with sufficient capital – for example in the case of recent Korean or Taiwanese immigrants to the United States. Indeed, it is with resources from within the ethnic community that enterprises are often established. Co-ethnic networks may provide capital for the initial establishment of a business, and credit during its operation, but they may also provide important information on how to navigate the bureaucratic intricacies of establishing a new business, market intelligence

Figure 13.2 A Filipino store in Toronto
Source: The authors.

on where opportunities are to be found, as well as the suppliers, employees, and customers that a new business needs. Box 13.2 relates the story of Korean convenience store owners in Ontario, Canada, and highlights the importance of these *ethnic resources*.

Such ethnic resources might amount to nothing, however, if the *opportunity structure* for establishing a business is not conducive. This relates to the opportunities and barriers that a host society presents. There may, for example, be government policies concerning the establishment of businesses in a particular sector. The general economic climate will also affect whether small businesses

Box 13.2 Ontario's Korean convenience stores

Korean immigration to Canada was minimal until new immigration laws in 1967 made the process less biased towards white European immigrants. Immigrants from South Korea started to arrive in the early 1970s, but it was in the 1990s that a significant immigration stream emerged. Of the 71,000 Koreans immigrants living in Canada by 2001, nearly half had arrived in the previous five years – reflecting a huge surge in immigration following the financial crisis in South Korea in the late 1990s. A remarkable 36 per cent of those in the labour force were self-employed (compared with 11 per cent for native-born Canadians) – reflecting the fact that nearly half of all Korean immigrants in the 1990s had entered Canada through the business immigrant programme. Many of these Korean entrepreneurs established video rental shops, restaurants, dry cleaners, and, in particular, convenience stores. By 2005, there were 3,200 convenience stores operated by Korean immigrants in the province of Ontario alone.

Why did so many Korean immigrants establish convenience stores? On one side of the equation, from the late 1980s onwards, a restructuring in the retail sector led several corporate owners of chains of convenience stores to sell off their retail outlets – hence a business opportunity presented itself at precisely the time when Korean business immigrants were arriving. It was also a business on a scale that could be operated by a family, required no specific technical know-how, supplied a modest but reliable daily cash income, and could be acquired for about the required investment under Canada's business immigration programme (C$300–400,000 depending on the place of settlement). But why Koreans in particular? In part, this reflects the relatively low levels of English language fluency among Korean immigrants – which made only certain kinds of business possible. But a large part of the explanation relates to the co-ethnic support networks created by the Korean immigrant community. As noted earlier in this chapter, when a particular group gains a foothold in a particular sector of the economy or the labour market, it can lead to deepening concentration as new arrivals are guided by previous immigrants into the field that they themselves have entered. In the Korean case, this was also institutionalized through the Ontario Korean Businessmen's Association (OKBA). The OKBA was established in 1973 and emerged to become a major factor in channelling new Korean immigrants into retail enterprises – providing information, credit, and wholesale services to convenience store owners. Less tangibly, the OKBA also created a social environment in which members could find mutual respect and solidarity – a reinforcement of self-esteem in a Canadian context of deprofessionalization and cultural difference (for more, see Kwak, 2002).

are able to take root and grow. And, the size of an ethnic community and its spatial distribution might determine whether businesses can be established that will serve co-ethnics alone.

In some cases, ethnic minority groups may be forced into certain kinds of businesses or sectors. In the early part of the twentieth century, Chinese immigrants in North America, for example, established restaurants and laundry businesses because they were largely excluded from all but the most menial of tasks in the waged labour force. This motivation for self-employment is often referred to as the *blocked mobility thesis*. In the Philippines, where Chinese traders had settled for hundreds of years, citizenship was often denied and legislation in the 1950s explicitly excluded non-citizens from land ownership or any form of retail business. This forced many ethnic Chinese into wholesaling, banking, and manufacturing industry, which, ironically, proved to be far more profitable and dynamic sectors over the subsequent decades (Yeung, 2004). The same has been true for Jewish immigrants in North America, and elsewhere – exclusion from land ownership and skilled worker guilds in Europe forced many Jewish entrepreneurs to take on roles as 'middleman' traders, shopkeepers, and money lenders.

By the late twentieth century, such institutionalized racism was less common, but ethnic differences (including, for example, linguistic and religious differences) were still a basis for the marginalization of some groups in immigrant-receiving societies. In addition, several major destination countries had put in place 'business immigrant' programmes that granted residency or citizenship to individuals with entrepreneurial experience and large amounts of capital to invest. In Canada and Australia, for example, many ethnic Chinese from Hong Kong and Taiwan took advantage of these programmes and were thereby obliged to establish businesses. While these schemes have brought a huge amount of capital into these host countries, the enterprises themselves have often been less successful. Although such business immigrants had to demonstrate previous aptitude and success in entrepreneurial activity, the reality is that many fail to succeed in a new cultural and institutional environment. In some cases this is because of linguistic limitations, but in others it reflects unfamiliarity with labour practices, tax codes, as well as the less definable cultural dimensions of doing business. Business acumen is, therefore, to some extent *place-bound* and success as an entrepreneur does not guarantee a universal ability to make enterprises work in a different context (a phenomenon that echoes the points made about national business cultures in Chapter 11).

Ethnic business clusters

We have noted some of the reasons that ethnic minorities have turned to self-employment and entrepreneurship, especially when arriving in a host country as new immigrants. In many instances, a further feature of these businesses is their congregation in *spatial clusters*, forming identifiable commercial districts such as a Chinatown or a Little India. It would seem that space therefore plays an important part in the emergence of an ethnic economy. This can be so in several different ways (here we draw on Kaplan and Li, 2006).

First, ethnic business clusters are usually located in the heart of residential neighbourhoods associated with the same ethnic group. The neighbourhood is likely a source for employees, who may have been attracted to the area in the first place because of employment opportunities in ethnic enterprises. The advantage of such employment for new immigrants is that barriers such as non-recognition of credentials, linguistic limitations, and discrimination are much less likely to present problems than in the mainstream labour market (as noted in Section 13.3). Where a business is specifically targeting co-ethnics (for example, a video rental or grocery store) then it also clearly benefits from this concentration of potential customers. Even where ethnic business clusters are not located in the midst of the ethnic residential concentrations, an identifiable neighbourhood becomes a natural destination for members of that group when seeking goods and services, or simply when they want to operate in their native language.

Second, there may be *agglomeration economies* (described in Chapter 5) that generate benefits for businesses locating close to their competitors. Restaurants, for example, might collectively benefit from proximity to grocery stores selling the types of ingredients that they require. They might also benefit from customers (especially those from other ethnic groups) who are attracted to the neighbourhood as a whole as much as an individual shop or restaurant within it. Large Chinatowns in New York, San Francisco, and Sydney, for example, are tourist attractions in their own right and locating within them means access to a wealthy tourist clientele. In the UK, the Indian and Pakistani restaurants of London's Brick Lane or Manchester's 'Curry Mile' in Rusholme similarly benefit from the agglomerations they form.

Third, ethnic business clusters enable enterprises to access the ethnic resources mentioned earlier. Networks that provide capital, know-how and market information are often based on face-to-face interaction, which requires spatial proximity. These networks have been shown to be critical in a variety of settings where ethnic minorities have successfully established businesses, and for them to work effectively, the clustering of related businesses is often a necessary pre-requisite.

Finally, a spatial enclave collectively forms a tangible basis for the nurturing and maintenance of an ethnic identity. The architecture, languages spoken, products sold, and even faces seen in an ethnic business cluster ensure that identities are reinforced and positively affirmed. In other words, to be Chinese in Chinatown is to be reminded of one's 'Chinese-ness', which in turn increases the likelihood that one will continue to demand Chinese goods and services. In this way, ethnic enterprises collectively benefit from their visible presence on the urban landscape in ethnic places.

Through a variety of processes, then, spatial proximity and the creation of ethnic places are important elements of an ethnic economy (see Figures 13.1 and

13.2, again). A further effect of spatial clustering may be the increasing *closure* of an ethnic economy. This refers to the extent to which an ethnic economy is conducted exclusively by, and for, members of the same ethnic group. At the furthest extreme, an *ethnic enclave economy* involves a collection of entirely *intra-ethnic* economic relations, with ownership, management, employees, customers and suppliers across a range of economic sectors all sharing the same ethnic identity. In reality, this situation is very rare. In most cases, ethnic businesses are made up of some combination of ethnic and non-ethnic participants in each of these roles. Indeed, the line between an ethnic and a mainstream enterprise is a blurred one. However, where the clustering of ethnic businesses coincides with a distinct pattern of residential segregation, so that an ethnic community can live, work, and shop in the same neighbourhood, then the possibilities of enclave formation increase.

There has been much debate concerning the desirability of spatial clustering and enclave formation among ethnic groups in major immigrant-receiving cities. On the positive side, the ethnic economy is seen as providing opportunities for new immigrants, access to particular market segments and an array of mutual support mechanisms. On the negative side, however, ethnic economies are seen as marginalized, providing poor wages and working conditions, and holding limited opportunities for expansion. These various positions are summarized in Table 13.2. It is, however, important not assume that the movie-induced stereotypes of an ethnic economy hold true in all cases. Informal contract arrangements, small-scale businesses, and low-skilled, low-paid work might still hold true in some places and for some groups, but a more complex picture must now be acknowledged. Many ethnic communities in the largest immigrant gateway cities now have sophisticated and formalized forms of business. In 1999, there were 23 Chinese-American banks headquartered in Los Angeles and while they might be less visible than the numerous grocery stores and restaurants, the financial services they provide are essential in the creation of an ethnic enclave economy (Li et al., 2002).

We should also note that ethnic economies have proven to be dynamic and mobile phenomena. In recent years, with a new generation of wealthier immigrants, older downtown ethnic neighbourhoods in some cities have been joined by new ethnic residential and commercial concentrations in the suburbs. Dubbed *ethnoburbs* by some researchers, these are often not marginalized spaces of ghettoization, but are thriving concentrations of ethnic business that depend to a large extent on the ability of ethnic banks to finance their development. In Los Angeles, the ethnic Chinese banks mentioned above have, through both mortgage lending and commercial loans, brought about an increased concentration of ethnic Chinese in particular neighbourhoods, including both the downtown Chinatown and the Chinese ethnoburb of Monterey Park. Similar dynamic processes are occurring in the San Francisco Bay area. In Toronto (as well Vancouver), suburban neighbourhoods have become the major destinations of new, and

Table 13.2 The two sides of ethnic enterprise

Dimension	Positive	Negative
Causes for entrepreneurship	Market opportunities Entrepreneurial initiative	Discrimination Blocked mobility
Networks and organizational structure	Ethnic resources Organizing capacity of ethnic groups Cooperation in raising capital	Excessive internal competition Dependence on networks Commodification of ethnicity
Networks and labour	Job opportunities for immigrants Efficient labour recruitment Advantages due to low-cost labour Returns to human capital	Containment and segregation of ethnic groups Poor pay rates and exploitative working conditions Low returns to human capital
Networks and markets	Access to protected markets Successful business transactions	Limitations of a closed ethnic market
General outcomes	Internal support mechanisms and social mobility	Exploitative labour relations and broader structural marginalization Spousal exploitation

Source: Adapted from Walton-Roberts and Hiebert (1997), Table 1.

often wealthy, immigrants from Hong Kong, Taiwan, and China. Toronto's two downtown Chinatowns are now relatively minor residential and business concentrations for the Chinese population, and the presence of other groups such as Vietnamese has increased in these areas. The primary centres of Chinese commercial activity are now the city's north-eastern suburbs, where numerous Chinese shopping malls have been developed (Figure 13.3).

In all these ways, we see ethnicity affecting the location of businesses in a way that is entirely unpredictable when economic analysis is blind to the importance of culture and ethnic difference. Ethnic bonds motivate businesses to cluster together, in part because ethnic bonds result in consumers and clients gravitating towards businesses that provide a familiar environment. But the very act of creating a cluster also strengthens those bonds as a place-based ethnic economy reinforces the sense of identity and difference among ethnic minorities.

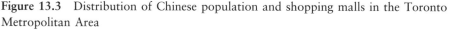

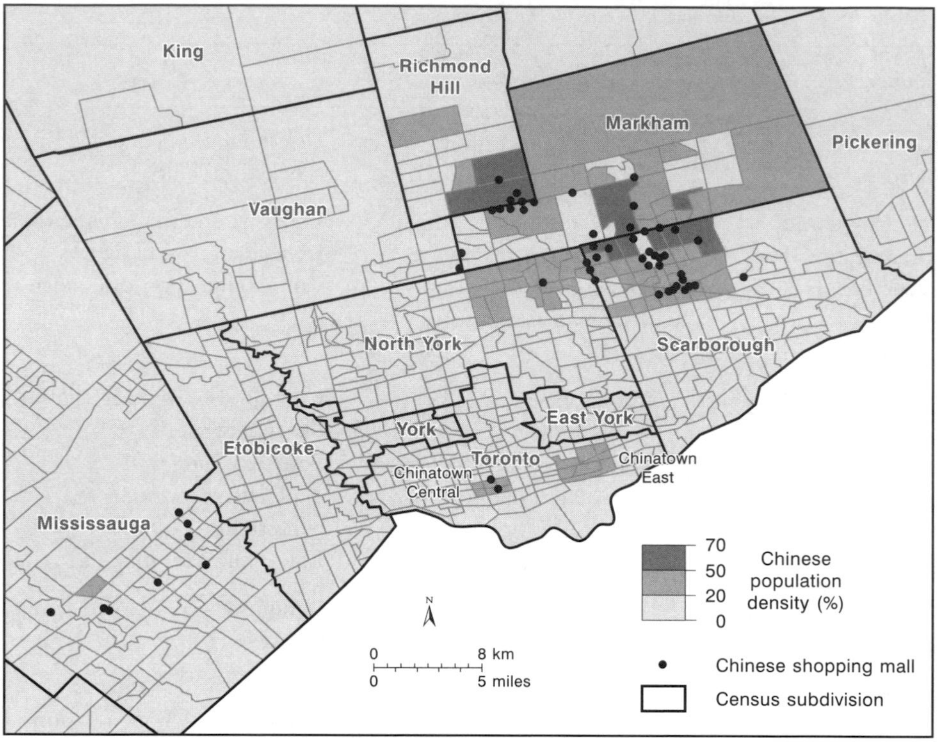

Figure 13.3 Distribution of Chinese population and shopping malls in the Toronto Metropolitan Area
Source: Lo (2006), Figure 6.1. With permission of Rowman & Littlefield Publishers, Inc.

13.5 The Economic Geographies of Transnationalism

We have so far seen the diverse ways in which ethnic enterprises are related to changing geographies at the scale of the city. Until recently, this was the spatial scale at which most geographical analyses would stop. Ethnic entrepreneurs and ethnic economies were understood as a part of the integration and settlement of immigrants in a host society. Over the past two decades, however, it has become increasingly evident that immigrants (of any ethnicity) do not simply arrive, settle, and start anew. Instead, they maintain close ties with their countries of origin. These ties are global in their reach and have been collectively labelled *transnationalism* (Box 13.3). In this section we will consider two forms of transnational ties that are commonly established by migrants between home and host countries. The first relates to the sending of *financial remittances* to family members left behind in their home countries. These funds often have significant economic effects at the national scale in the receiving countries, but

Box 13.3 Transnationalism

Since the early 1990s, the notion of transnationalism has been used to describe the multiple linkages that immigrants maintain with their places of origin. These linkages are not necessarily new. Immigrants have always maintained social, cultural, economic, and political linkages with their homelands. But in recent years several developments have intensified these linkages. First, the sheer growth in temporary and permanent migration, and the diversity of countries involved, has increased the total numbers of people living outside the country of their birth – hence the scale and scope of transnational linkages has multiplied. Second, technological developments have greatly facilitated the maintenance of strong ties: inexpensive telephone calls, email and text messaging, and even discount airfares, have all enabled relationships to be continued across global space with an intensity not previously possible. Third, institutional structures increasingly recognize and even encourage transnational activities – overseas investment by diasporas is positively encouraged by many governments, and is increasingly backed up with provisions for dual citizenship, absentee voting, and state support to maintain the bonds of overseas citizens. The result is an increasingly dense network of ties based on migration, flows of financial remittances and commodities, and flows of information. In all of these ways, then, contemporary transnationalism is not a new process, but a phenomenon that has increased dramatically in both its extent and its intensity.

they may also transform the lives of households who receive them. The second relates to the development of *transnational business practices* that are rooted in ethnic bonds and cultural practices.

Transnational remittances

All migrants maintain ties of one sort or another with their home country, and for many there will be a *financial* dimension to this linkage. Temporary contract workers in particular maintain strong economic ties with their place of origin by sending money back home. Globally, workers' remittances in 2004 amounted to $160 billion – almost the same as total Foreign Direct Investment flows (US$166 billion). Remittances are becoming an increasingly important source of income in many developing countries, as Table 13.3 indicates. While the absolute levels of remittance income are highest in India (US$22 billion in 2004), China ($21 billion) and Mexico (US$18 billion), the table shows the economic significance of remittances for many smaller nations on every continent. In all

Table 13.3 Top twenty remittance-receiving countries (as a percentage of GDP, 2004)

Country	GDP ($millions)	Remittances as Share of GDP (%)
Tonga	213	31.1
Moldova	2,595	27.1
Lesotho	1,312	25.8
Haiti	3,530	24.8
Bosnia and Herzegovina	8,533	22.5
Jordan	11,515	20.4
Jamaica	8,865	17.4
Serbia and Montenegro	23,997	17.2
El Salvador	15,824	16.2
Honduras	7,371	15.5
Philippines	84,567	13.5
Dominican Republic.	18,673	13.2
Lebanon	21,768	12.4
Samoa	375	12.4
Tajikistan	2,073	12.1
Nicaragua	4,555	11.9
Albania	7,590	11.7
Nepal	6,707	11.7
Kiribati	62	11.3
Yemen, Rep.	12,834	10

Source: Adapted from World Bank (2006: 90), Figure 4.1.

20 countries listed, remittances account for more than 10 per cent of the home country's economy. Across the developing world, remittances are now more than double the amount provided by official development assistance (i.e. aid) from all sources (Figure 13.4).

Clearly remittances have an effect in reducing poverty in developing countries. Moreover, remittance flows are in most cases much more immune to the vagaries of economic and political uncertainty than investment flows – especially in countries such as the Philippines where money is being sent from all around the world. But the benefits of remittances are a subject of some debate. Critics point out that it is never the poorest who are able to migrate and so remittances may increase inequality in migrant sending societies. Remittances are also often used for consumption (and often imported) goods, or for services such as healthcare and education, not for productive investment. They may push up the cost of land, housing, good quality education, and other commodities, such that migration becomes a necessity in order to enjoy a comfortable quality of life. A high volume of remittances may also push up the exchange rate of the country's currency, therefore making its exports more expensive, and hence further

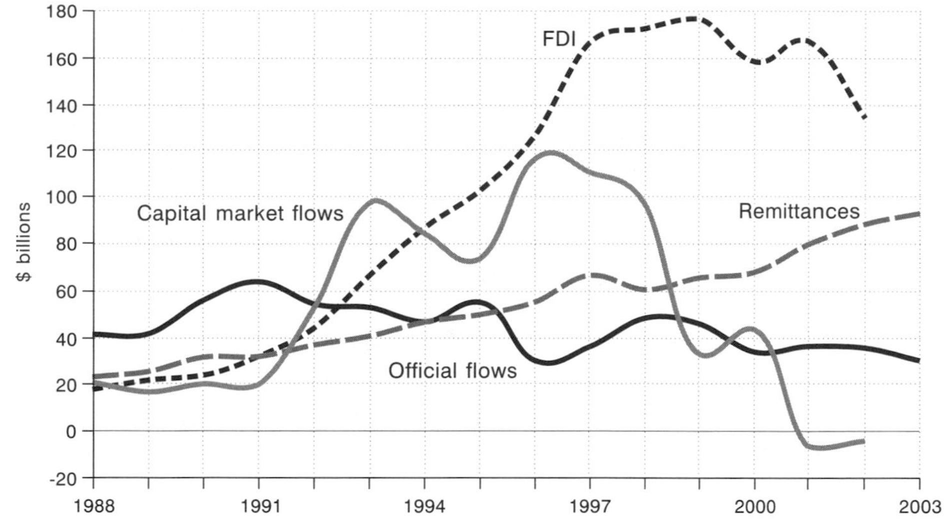

Figure 13.4 Resource flows to developing countries (in billions of dollars)
Source: Redrawn from World Bank (2004: 170), Figure A.1.

deepening the dependence on migration. These, of course, are only the economic costs – to generate remittances, especially by temporary workers, families must often be oceans apart, so that children are separated from parents and spouses from each other, creating a very human cost.

There is, however, a counter-argument. Where remittances are used for the education and healthcare of children, they are fostering the long-term well-being (and productivity) of the workforce (quite apart from the less economic and more humane benefits). When remittances are used for immediate consumption needs, they are circulated among local businesses and can therefore have multiplier effects on the local economy. In some cases, collective remittances by hometown associations will develop public infrastructure such as potable water projects, health clinics, orphanages, and schools. Remittances may also have positive cultural impacts – financing religious and ritual activities and therefore keeping traditions alive; or enhancing the power of women in households where they become the prime breadwinner due to their overseas work.

Transnational ethnic enterprises

Aside from household expenditures, migration also gives rise to a transnational variety of ethnic entrepreneurship. These are not the powerful transnational corporations discussed in Chapter 8, but generally small enterprises that nevertheless transcend national borders. Four specific types of transnational migrant enterprises have been identified (Landolt et al., 1999):

- *circuit* firms provide services transporting or transmitting goods and remittances, or in facilitating the migration process itself, for example as recruitment agencies;
- *cultural enterprises* import and sell cultural goods such as magazines, music, and videos from the 'home country';
- *ethnic enterprises* supply goods such as food and clothing from 'home' to both immigrant and non-immigrant urban markets;
- *return migrant micro-enterprises* are established by returnees to their home country and utilize contacts made while overseas.

In Germany, in recent years, Turkish enterprises of all of these kinds have expanded prolifically, changing the relationship between Turks and Germans, and between Turks and their places of origin in Turkey. Box 13.4 examines this phenomenon in more detail.

Box 13.4 Ethnic Turks in Germany

Turks first migrated to Germany in the early 1960s as contract labour migrants, or 'guest workers' (*Gastarbeiter*). Many stayed and they now constitute by far the largest ethnic minority population in Germany (and one of the most significant in any European country), numbering over 2 million. Turkish businesses are now a common sight in the landscapes of German cities. In Berlin alone it is estimated that there are 5–6,000 German-Turkish enterprises.

Initially, businesses were established in the 1960s to cater to the demand for Turkish services and products among guest workers – circuit firms and cultural enterprises such as travel agencies, translation firms, restaurants and cafes. By the 1970s, cultural enterprises had expanded to provide clothing, music, groceries – food and grocery stores still constitute over 60 per cent of all German-Turkish enterprises.

By the 1990s, however, the vast majority of businesses were not just drawing upon the Turkish community but also had ethnic Germans as their suppliers and customers. Almost one-third had non-Turkish employees. Thus, Turkish entrepreneurship had shifted from solely circuit or cultural enterprises and into ethnic enterprises, catering to increasing demand by Germans for 'ethnic' food and other products. We also now see the emergence of transnational return migrant enterprises as German-Turks invest in businesses in Turkey. These have included clothing manufacturers established in Turkey to produce for the German market, and German-Turkish involvement in the tourism business in Turkey. For more, see Pecoud (2002).

In recent years, the business practices associated with transnational ethnic firms have received increasing attention. The economic practices of ethnic Chinese entrepreneurs have been a particular source of fascination, largely because the rise of China itself as an economic force has been paralleled by the great success of ethnic Chinese communities around the Pacific Rim, and elsewhere across the world. From Sydney, to Singapore, to San Francisco, ethnic Chinese immigrant communities (or, more accurately, some segments within them) have thrived economically, in part based on the global networks forged by bonds of co-ethnicity (Yeung, 2004).

There are a number of distinctive business practices that seem to prevail in such global networks that are not always found in the mainstream Anglo-American market economy and which appear to have been especially well adapted to competition in the contemporary global economy. These practices emphasize the importance of social and cultural connections or relationships in initiating and cementing what is otherwise a business relationship. This has become known as Chinese, or Confucian, or '*guanxi*' capitalism – referring to the Mandarin Chinese word for 'relationship'. Five characteristics in particular are seen as competitive advantages of ethnic Chinese firms:

- Family ties are a significant component of business management. Even some very large ethnic Chinese businesses are controlled by sons, sons-in-law, brothers, and nephews of a founding 'father' figure (and such managerial posts are, on the whole, the domain of male relatives alone; see Chapter 11 for an example of Li Ka-shing from Hong Kong). The advantage of family ownership is that it allows quick decision-making even across a transnational business organization.
- Business transactions outside of the family circle will often be based on other forms of cultural affinity – membership of the same clan, place-based identities, and linguistic commonality. There is, then, a social basis to business ties that cements the business relationship into one of mutual trust and reciprocity that goes well beyond a purely contractual relationship. Where joint business ventures with foreign partners are involved, this creates a reliable relationship where legal and contractual agreements may not be enforceable. This has been the basis for a great deal of the investment by Taiwanese and Hong Kong manufacturers in mainland China.
- Both familial and cultural relationships point to the importance of reputation and trust in business practices. These might apply to relationships between competitors, between contractors and suppliers, and between employers and employees. The bond of trust in business relationships is especially effective in ensuring that contracts are honoured even when partners are oceans apart – a relationship based on trust is far more easily 'stretched' across global space than one based on legal contracts.
- The ability to raise capital for new business ventures through informal networks creates a great deal of flexibility. Instead of going through formal

banking or stock market channels to raise money, a new venture can be financed through loans from members of the same co-ethnic network. In ethnic Chinese communities such mutual aid has been a long tradition, largely on account of the discriminatory environments in which they found themselves and the lack of access to formal credit channels. In a contemporary transnational business operation, such sources of credit make the raising of capital for new ventures both relatively easy and inexpensive.

- The exchange of market intelligence: relationships in networks can be cemented not just through business partnerships but also through binding indebtedness – not of a financial kind, but through the exchange of gifts and favours, including information. A cultural understanding of the 'gift economy' then dictates that a gift can never be exactly repaid, thus establishing an ongoing relationship.

In this section we have seen how familial or co-ethnic bonds can operate at scales much greater than the neighbourhood scale of the ethnic economy. Migrant remittances represent one form of transnational linkage with households in migrant-sending countries, often with profound implications both for both individual recipients and national economies. Transnational ethnic enterprises, on the other hand, represent a stretching of the same bonds and ethnic resources that create urban business clusters. In some cases, particularly the example of ethnic Chinese business networks, these bonds have proven to be well adapted to the need for flexible and reliable business partnerships in the contemporary global economy.

13.6 The Limits to Ethnicity

We have sought in this chapter to highlight the importance of ethnicity in economic life at multiple geographical scales. In closing, however, it is important to note some of the inherent limitations that exist in using ethnicity as a lens through which to understand various dimensions of the economy. First, in discussing ethnicity there is always an inclination to create groups and categories and to assume that they are both meaningful and homogeneous. In other words, it is often assumed that there is something *essential* to ethnicity. In reality, however, ethnicity is a complex and contradictory phenomenon and we should not assume that it predicts any particular forms of economic practice or exhibits any essential characteristics (see also Chapter 11, Section 11.6). This point can be illustrated through an example drawn from one of the Chinese ethnoburbs mentioned earlier. In the Toronto suburb of Richmond Hill, a contentious dispute emerged in the 1990s over a proposed new shopping mall that would be modelled on a Hong Kong retail environment – with a large numbers of restaurants, the enclosure of shops in a climate controlled environment, and extensive parking facilities – all of which set it apart from neighbourhood shopping facilities in the area. Proponents and opponents of the mall were caught in a tense standoff over the development, which eventually went ahead with some design modifications. What is of particular interest, however, is that both proponents and opponents were ethnic Chinese-Canadians, some of whom wanted to maintain the character of the neighbourhood, while others favoured the new development. There are, then, dangers in assuming that an ethnic group can be defined homogeneously and essentially in its economic preferences (see Preston and Lo, 2000).

The same caveat is needed when we look at statistical data on the employment experiences of particular ethnic groups – ultimately there is no 'average' experience for Filipinos, Koreans, Ghanaians etc. in the urban labour market. Such groups often exhibit as much diversity as distinctiveness. In 2001, for example, a total of 12,903 immigrants arrived in Canada from the Philippines. Almost 3,000 arrived as 'skilled workers', while over 2,000 were granted permanent resident status through the Live-In Caregiver Programme mentioned earlier in this chapter. Another 33 arrived under business immigrant programmes. Meanwhile, just over 1,000 were under the age of 5, and would therefore receive all of their formal schooling in Canada. All are ethnically Filipino immigrants, and yet the circumstances of their arrival alone will have huge impacts on their experiences of economic life in Canada. There is, in short, no such thing as an average Filipino immigrant. There is a danger that using this ethnic category to group or essentialize diverse individuals will obscure as much as it reveals.

A further point of caution is that ultimately other factors can 'trump' ethnicity in the conduct of business transactions. For example, there exist extensive linkages between the Hsinchu high technology region in Taiwan and Silicon Valley in California, based on co-ethnic networks. But while co-ethnic bonds may facilitate linkages such as subcontracting and sharing of information, many entrepreneurs remain cautious with respect to bonds of co-ethnicity (Hsu and Saxenian, 2000). More important are the technological and economic logics for collaboration, rather than the cultural logics that supposedly shape '*guanxi* capitalism'. *Guanxi* is not the dominant organizational principle behind such relationships. The president of an electronics manufacturer in Taiwan who had recruited many skilled Taiwanese workers from Silicon Valley puts the case plainly:

> Of course we look for qualified Chinese talents, and in most time, it is easier to get information about the candidate's reputation if they are Chinese. But we do not target on Chinese in itself. We search for qualified engineers. That's our top priority in the evaluation process. It is stupid to hire a guy just because he is in my circle. We hire people based on their competence, not their guanxi or ethnicity. Guanxi helps us to get the right people easily, but it does not mean we exploit the advantage of guanxi itself. The key issue here is that all the people we hire are professional engineers, and they happen to be Chinese.
>
> (Hsu and Saxenian, 2000: 2002)

Thus, *guanxi* was the lubricant in the system, rather than the system itself. Ultimately, technological competencies and market-based rationality were the determining factors. This reminds us that we must not replace a simple economic logic with one that over-emphasizes cultural processes. Cultural and social processes are integral to how economic processes work, but they are not necessarily determining.

13.7 Summary

This chapter has explored the ways in which ethnic identity intersects with economic practice, and the ways in which space is integral to this intersection. We should now recap some of these geographical arguments. In the case of ethnically diverse *labour markets*, we saw the sorting that occurs whereby certain ethnic groups get matched to certain kinds of work. This phenomenon of segmentation was explained on the basis of human capital, institutions, discrimination, and home-work linkages. In each case, the spatiality of the process was critical. Qualifications and professional experience are frequently devalued as they travel across jurisdictional boundaries, while skills may also be specific to particular cultural contexts. Thus the localized institutional and cultural specificity of labour markets is an important basis for understanding segmentation, especially among those such as migrants who move between them. Equally, we cannot dismiss the role that stigmatization and discrimination play in segmentation, and this is frequently spatialized, with particular places seen as producing stereotyped workers. An explicitly spatial argument also exists in relation to home-work linkages, where residential segregation can lead to occupational segmentation.

In exploring *ethnic businesses* we saw the significance of geographical processes. For many businesses, the establishment of ethnic concentrations in certain parts of major cities allows an economy to develop that draws upon co-ethnic financing, customers and employees. The emergence of ethnic places in the city also reinforces a sense of ethnic identity. But ethnic neighbourhoods are not static. In many cases, new ethnic business clusters have emerged, often reflecting the greater wealth of more recent immigrants. Driving these new suburban developments have been ethnic financial institutions, highlighting the fact that ethnic businesses are far more diverse and sophisticated than the popular image of grocery stores and restaurants.

We then examined the *transnational dimensions* of the economic connections established by migrants. Remittances between households, for example, which have multiplied dramatically in recent years, can have significant effects on economies and households where they are received. We also saw how transnational business enterprises can bind together different local economies in ways that are rather distinct from the patterns of transnational corporations described in Chapter 8. A celebrated feature of these enterprises has been their supposedly ethnic quality, making them distinct from mainstream corporate models. In closing, however, we pointed out the pitfalls in placing too much emphasis on ethnicity as an explanatory factor in understanding economic experiences and practices.

Further reading

- For further geographical analysis of ethnic differentiation in North American urban labour markets, see Ellis and Wright (2000) and Hiebert (1999).
- Much of the research on ethnic entrepreneurship and ethnic economies has been conducted by American sociologists. Roger Waldinger, Alex Portes, Ivan Light, Edna Bonacich and their various collaborators have been especially influential. More recently, geographers have also paid increasing attention. A collection of geographical interpretations can be found in Kaplan and Li (2006).
- On remittance economies in Mexico, see Conway and Cohen (1998), and in India, see Walton-Roberts (2004).
- For a collection of geographical essays on transnationalism, see Jackson et al. (2004).
- For some recent geographical work on the ethnic Chinese diaspora and their transnational business networks, see Yeung and Olds (2000) and Ma and Cartier (2003).

Sample essay questions

- How does ethnicity influence the spatial organization of labour markets?
- What are the main factors accounting for the clustering of ethnic business in major gateway cities?
- Why is ethnicity a relevant explanation for transnational economic activities?
- What are the dynamic changes that ethnic economies experience in today's globalizing world economy?
- Describe and explain how ethnicity works in your own economic life.

Resources for further learning

- http://users.fmg.uva.nl/jrath/imment.htm: Jan Rath, a researcher in Amsterdam, has an extensive website relating to ethnic entrepreneurship.
- http://www.eeoc.gov/stats/jobpat/jobpat.html: the US Equal Employment Opportunities Commission has statistics for all US cities and states along the

lines discussed for Los Angeles in Section 13.3. http://www.cre.gov.uk/research/statistics_labour.html: data on Britain are available from the Commission for Racial Equality.

- http://www.statcan.ca/english/Estat/licence.htm: Statistics Canada's E-stat system provides extensive data on migration, ethnicity and labour markets.
- http://www.migrationinformation.org/: the Migration Policy Institute, and its Migration Information Source, has extensive information on global migration, transnationalism and remittances.
- http://www.wceckorea.org/en/wcec2005/about_wcec.asp: the website of the 2005 meeting of the biennial World Chinese Entrepreneurs Convention (WCEC) provides an interesting window onto the activities of the international Chinese business community.

References

Bauder, H. (2001) Culture in the labor market: segmentation theory and perspectives of place, *Progress in Human Geography*, 25(1): 37–52.

Conway, D. and Cohen, J. (1998) Consequences of migration and remittances for Mexican transnational communities, *Economic Geography*, 74(1): 26–44.

Ellis, M. and Wright, R. (2000) The ethnic and gender division of labor compared among immigrants to Los Angeles, *International Journal of Urban and Regional Research*, 24(3): 567–82.

Hiebert, D. (1999) Local geographies of labor market segmentation: Montréal, Toronto, and Vancouver, 1991, *Economic Geography*, 75(4): 339–69.

Hsu, J.Y. and Saxenian, A. (2000) The limits of guanxi capitalism: transnational collaboration between Taiwan and the USA, *Environment and Planning A*, 32(11): 1991–2005.

Jackson, P., Crang, P. and Dwyer, C. (eds) (2004) *Transnational Spaces*, London: Routledge.

Kaplan, D. and Li, W. (eds) (2006) *Landscapes of the Ethnic Economy*, Lanham, MD: Rowman and Littlefield.

Kwak, M.J. (2002) Work in family businesses and gender relations: a case study of recent Korean immigrant women, Master's thesis, Department of Geography, York University, Canada.

Landolt, P., Autler, L. and Baires, S. (1999) From hermano lejano to hermano mayor: the dialectics of Salvadoran transnationalism, *Ethnic and Racial Studies*, 22(2): 290–315.

Li, W., Dymski, G., Zhou,Y., Chee, M. and Aldana, C. (2002) Chinese American banking and community development in Los Angeles County, *Annals of Association of American Geographers*, 92(4): 777–96.

Lo, L. (2006) Changing geography of Toronto's Chinese ethnic economy, in D. Kaplan and W. Li (eds) *Landscapes of the Ethnic Economy*, Lanham, MD: Rowman and Littlefield, pp. 83–96.

Ma, L.J.C. and Cartier, C. (eds) (2003) *The Chinese Diaspora: Space, Place, Mobility and Identity*, Boulder, CO: Rowman and Littlefield Publishers.

Pecoud, A. (2002) 'Weltoffenheit schafft Jobs': Turkish entrepreneurship and multiculturalism in Berlin, *International Journal of Urban and Regional Research*, 26(3): 494–507.

Pratt, G. (2004) *Working Feminism*, Philadelphia, PA: Temple University Press.

Preston, V. and Lo, L. (2000) Asian theme malls in suburban Toronto: land use conflict in Richmond Hill, *The Canadian Geographer*, 44: 86–94.

Preston, V. and McLafferty, S. (1999) Spatial mismatch research in the 1990s: progress and potential? *Papers in Regional Science*. 78: 387–402.

Walton-Roberts, M. (2004) Returning, remitting, reshaping: non-resident Indians and the transformation of society and space in Punjab, India, in P. Jackson, P. Crang, and C. Dwyer (eds) *Transnational Spaces*,. London: Routledge, pp. 78–103.

Walton-Roberts, M. and Hiebert, D. (1997) Immigration, entrepreneurship, and the family: Indo-Canadian enterprise in the construction industry of Greater Vancouver. *Canadian Journal of Regional Science*, 20(1–2): 119–39.

World Bank (2004) *Global Development Finance 2004: Harnessing Cyclical Gains for Development*, New York: World Bank.

World Bank (2006) *Global Economic Prospects 2006: Economic Implications of Remittances and Migration*, New York: World Bank.

Yeung, H. (2004) *Chinese Capitalism in a Global Era: Towards Hybrid Capitalism*, London: Routledge.

Yeung, H. and Olds, K. (eds) (2000) *Globalization of Chinese Business Firms*, New York: Macmillan.

INDEX